MEDIEVAL ISLAMIC MAPS

MEDIEVAL ISLAMIC MAPS

An Exploration

Karen C. Pinto

The University of Chicago Press / Chicago and London

Karen C. Pinto is assistant professor of Islamic and Middle Eastern history at Boise State University.

The University of Chicago Press, Chicago 60637
The University of Chicago Press, Ltd., London

Printed in China

25 24 23 22 21 20 19 18 17 16   1 2 3 4 5

ISBN-13: 978-0-226-12696-8 (cloth)
ISBN-13: 978-0-226-12701-9 (e-book)
DOI: 10.7208/chicago/9780226127019.001.0001

Library of Congress Cataloging-in-Publication Data

Pinto, Karen C., author.
Medieval Islamic maps : an exploration / Karen C. Pinto.
pages cm
Includes bibliographical references and index.
ISBN 978-0-226-12696-8 (cloth : alk. paper) — ISBN 978-0-226-12701-9 (e-book) 1. Cartography—Islamic countries—History. 2. Geography, Arab. I. Title.
GA221.P56 2016
912.092'21767—dc23

2015017867

⊚ This paper meets the requirements of ANSI/NISO Z39.48-1992 (Permanence of Paper).

IN LOVING MEMORY OF MY PARENTS,

ADELE BERTHE PINTO
(1928–2008)
AND
LT. COL. FELICE GEORGE PINTO
(1918–1984)

عِلم نے مجھ سے کہا عِشق ہے دیوانہ پن
عِشق نے مجھ سے کہا عِلم ہے تخمین وظن
عِلم ہے پیدا سوال، عِشق ہے پِنہاں جواب
عِلم ہے اِبن الکتاب، عِشق ہے اُمّ الکتاب!

Knowledge said to me, Love is madness;
Love said to me, Knowledge is estimation and presumption.
Knowledge is born a question, Love is the hidden answer;
Knowledge is Son of the Book, Love is Mother of the Book!

ALLAMA IQBAL, *'Ilm wa 'Ishq* (Knowledge and Love)

It is seeing which establishes our place in
the surrounding world; we explain that
world with words, but words can never undo
the fact that we are surrounded by it.
The relation between what we see and what
we know is never settled.

JOHN BERGER, *Ways of Seeing*

Contents

Note on Transliteration

A book involving the admixture of Arabic, Persian, and Ottoman transliteration systems is, as I have discovered, no easy matter. The standard IJMES system, which I employ throughout this book, does not fit Ottoman transliteration well. Thus, Mehmet becomes Meḥmed but *Tārīḫ-i Hind-i Gharbī* (History of the West Indies) cannot be changed to the Arabic form of *Tārīkh*. Similarly, Loḳmān looks odd when transliterated according to the Arabic system as Luqmān. The problem is compounded by modern Turkish, which uses, for example, "c" for "j." Thus, *Cerrāḥiyyetü'l-Ḥāniyye* (Imperial Surgery) should be transliterated as *Jerrāḥiyyet* according to the IJMES system, but no one would be able to locate the book under that spelling. For this reason, the transliteration of Ottoman Turkish under the IJMES scheme cannot be foolproof. The case of Persian is easier. I apply the IJMES transliteration convention. Instead of the Persian spelling Iskander, for instance, I use Iskandar according to the Arabic convention. In order to preserve the narrative flow I have removed the prefix "al-" from the start of Arabic last names. Thus Ṭabarī instead of al-Ṭabarī, Ma'mūn instead of al-Ma'mūn, and so on. Exceptions to this are quotes and full names.

Place-names present especially complex choices: Does one list Mecca according to common practice or according to the correct transliteration of Makka? Medina or Madina? More vexing is the correct use of names for which a local equivalent has become popular. Seville versus Sevilla is a case in point. Troublesome is the spelling for the East African tribe that forms the core of chapters 8 and 9: Buja according to Arabic orthography or Beja according to common practice? I decided in favor of common practice. So I use Mecca, Medina, Sevilla, and Beja. I transliterate only those place-names that I judge as being not well known or without an Arabic alternative—for example, Ifrīqiya, Zaghāwa, and Mafāza al-Buja.

In cases where usage of a word is common, I do not list it in its transliterated form (e.g., "sultan" not "sulṭān"). Exceptions to this are references to the Qur'ān and ḥadīth. I also elected to not transliterate dynastic names. Outside of what falls into the domain of common practice, I chose to transliterate all Arabic, Persian, and Turkish words, names, and book titles. I may have missed a few and for this the error lies with me alone.

In short, I use a hybrid IJMES system in consultation with the method adopted by the associate editor of book 1 of the second volume of *The History of Cartography*, Ahmet Karamustafa, who also had to wrestle with the transliteration scheme of three different languages and the place-name dilemma and developed brilliant solutions.

All translations are my own unless otherwise noted.

1

Introduction
Ways of Seeing Islamic Maps

Scattered throughout collections of medieval and early modern Arabic, Persian, and Turkish manuscripts are thousands of cartographic images of the world and various regions.[1] The sheer number of these extant maps tells us that from the thirteenth century onward, when copies of these map manuscripts began to proliferate, the world was a much-depicted place. It loomed large in the medieval Muslim imagination. It was pondered, discussed, and copied with minor and major variations again and again, and all with what seems to be a peculiar idiosyncrasy to modern eyes. The cartographers did not strive for mimesis (imitation of the real world). They did not show irregular coastlines even though some of the geographers whose work includes these maps openly acknowledge that the landmasses and their coastlines are uneven.[2] They present instead a deliberately schematic layout of the world and the regions under Islamic control.

These images employ a language of stylized forms that make them hard to recognize as maps. Scholars of Islamic science and geography often ignore and belittle these maps

Fig. 1.1. Classic KMMS world map, "Ṣūrat al-Arḍ" (Picture of the World), from an abbreviated copy of al-Iṣṭakhrī's *Kitāb al-masālik wa-al-mamālik* (Book of Routes and Realms). 589/1193. Mediterranean. Gouache and ink on paper. Diameter 37.5 cm. Courtesy: Leiden University Libraries. Cod. Or. 3101, fols. 4–5.

on the grounds that they are not mimetically accurate representations of the world.[3] What these scholars miss is that these schematic, geometric, and often symmetrical images of the world are iconographic representations—"carto-ideographs"—of how medieval Muslim cartographic artists and their patrons perceived their world and chose to represent and disseminate this perception.[4]

On the surface it seems that these often elaborately illuminated nonmimetic cartographic works, employing pigments made from precious metals and stones, must have been produced for the elite literati of medieval Islamic society such as the commissioners/patrons, collectors, copyists, and high-status readers of the geographic texts within which these maps are found. This conclusion ignores the easy-to-replicate nature of these schematic images, which would have enabled students visiting the libraries of sultans, amirs, and other members of the ruling elite to transport basic versions of these carto-ideographs back to the people of their villages and far-flung areas of the Islamic world.

As the scholar who pioneered the deconstructive analysis of maps in the history of cartography, J. B. Harley, points out: "What constitutes a text is not the presence of linguistic elements but the act of construction so that maps, as constructions employing a conventional sign system, become texts."[5] Nowhere is the role of map as text better played out than in the realm of medieval and early modern mapping where, due to the developing skills of the cartographers, there was less concern for scientific exactitude and greater interest in the artistic rendering of thought. See, for example, the mid-sixteenth-century representation of Europe from S. Münster's *Cosmographia* (fig. 1.2). To simply dismiss this map as an inaccurate representation of Europe and therefore an invalid source is to miss the point. It is reflective of a much deeper sociocultural, historical, and political context that needs to be read and interpreted. Are the metaphorical coastlines reflective of an emerging consciousness of European supremacy? Can they not be read as a sort of "pre-colonial colonialism" with Spain and Portugal leading the way, assisted by the arm of Italy and the pivotal island of Sicily on one side and by the staff of the North Sea formed by the Danes, the English, and the Scots on the other? The Münster map demonstrates how cartography can be used to navigate the medieval and early modern imagination.

The practice of mapping, representing the world that surrounds us, is not a new one. People have been making maps, in sand, in rock, with pebbles, on the bark of trees, since time immemorial.[6] We have grown so accustomed to our modern maps and sophisticated satellite images from space that most of us rarely think of what it must have been like to determine the shape and geopolitical features of the world prior to the existence of chronometers and space technology. It is hard to relocate one's perspective to that of the medieval Islamic cartographers who had only the assistance of mountains, masts of ships, travelers' accounts, cosmographic myths, and rudimentary astronomical equipment to guide them. This book aims to guide the reader through the thicket of Islamic maps by presenting three different ways of seeing a selection of them.

The plethora of extant copies produced all over the Islamic world, including India, testifies to the enduring and widespread popularity of these medieval Islamic cartographic visions. For no less than six centuries (eight centuries if we include the nineteenth-century subcontinental examples), these images were perpetuated primarily in a standardized geographic series, sometimes referred to as the Islamic Atlas or the Balkhī school of mapping. I have chosen to disregard this misnomer and have opted instead for a new acronym: the KMMS mapping tradition. I base this acronym on the title of the genre's most widely disseminated version: Iṣṭakhrī's *Kitāb al-masālik wa-al-mamālik* (KMM) (Book of Routes and Realms).[7] The "S" appended to the acronym stands for *ṣūrat* ("picture" in Arabic)—that is, those KMM geographical manuscripts that are accompanied by map images.[8]

My aim in this book is twofold. First, I will introduce readers to the KMMS maps.

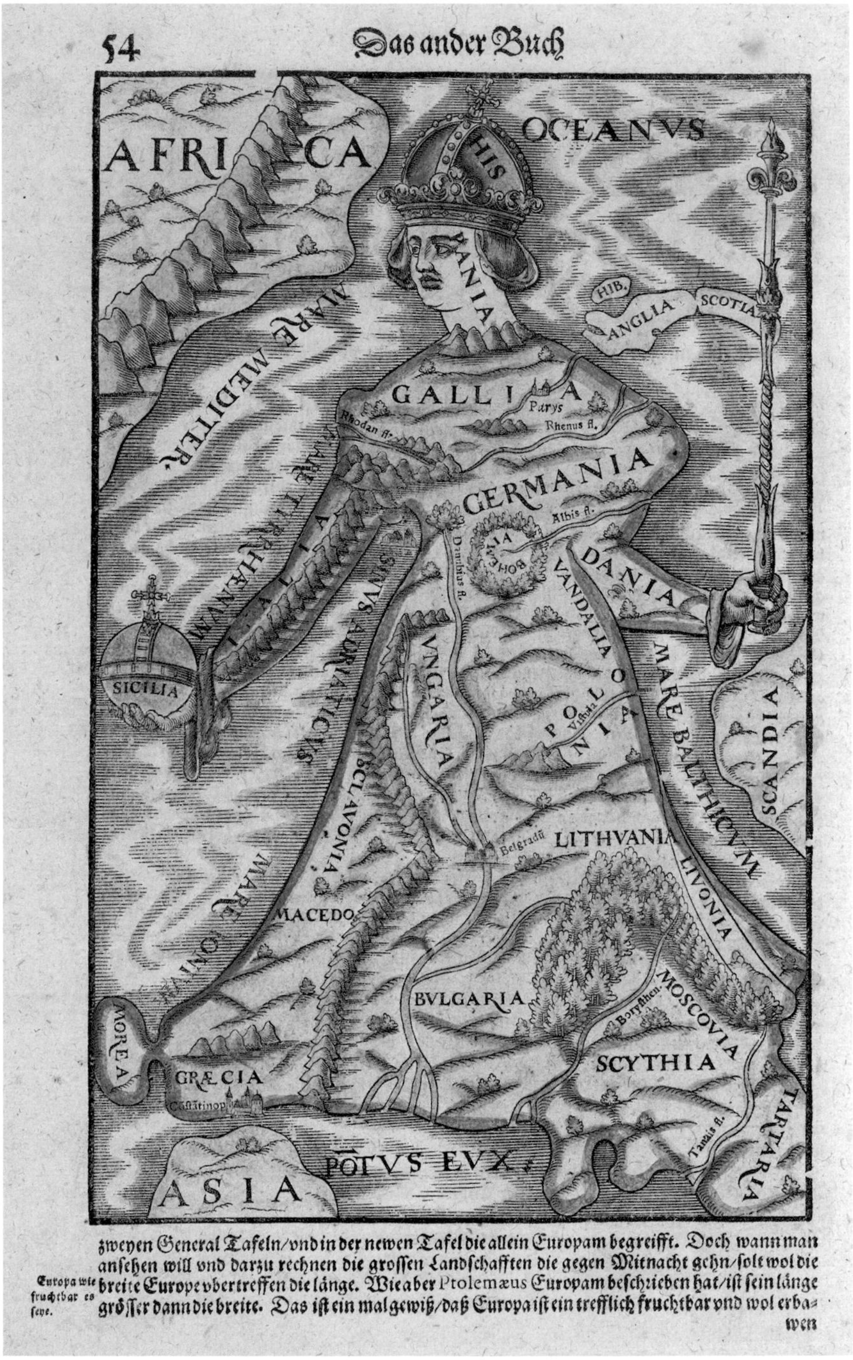

Fig. 1.2. Europe as queen. 1550–1570. Hand-colored woodcut print on paper. 26 × 17 cm. S. Münster, *Cosmographia*. Courtesy: Special Collections, The University of Texas at Arlington Library, Arlington, Texas. 2007-357.

Second, I will demonstrate my own approach to reading these maps and how this approach can be employed to expand the boundaries of Islamic history and the history of cartography. In this book, I present a tripartite approach to the KMMS maps of the world. Each part is an interpretive method unto itself and can be categorized respectively as iconography, context, and patronage. Examination from these points of view places the study of medieval Middle Eastern maps within modern and postmodern theoretical paradigms. Taken together, these three analytical modes function like the legs of a tripod, supporting new conclusions about Islamic history and Islamic maps.[9]

Objectives and Structure of This Book

In this book I explore ways of revising Islamic history using this tripartite approach to medieval Islamic maps. I examine the maps from the point of view of iconography, specifically of one form, the Encircling Ocean; I analyze the way imagination is mapped through a contextual study of a particular place, the lands and deserts of the Beja; and I show via evidence of patronage how the maps can be used as an alternate means of biography.

In chapter 2, "A Look Back," I survey the work that has been done on the Islamic mapping tradition and discuss the growing role of maps as wellsprings of historical information from material culture.

Chapter 3, "A Sketch of the Islamic Mapping Tradition," provides essential background to understanding the tradition within which the maps are encased. This is the beginning of a familiarization process that will acquaint readers with these relatively unknown images.

In chapter 4, "KMMS World Maps Primer," my goal is to make the maps "legible" to the general reader. I discuss the forms that make up the classical Islamic KMMS map of the world and elaborate the matrix of places and spaces that overlay each world map. A reading of this chapter will enable readers to follow the later discussion and deeper reading of the maps that I demonstrate in this book, showing how the maps can be used as alternate gateways into Islamic history and the history of cartography.

Chapter 5, "Iconography of the Encircling Ocean," is the first of three chapters that focus on the Encircling Ocean form in the world maps. The goal of these chapters collectively is to conduct an iconographic exploration of the Encircling Ocean form, establishing the KMMS maps in their place within a tradition of great breadth in both time and space. This chapter begins this task by offering a broad look at encircling metaform occurrences across a multitude of ancient, medieval, and early modern mapping traditions.

Chapter 6, "Classical and Medieval Encircling Oceans," considers Greco-Roman and subsequent European cartography from the perspective of the Encircling Ocean form. The chapter works forward in time to medieval maps of European origin, culminating with a remarkable example of intersection between medieval Islamic and European maps.

Chapter 7, "The Muslim Baḥr al-Muḥīṭ," details medieval Islam's understanding of the Encircling Ocean from religious and cartographic perspectives, and follows the progression of the form's symbolic evolution from a marker of unknown terrors to a sign of both earthly and divine power. This chapter concludes the iconographic study of a quintessential feature of the KMMS maps.

Chapter 8, "The Beja in Time and Space," is the first of two chapters pondering a curious anomaly that occurs in every medieval Islamic map of the world. Located on the eastern flank of Africa is a double-territorial ethnonym for an obscure East African tribe: the Beja. Mention of them in medieval Middle Eastern historiography is rare and, at best, superficial, yet no Islamic map from the eleventh to the nineteenth century leaves them out. The question I raise in this chapter is, quite simply, who were the Beja?

Chapter 9, "How the Beja Capture Imagination," builds on the answer to chapter 7's query by asking another question. Knowing who the Beja are, we are led to wonder, why are they absent in Islamic historiography yet ever present on the KMMS world maps? The surprising answers help us to understand how capturing of the imagination of the cartographer affects what makes it onto a world map. This chapter completes my contextual reading of the Beja ethnonym.

Chapter 10, "Meḥmed II and the Ottoman Cluster," is the first of three chapters in which I conduct a case study approach to the unstudied subject of patronage of Islamic cartography. I employ a Schamaesque approach to reading a set of classical Islamic KMMS maps from the period of Meḥmed II, known as Fātiḥ, "the Conqueror," because of his conquest of Constantinople in 1452. This chapter begins with a brief survey of work already done on the role of patronage in cartography, and moves on to a consideration of the character of a particular patron, namely, Meḥmed II. I offer an alternate view of Fātiḥ.

In chapter 11, "The KMMS Ottoman Cluster," I look specifically at the identifying characteristics of the maps I refer to as the Ottoman cluster, embedded within the larger set of KMMS maps. Via the Ottoman cluster reading, I establish that all copies are a product of a particular time and a particular milieu and that they therefore can be read to reveal the sentiments of both the illustrator and his patron.

Chapter 12, "Source of the Ottoman Cluster," examines the probable origin of the cluster both in terms of the images used and the reason for their use. In this last of the three chapters on these maps, I point out a key manuscript and consider the impact of this work on Meḥmed II. By making a connection between patronage and propaganda, I show how a set of KMMS map manuscripts can be used to provide insights into map audience, patronage, and politics in fifteenth-century Anatolia.

These chapters are intended to be illustrative of a new approach to Islamic cartography. This book does not set out to provide a definitive reevaluation of the whole Islamic mapping tradition. Rather it presents a series of scenarios that illustrate new ways and alternate methodologies for addressing the rich Islamic cartographic heritage. The hope is that this study will pave the way to a major reevaluation of the use of visual sources as an approach to Islamic history and enrich the field of the history of cartography.

2

A Look Back

Islamic cartography remains an under-studied field. The problem lies in the resistance of Islamic historians to the study of images and new theoretical approaches. Iconographic analysis of images is unusual outside of Islamic art history, and within the domain of art history, the study of images and objects that do not conform to the strict definition of miniature art and architecture is also rare. As a consequence, the illustrations in manuscripts of science and geography have fallen through the cracks between specialties. Discussions focusing on cartography are few and those that exist focus on veracity, authentication, chronology, and the description and classification of the cartographers into specific schools. Compositional analysis and interpretation of the images is rare. The emphasis instead is on mimesis and the contributions of the most mimetic traditions at the expense of the bulk of the KMMS tradition.[1]

The texts of the medieval Islamic geographers have been a favorite topic of Orientalists since the mid-eighteenth century, in contrast. Countless volumes of meticulous editions and translations line the shelves under the 893s in the old Dewey decimal system. Numerous accounts relate the story of the rise of Muslim sciences and world geography; the rapid progress of the sciences during the early Abbasid years under the active patronage of the Abbasid caliphs Manṣūr, Hārūn al-Rashīd, and Ma'mūn; and the frenzied pace of translation of Greek and Indian sources.[2] Translation, extraction of useful "facts," and classification dominated the corpus of scholarship on the medieval Islamic geographers until the late twentieth century.[3]

There are exceptions to the Orientalist study of Islamic geographers and fortunately for us these exceptions are on the increase.[4] Foremost among these are the four-volume seminal work of André Miquel, *La géographie humaine du monde musulman jusqu'au milieu du 11e siècle*; the chapter on geography in Paul Heck's *The Construction of Knowledge in Islamic Civilization*; Adam Silverstein's essay "Medieval Islamic Worldview" in the *Geography and Ethnography* volume and his work on Islamic geographers in *Postal Systems of the Pre-Modern Islamic World*; the chapter "Autopsy of a Gaze" in Houari Touati's *Islam and Travel*; Travis Zadeh's *Mapping Frontiers across Medieval Islam*; and Zayde Antrim's *Routes and Realms*. Though these are exposés of vastly differing length, they present some of the best and most innovative arguments on the works of the Islamic geographers to date.

In keeping with the broad, sweeping *longue durée* approach of the French Annales school of history, Miquel takes on the Islamic geographical tradition in what is the most comprehensive study to date.[5] His four-tome classic studies in depth everything from the mathematical end of the tradition to air, plants, routes, marvels, and the divisions (and indivisibility) of the earth. The guiding beacon for Miquel is the Muslim humanistic principle of *adab* (a concept that governs a range of meanings in Muslim culture and literature, including etiquette, style, taste, and good manners).[6]

Heck's work is significantly shorter and geography is only an oblique investigation as it pertains to his central inquiry on *Qudāma ibn Jaʿfar* and his *Kitāb al-Kharāj wa-ṣināʿat al-kitāba* (Book of the Land-Tax and the Craft of Writing). Heck does us the service of burying once and for all the outdated argument of the Iraqi versus Balkhī schools. In addition, he challenges Miquel's *adab*-based approach. Heck asserts instead that administrative considerations are the raison d'être of the early Muslim geographers, who were also state officials. He argues that their primary concern was to create an encyclopedic manual to guide administration of Islamic lands. Thus, taxation and the upkeep of roadways receive priority, as does the central domain of Islam, at the expense of travelers' reports on more distant places.

In *Postal Systems of the Pre-Modern Islamic World*, Silverstein continues Heck's trajectory to discuss the role that the geographies played administratively in outlining and supporting the official information system of postal (*barīd*) routes. He casts the *masālik wa-al-mamālik* (routes and realms) tradition as "the genre of caliphal itineraries" and the closest one can get to a postal manual. He sees Ibn Khurradādhbih's work as akin to the *Peutingerian Table* or the *Parthian Stations*.[7] In "Medieval Islamic

Worldview," Silverstein concurs with my findings that the Islamic geographic tradition was heavily influenced by ancient, pre-Islamic views of the world, especially Hellenistic, Sassanid, Mesopotamian, and Jewish. Silverstein argues that a specifically "medieval Islamic" geographical vision does not emerge until the tenth century and that when it does it drops its ancient baggage and ceases to represent a "*world*view."[8] While this may be true for the geographical work of Muqaddasī, the plethora of world maps that proliferate from the eleventh century onward, as demonstrated in the course of this book, flies in the face of Silverstein's conclusion.

Taking as his leaping-off point that medieval Muslim scholars were "mad for travel," Touati takes a completely different and novel approach. He establishes how travel came to be the central pivot of the medieval Islamic scholarly world and not only for religious and administrative concerns. For Touati, the primary purpose of travel for medieval Muslims was not administrative nor was it for seeking out "otherness"; rather it was for the purpose of creating the "sameness" of a single *mamlaka* (realm) from the eighth century onward in which Muslims could feel at home anywhere.[9] Through this lens, Touati sees the geographers as active travelers determined to create a narrative for a single unified Islamic world. He argues that from the late ninth century onward *ʿiyān* (seeing) came to take precedence over *samāʿ* (hearing) and that this affected the geographers, in particular Yaʿqūbī, Muhallabī, Iṣṭakhrī, Ibn Ḥawqal, and Muqaddasī, for whom the "primacy of the eye" made the voyage a necessary rule of conduct for writing a geography.[10] In doing so, Touati tries to dispel, though not always convincingly, the armchair scholar label with which many of the geographers have been branded.

Through a remarkably detailed and thorough investigation of the tale of Sallām the Interpreter and his mid-ninth-century ʿAbbasid caliphal-ordered mission to uncover the mysteries of Gog and Magog and the wall that Alexander the Great supposedly made to lock them out of the inhabited world, Zadeh takes us on an extraordinary and fascinating in-depth tour of everything from the role of the translator and the primacy of the discourse of *ʿajāʾib* (the fantastic) to the reception of the tale by other Muslim geographers and how it morphed later through the hands of Orientalist scholars. Zadeh deftly uses this slice in *Mapping Frontiers* to expose the nuances of medieval Islamic geography from the microcosmic to the macro with a special emphasis on the *Routes and Realms* work of one particular geographer in whose work Sallām's tale first appears: Ibn Khurradādhbih's *Kitāb al-masālik wa-al-mamālik* (Book of Routes and Realms), composed sometime in the latter half of the ninth century. Zadeh's detailed research on the work of this early geographer and the manuscript copies of his work along with the way in which the published editions obscure the manuscript variations is monumental and points the way for future work on the Islamic geographers: pick a single thread and follow it through to the present.[11]

Antrim's *Routes and Realms* focuses on the discourse of place, in particular land

and belonging in the premodern Islamic world. Antrim studies home/homeland, city, and region to prove that "attachment to land" (*arḍ/waṭan*) mattered. Antrim conducts a sweeping overview of medieval Arabic literature from anthologies, topographical histories, and religious treatises to travel literature, geographies, and maps. Specifically, she examines the geographical literature from the perspective of the three key cities of Mecca, Jerusalem, and Baghdad, and looks at what the geographers have to say about climatic divisions (Hellenistic *aqālīm* versus Persian *kishwar*) and regional boundaries. Antrim coins a new term, "citational performances," for referring to the practice of scholars copying material from past work and claims that this practice enhances "the legibility of the cities" because it brings with it "the ring of the familiar and the authoritative." She proves that the tenth-century geographer Muqaddasī cites passages from earlier geographers on Jerusalem instead of calling upon his own native knowledge, and in doing so, she counters the views of Touati.[12]

Although maps are a part of this discourse, none of these authors address them at length. Silverstein is not enamored of them at all as evinced by the only comment that he makes on the subject: "They are not remarkable for what we would think of as precision but serve as a reminder of the rough outline of the known world surrounded by the ocean sea."[13] Seeking to determine how Sallām's ekphrastic account translates into the "cartographic being," Zadeh addresses a small selection of world maps. Although his survey is cursory and incidental to the larger purpose of tracking Gog and Magog, Zadeh is insightful on "the power of geographical discourse to collapse the vast distance of space and time before the eyes of the readers and viewers who behold the wonders of the world in the highly transportable capital of mimetic reproduction."[14] For Touati cartography is but a small part of the spectacle of the world that geography feeds and the love of travel motivates.[15]

In the context of her analysis of regionalism as a category of the discourse on place, Antrim examines the construction of boundaries and the location of some cities on a smattering of maps from the Balkhī school (three of the Arabian Peninsula, one of the Mediterranean, two of Fars, and one of Kirman) from the late eleventh and the mid-fifteenth centuries.[16] Nowhere does Antrim list the dates of the manuscripts from which these maps are culled, so readers are not aware of the four-century gap in their construction. Nor does she account for variations in place and space over time. Instead Antrim claims that these maps have "a strong relationship with the tenth-century originals despite the collaborative process by which they were produced and reproduced over the centuries," even though the tenth-century originals are no longer extant.[17]

Miquel devotes the most space to analyzing the KMMS maps in a section of his chapter "La terre indivise." But he too falls prey to the oversight of provenance and dating and does not provide information on the manuscripts whose maps he studied. Nor does he address the internal variations of form and place and the range of the tradition from the eleventh to the nineteenth century. Instead his is a vague and impressionistic overview of the maps from one manuscript that he does not even identify.[18] For extensive visual images of the maps, we still have to rely on two outdated classics in the field.

I speak here of the two major efforts in the first half of the twentieth century: Konrad Miller's six-volume *Mappae Arabicae* (1926–1931) and Youssouf Kamal's five-volume *Monumenta Cartographica Africae et Aegypti* (1926–1951). Both Miller and Kamal focus on cataloguing and reproducing cartographic material instead of analyzing its content. They make significant errors by dating the maps as contemporaneous to the original cartographers to whom the maps were attributed rather than according to the date of the manuscript copy. This is a serious oversight since not a single KMMS manuscript is extant from the time of the original authors. This results in the misattribution of mid-nineteenth-century copies to the tenth century.[19] Kamal's work focuses on ancient Greek and medieval European carto-geography, paying only secondary attention to the Islamic maps and their geographical texts. Miller, on the other hand, focuses exclusively on the Islamic mapping tradition. His *Mappae Arabicae* is to date the only available extensive reprint of the Islamic maps.[20] Sadly, his valiant attempt to bring this material to light suffers from serious dating errors, as already mentioned, as well as inaccurate Arabic transcriptions. Last but not least, the black-and-white reproductions conceal the significance of color.

Subsequent to Miller and Kamal, the Belgian scholar J. H. Kramers published a series of articles on the subject. In 1932 he published "La question Balḫī-Iṣṭaḫrī-Ibn Ḥawḳal et l'Atlas de l'Islam," devoted to the question of the original author of the KMMS cycle of maps. In this article, Kramers includes six pages on the picture cycles from four KMMS manuscripts detailing the variations between maps of the world, the Persian Gulf/Indian Ocean, the Maghrib, Egypt, and the Mediterranean. This represents the most extensive iconography ever conducted on these maps until my work. In 1938 Kramers penned the first *Encyclopaedia of Islam* entry on *djughrāfiyā* in which he briefly mentions the maps. Also in 1938 Kramers revised Michael Jan de Goeje's Arabic edition of Ibn Ḥawqal's *Kitāb ṣūrat al-arḍ* (Book of the Picture of the Earth), which he translated into French in 1964.[21] To Kramers's credit, he was the first scholar to recognize the integral nature of the maps and text, so that in his revised Arabic edition of Ibn Ḥawqal's text and in his French translation of the same, he reproduced the maps along with the texts and thus broke with the pattern of his Orientalist predecessors who consistently excluded the maps from their editions of the geographical texts. He held back, however, from analyzing the images and criticized them as simplistic throwbacks to an ancient heritage that had forgotten its "superior" Greek roots.[22]

Kramers's Austrian counterpart Hans von Mžik edited and published a facsimile of the earliest extant medieval Islamic geographical manuscript, Khwārazmī's *Kitāb ṣūrat al-arḍ* (Book of the Picture of the Earth), with its four intriguing maps.[23] In a separate monograph Mžik published a copy of the KMMS maps, a unique Iṣṭakhrī copy housed at the Austrian National Library.[24] Although the maps in this manuscript are among the most schematic and imaginative of the KMMS repertoire, Mžik avoids any pictorial analysis. Instead he devotes 127 pages to the laborious identification of place-names, missing completely the hallmark of this manuscript: the significance behind its misplacement and confusion of sites.[25]

Maqbul S. Ahmad revised Kramers's *djughrāfiyā* article for the second edition of the *Encyclopaedia of Islam* and added a separate entry on maps, *kharīṭa* (the modern Arabic term for "map"). Ahmad repeats the unchanged nineteenth-century narrative of how Arab maps can be traced back to Greek, specifically Ptolemaic influence, and how they began under Abbasid patronage in Baghdad. Ahmad is, however, insightful about the KMMS mapping tradition when he notes that it "exercised a deep influence on later cartographers and became the most popular style of cartography in the Islamic world."[26]

Around the same time as Maqbul Ahmad's contribution, C. F. Beckingham published a singleton article on the Islamic mapping tradition devoted to an in-depth examination of the depiction of Italy on one of Ibn Ḥawqal's maps of the Mediterranean. He proves Kramers and Wiet wrong in one of their identifications and makes the startling discovery that the late twelfth-/early thirteenth-century geographer Yāqūt al-Ḥamawī relied on Ibn Ḥawqal's map *not* text for his own opus magnum (*Mu'jam al-buldān*/Compendium of Countries). Insightfully, Beckingham asserts that "when Yāqūt writes '*qāla 'bn* [*sic*] *Ḥauqal*' ['said Ibn Ḥawqal'] he may mean no more than 'Ibn Ḥauqal's map shows.' Yāqūt must in fact have regarded the maps as an integral part of the book, as indeed they are."[27]

In the late 1980s, Fuat Sezgin published his first monograph on the subject of Islamic mapping, *The Contribution of the Arabic-Islamic Geographers to the Formation of the World Map*. This work had the potential to make a significant contribution because it examined the important under-studied subject of the connections between Islamic and European (both medieval and Renaissance) mapping. Instead of analyzing the Islamic maps for their own unique contributions, Sezgin views them as an expansion on the Ptolemaic mapping tradition. He is determined to prove that mimesis in European cartography derives entirely from Islamic models and that the Muslim mapmakers were the first to employ graticules.

In particular, Sezgin asserts that the no longer extant mid-ninth-century silver globe commissioned by the Abbasid caliph Ma'mūn (r. AH 197–218/813–833 CE) was an extremely sophisticated model of the world that paralleled early Renaissance maps and therefore influenced the development of mimetic Renaissance cartography.[28] Sezgin attempts to prove this untenable thesis by assembling a host of late thirteenth- and fourteenth-century Islamic and European maps as examples without fully explaining the parallels. He argues that the sophisticated maps with grids found in a late fourteenth-century manuscript of Ibn Faḍl Allāh al-ʿUmarī are an exact representation of the missing ninth-century Ma'mūnid globe miraculously reappearing unchanged five centuries later![29] At no point does Sezgin account for how ʿUmarī alone managed to preserve visual examples of the Ma'mūnid map in striking contrast to the bulk of the surviving corpus of Islamic maps.[30]

At the turn of the present century, Sezgin followed up on his effort of the late 1980s with an expanded three-volume study on Islamic mathematical geography and cartography and its contribution to European cartography. The gist of this expanded set is exactly the same as his earlier single volume only in many more words.[31] Sezgin's aim in both projects is to prove that Muslims scholars under the patronage of the Abbasid caliph Ma'mūn had created in the ninth century an extremely mimetic rendition of the world map rivaling and influencing even eighteenth- and nineteenth-century European models. To this end Sezgin throws hundreds of maps at his reader without explaining the visual connections.[32]

In his single-minded quest for mimesis and scientific accuracy and in his inability to grasp that mapping is part of a continuum, Sezgin makes no room for the prolific KMMS world mapping tradition and the significance of its influence on early modern European mapping. Instead, Sezgin, who is deeply steeped in antiquated positivist, nineteenth-century Rankean thinking, sidelines the bulk of the KMMS tradition as "a simple picture, a bottle, a circle or semi-circle, a few straight or curved lines form[ing] the framework."[33] Had Sezgin not placed exclusive reliance on the mythical Ma'mūnid globe and his quest for nineteenth-century mimesis in the ninth century, he would have been able to establish the Renaissance mapping connections that he so urgently seeks. Issues of partisanship aside, one must not ignore Sezgin's immense service to the history of Islamic geography and cartography by collecting and reprinting many crucial articles, translations, and Arabic texts in his *Islamic Geography* series, which at last count was up to volume 318.[34]

Until the mid-twentieth century, the Islamic mapping tradition received only passing, and mostly derogatory, attention within the mainstream of history of cartography studies.[35] This trend was partially reversed by Leo Bagrow's 1951 classic *Die Geschichte der Kartographie* (The History of Cartography). This ten-page exploration of the subject highlights the work of Khwārazmī and Idrīsī but ignores the KMMS tradition. Bagrow is to be lauded for incorporating information on the history of Islamic cartography at a time when it was ignored by the mainstream of the history of cartography.

The ongoing and seminal revision of the field in *The History of Cartography* series, under the editorship of the late J. B. Harley and David Woodward (pioneers in the movement to expand the frontiers of cartographical investigation), has significantly altered the gaze of historians of cartography. Volume 2, book 1, *Cartography in the*

Traditional Islamic and South Asian Societies (hereafter *HC 2.1*), presents the most comprehensive and theoretically innovative accounts of Islamic cartography to date. It opens avenues to other, equally valid forms of cartography such as cosmographical diagrams, Qibla charts and instruments, celestial maps, and even the crucial Muslim contributions to the art of geodesy.[36] In addition, the assistant guest editor of the Islamic portion of the volume, Ahmet Karamustafa, provides a stimulating introduction in which he raises thought-provoking—albeit unanswered—questions on topics such as the relationship between the maps and the texts, their linkage to the Islamic tradition of illustrated manuscripts, and questions regarding audience and patronage. Karamustafa's personal contribution on cosmographical diagrams is a welcome foray into a hitherto ignored tradition.

This volume also features an important article on Qibla maps coauthored by David King, who has devoted his scholarly career to tracing and studying Qibla maps and establishing the importance of the Qibla in Islamic architecture—specifically mosque construction.[37] Since his contribution to *HC 2.1*, King has gone on to propose the innovative thesis that astrolabes should be seen as world maps. In 1989 and 1995, King found two seventeenth-century Safavid scientific instruments (resembling astrolabes) engraved with world maps of a kind previously unknown to the history of cartography that go back to the work of the eleventh-century scholar Bīrūnī or even Ḥabash al-Ḥāsib (Bīrūnī's predecessor by two hundred years). In his 1999 book *World-Maps for Finding the Direction and Distance to Mecca*, King theorizes that the technology for making these instruments was revived in Timurid times and continued by Safavid scientists.[38]

The three articles by Gerald R. Tibbetts in *HC 2.1* on the central topic of medieval Islamic mapping leave much to be desired, however. They are a reiteration of earlier secondary literature on the subject. Tibbetts focuses on veracity, authentication, description, and the classification of the cartographers into the problematic Iraqi versus Balkhī schools—successfully dismantled by Heck.[39] In his first article, "The Beginnings of the Cartographic Tradition," Tibbetts relates the familiar story of the rise of science in the Muslim world. In his second piece, "The Balkhī School of Geographers," he employs a confusing classification scheme based on J. H. Kramers's seventy-year-old criticism of Konrad Miller's eighty-year-old work.

Through my own subsequent perusal of the original manuscripts, it has become clear that Tibbetts did not reexamine the maps and their manuscripts in situ. As a result, many manuscripts are not correctly dated or are erroneously listed as including or not including maps. Tibbetts does attempt a limited iconographic exercise, but even for this he relies on the seriously outdated work by Kramers.[40] The result is that Tibbetts develops a flawed stemma for the KMMS manuscripts. In his third and final article, "Later Cartographic Developments," Tibbetts discusses a wide variety of pieces without linking them to the larger Islamic mapping tradition or explaining their development. Some of the manuscripts in this chapter, such as the Bīrūnī copy of *Kitāb al-tafhīm*, Tibbetts dates to the lifetime (420/1029) of the author. I have personally reexamined this manuscript and determined that it is not from the lifetime of Bīrūnī; rather it is a copy several centuries removed from the original. Tibbetts does not address the meanings behind the variations that crept into the maps via the later copyists or the thought-provoking questions that the guest editor, Ahmet Karamustafa, raises in his introduction. Still it must be said that in spite of these flaws Tibbetts's articles are the most up-to-date and comprehensive reference on the subject of Islamic cartography available for those keen to identify the location of the maps and relevant secondary source work.[41]

Ralph Brauer's 1995 *Boundaries and Frontiers in Medieval Muslim Geography* warrants mention. Brauer delves into the semantics of the two words in Arabic for "boundary/frontier," *ḥadd* and *thughūr*. Enlisting the support of medieval Islamic maps, and geographical texts, as well as modern maps, to prove that *ḥadd* was a reference to internal boundaries within the Islamic world and that *thughūr* was a reference to the frontiers that separate Muslim lands from those of non-Muslims, Brauer produced what is to date one of the most innovative studies involving medieval Islamic carto-geography.[42]

In 2001 Jonathan Bloom published a seminal study called *Paper before Print*. As part of his discussion on the impact of paper on the Islamic world from the eighth century onward, Bloom examines maps. His thirteen-page overview of the Islamic mapping tradition is the best short introduction on the subject to date. Bloom recognizes that the maps were not meant for mass consumption or as aides for navigation. Among his insights are telling quips, such as Idrīsī's maps were probably so original "because as a novice in the field, he was unencumbered by generations of cartographic and geographical scholarship."[43] Rare is the Islamic art historian who examines maps. This is a welcome addition on many fronts.

At the turn of the twenty-first century, Emilie Savage-Smith, an expert on Islamic medical manuscripts, discovered a new manuscript with maps called the *Book of Curiosities*. This resulted in an extensive website with translations, a series of articles, and a jointly edited book with Evelyn Edson, *Medieval Views of the Cosmos*.[44] Yossef Rapoport, Savage-Smith's former assistant on the Bodleian's *Book of Curiosities* project and expert on medieval Islamic divorce law, published a series of articles, "View from the South" and "Reflections of Fatimid Power," that are insightful studies on the maps in the *Book of Curiosities*.[45]

In 2005 Brill published Cyrus Alai's *General Maps of Persia: 1477–1925*. Alai's primary interest is modern maps.[46] As such, his book touches briefly on a few Iṣṭakhrī maps without insight. Sonja Brentjes steps in to fill the void with her 2009 *International Encyclopedia of Human Geography* entry, "Cartography in Islamic Societies." In this short but far-ranging piece, Brentjes goes from the medieval Islamic mapping tradition to twentieth-century surveying and cartography in Iran. Along the way, she covers all forms of mapping from KMMS models to hajj and Qibla maps to Ottoman examples and the debate about the earliest maps to use grids.[47] Testament to the rejuvenated interest in the subject, Ahmad Nazmi tackles the geographical corpus in *The Muslim Geographical Image of the World in the Middle Ages*. Nazmi presents a lucid overview of the subject with a useful listing of the key sources. It is unfortunately mired in the nineteenth-century "naturalistic" Rankean mode of thinking. He fails to take into consideration contemporary theoretical literature and makes some untenable arguments about the material. Nazmi's limited understanding of the maps is demonstrated in a small and confused final chapter.[48]

Of late the most significant contributions to the study of Islamic geography and maps have been made in the Francophone world by Christophe Picard and Jean-Charles Ducène. Picard has written extensively on the Mediterranean between the Latin and Islamic worlds and the Mediterranean's Umayyad heritage as well as on the Muslim Atlantic. He does not, however, address Islamic cartography at length. Instead Picard works primarily with the Arabic geographical and historical texts and legal documents.[49] Ducène, on the other hand, has written a number of articles on the full range of Islamic cartography, from discovering a new nineteenth-century Iṣṭakhrī manuscript and querying the title of KMMS works to analyzing the Nile in Ibn Ḥawqal manuscripts, Indian Ocean islands in medieval Arabic sources, the depiction of Africa in KMMS maps, France in Idrīsī's works, the relationship between Sufism and Islamic cosmography, and the unique circular map in Ibn al-Qāṣṣ's seventeenth-century manuscript. His books, however, focus on full-length translations of geographical texts such as that of Bakrī.[50]

Andreas Kaplony, a specialist in Arabic papyri, has devoted considerable attention to studying and understanding a very unusual world map by Kāshgarī showing the layout of Turkic tribes in Central Asia. His work is best encapsulated in the innovative essay "Comparing al-Kāshgarī's Map to His Text: On the Visual Language, Purpose, and Transmission of Arabic-Islamic Maps." This article is part of a book that Kaplony jointly edited with Phillipe Forêt in 2006, *Journey of Maps and Images on the Silk Road*, which includes other articles on Islamic mapping.[51] Subsequently, the Korean scholar Hyunhee Park has opened the door to Sino-Islamic mapping comparisons with her book *Mapping the Chinese and Islamic Worlds*, and Pınar Emiralioğlu briefly addresses medieval Islamic influences in *Geographical Knowledge and Imperial Culture in the Early Modern Ottoman Empire*.

Since the publication of *HC 2.1*, Islamic maps have received regular mention in general coffee table books on the history of cartography, such as Peter Barber's 2005 *Map Book* and Jeremy Harwood's 2006 *To the Ends of the Earth: 100 Maps That Changed the World*, as well as in collections of theoretical essays, such as the volume edited by James Akerman, *Maps: Finding Our Place in the World*. In the 2012 *A His-*

tory of the World in 12 Maps, Jerry Brotton includes one Islamic map—predictably an Idrīsī's world map.[52] Book-length manuscripts devoted exclusively to the subject of Islamic maps are still rare even though book covers sporting a striking map from the medieval Islamic mapping tradition have become all the rage.[53]

Trends in Map Theory and History

History of cartography, a young field like that of Islamic art history, remained mired in the positivist mode of analysis until the early 1980s when J. B. Harley led a charge to alter the mode of understanding and viewing maps.[54] Invoking the work of Roland Barthes, Robert Darnton, Jacques Derrida, Terry Eagleton, Michel Foucault, E. H. Gombrich, Frederic Jameson, Thomas Kuhn, W. J. T. Mitchell, Erwin Panofsky, Edward Soja, Yi-Fu Tuan, Hayden White, and other cutting-edge modern and postmodern theorists, Harley pushed at the frontiers of traditional map analysis by arguing for an interdisciplinary, metatheoretical approach.[55] In "Deconstructing the Map," Harley brazenly accuses historians and practitioners of cartography of being stuck "in a tunnel created by their own technologies without reference to the social world . . . unable to situate their maps within the discourse of cartography."[56] Arguing for a deconstructive approach that reads between the lines of the map just like a text, Harley asserts that it is only through this mode that we will begin to see that "cartographic facts are facts only within a specific cultural perspective."[57] Harley was interested in the links of social power that enable "the way in which maps become active agents, images taking on a social life of their own, helping to impose the reality they pretend to mirror."[58]

The opening of the study of historical cartography to new forms of discourse brought about an explosion of innovative literature. Harley and David Woodward collaborated to prepare a revamped reference for the field in the form of the ongoing *History of Cartography* series (six volumes at present). In volume 1, *Cartography in Prehistoric, Ancient, and Medieval Europe and the Mediterranean*, the editors established a new definition for "map" that has since been adopted widely as the working definition of what constitutes a map: "Maps are graphic representations that facilitate a spatial understanding of things, concepts, conditions, processes, or events in the human world."[59]

HC 2.1, discussed earlier, broke new ground for Islamic and South Asian cartography, as did the subsequent volumes *Cartography in the Traditional East and Southeast Asian Societies* (vol. 2.2, 1995); *Cartography in the Traditional African, American, Arctic, Australian, and Pacific Societies* (vol. 2.3, 1998), and, most recently, *Cartography in the European Renaissance* (vol. 3, 2007). The latter was published posthumously under the stewardship of Matthew Edney following the untimely deaths of first J. B. Harley and then David Woodward.

Since Harley's death, Christian Jacob has taken over as one of the leading theorists in the field. His work focuses primarily on the analysis of space in the classical Greek and Hellenistic cartographical imagination, although his questioning spans diverse subjects such as what constitutes a map, maps of the moon, the power of maps, and metacartography.[60] Other innovative exposés in the history of cartography include Mark Monmonier's *How to Lie with Maps*,[61] Peter Gould and Rodney White's cognitive investigations in *Mental Maps*, Denis Wood's *The Power of Maps*, Geoff King's *Mapping Reality*, David Turnbull's imaginative interpretation of aboriginal maps in *Maps Are Territories*, and Denis Cosgrove's corpus.[62] These are but a few of the epistemologically groundbreaking inquiries that revise our understanding of maps and the history of cartography.[63]

Along with a re-inquiry into the epistemology of maps, interdisciplinary attention to maps has also been growing. The architectural historian Lucia Nuti, for instance, uses maps to assess the vision of the Renaissance as represented in architectural plans, and the art historian Karl Whittington examines the legacy of Opicinus de Canistris from the perspective of the medieval cartographic imagination.[64] In *The Self-Made Map*, Tom Conley seeks out the cartographic impulse in early French Renaissance literature. He argues that a new sense of self, in relation to space, emerged from the explosion of mapping that followed the discovery of the New World. This in turn generated a unique form of "cartographic writing" that makes of Cervantes, for example, "a cartographer of both the post-medieval Mediterranean as well as the pastoral novel."[65] Scholar of English Daniel Birkholz has taken the study of maps to new heights with his provocative and deservedly prize-winning whodunit on two long-lost medieval English *mappaeregni*.[66] In *Eccentric Spaces*, Robert Harbison takes the reader on a voyage through topographical and architectural fictions and discusses maps as miniatures of the mind. Postmodern geographers such as James Duncan and Derek Gregory look to landscape painting and modern social theorists for maps of the imagination, Martin Lewis and Kären Wigen conduct a metageographic critique of the myth of continents, and the phenomenologist Edward Casey opens up dialogue between maps and landscape paintings in *Representing Place*.[67]

Although the domain of history has been slow to deny the primacy of written text, the use of maps and other visual images under the rubric of material culture has been on the rise since the visual turn of the 1990s. Breaking history's hallowed textual ground with works such as *The Embarrassment of Riches* and *Landscape and Memory*, Simon Schama wrote richly textured vignettes around a variety of images—mostly paintings, some sketches and photographs, as well as the occasional map—supplemented by textual sources, to serve up sumptuous historical banquets.[68] The enormous popularity of Schama's use of visual images to generate exciting speculative history has paved the way to greater acceptance of this approach. Famed for her work in medieval European religious history, Caroline Bynum harnesses imaginative religious imagery of the body of Christ, the female body, death, and resurrection to buttress metaphors as well as theological and philosophical discourse in *Fragmentation and Redemption* and *Resurrection of the Body*.

Veteran Islamic historian Richard Bulliet employs a wide range of visual images and examples of material culture to document the evolution of camel saddles and the disappearance of the wheel in the medieval Middle East in his seminal study *The Camel and the Wheel.*[69] Jere Bacharach, author of *Islamic History through Coins*, is well known for his keen eye for Islamic material culture and its impact on Islamic history. Carole Hillenbrand's monumental study *The Crusades: Islamic Perspectives* interweaves image and text to create a richly textured understanding of the period. In *The Ottoman Age of Exploration*, his pioneering bid to prove that the Ottomans actively partook in the Age of Exploration, Giancarlo Casale mines data from maps and images alongside textual sources. In general, though, historians of the Middle East have been slow to adapt to the visual turn, caught up as they still are with the struggle over the linguistic one.[70]

The closest parallel to my own work is that of historian Sumathi Ramaswamy, who uses maps and visual material as a gateway to history. Comparing nineteenth-century European and twentieth-century Tamil efforts to map the continental landmass of Lemuria, Ramaswamy argues against the existence of place. She examines the way in which cartographic conversion from the "paleo-space" of the European colonial imagination gives way to the "lived place" of the postcolonial Tamil imagination in *Lost Land of Lemuria.*[71] Ramaswamy's recent work, *The Goddess and the Nation*, picks up where she left off with "Lemuria" by analyzing the similarity of form between the late nineteenth-century Indian goddess Bharat Mata (Mother India) and the shape of the modern state of India. She regales her readers with a convincing panoply of maps and maplike images.

Above all rises the work of global cultural historian Felipe Fernández-Armesto, who harnesses maps and other visual data and objects of material culture along with anthropological, archaeological, and sociological data for his expansive world history projects that range from *A History of Food* to *Pathfinders*. In the latter, Fernández-Armesto uses maps to understand the role that pathfinders played in the divergences and convergences that brought us from a globalized Stone Age to our present era.[72]

Michael Wintle's book *The Image of Europe* is an indicator that the use of maps as part of the visual historical record is gaining ground. Examining conceptions of Europe from the classical period through the twenty-first century, Wintle reveals European anxieties of self-definition. Through his innovative use of Chinese merchant shipping routes on the Selden Map to show that the rise of London in the mid-sixteenth century owed as much to East Asian trading networks as it did to the rise of Atlantic settlements, Robert Batchelor has given us one of the most innovative histories involving a map. His book *London* is destined to be an "off-the-map" classic.

The interest in textualizing maps is growing exponentially as more people realize their value for revealing a hidden angle of historical perception of place and space embedded within the constructs of time, milieu, and patronage. It is in this context that I insert an exploration of medieval Islamic cartography in the hope that it opens up new avenues for working with this rich body of material culture.

3

A Sketch of the Islamic Mapping Tradition

In keeping with the general predilection to evaluate maps according to their accuracy, the best-known examples of Islamic maps are those that are the most mimetic. Famous for precisely this reason is the work of the twelfth-century North African cartographer Sharīf al-Dīn al-Idrīsī (d. AH 560/1165 CE), whom the Norman king Roger II (1097–1154) commissioned to produce an illustrated geography of the world: *Kitāb nuzhat al-mushtāq fī ikhtirāq al-āfāq* (Book of Pleasant Journeys into Faraway Lands).[1] There is no question that the maps that accompany the copies of Idrīsī's manuscript are extremely detailed representations of the world, well ahead of their time (fig. 3.1). Not only are the Idrīsī maps ranked among the most mimetic world maps of the later Middle Ages; they also include detailed regional maps that show an astounding depth of understanding of the topography of the greater Mediterranean region.

One of the lingering issues with Idrīsī's work, however, is that there are no extant examples of his work from his time period of the twelfth century. The earliest extant Idrīsī man-

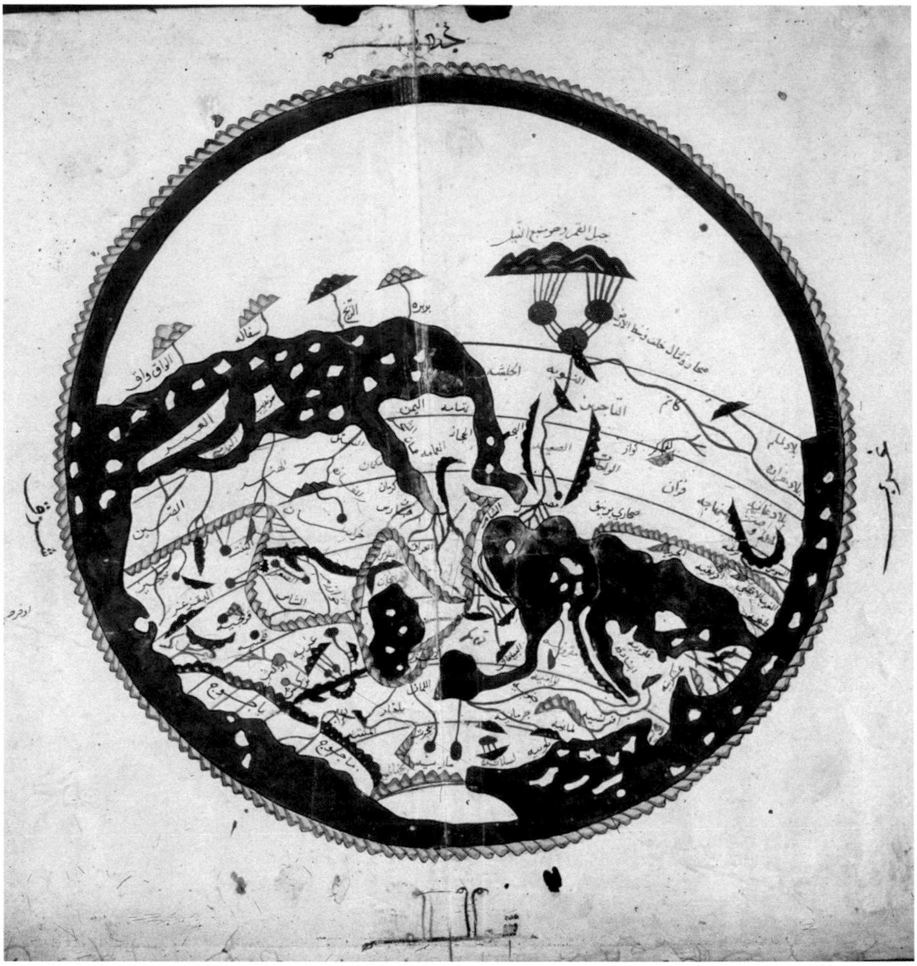

Fig. 3.1. Typical al-Idrīsī-type mimetic map of the world. 960/1553. Gouache and ink on paper. Diameter 23 cm. Courtesy: The Bodleian Library, University of Oxford. Ms. Pococke 375, fols. 3b–4a.

uscripts are from the fourteenth century.[2] The other issue with Idrīsī's work is that it is a unique tradition not representative of the bulk of medieval Islamic mapping. Thus maps that are attributed to him cannot be used as a source of insight into the cartographic worldview of medieval Muslims. Idrīsī's work can be used to illumine the worldview of the milieu surrounding Roger II in Norman Sicily with insight into the North African—specifically Tunisian—ambit. Even with this there are issues since the four extant Idrīsī manuscripts are productions of more than a century after his death, well after the period of Islamo-Norman scholarly interaction. We cannot, therefore, ascribe with surety the worldview expressed in these copies as being representative of the Norman-Muslim world of the twelfth century. As outstanding as Idrīsī's work is for inserting a heightened cartographic mimesis into the late medieval Islamic mapping repertoire, it needs to be addressed with these cautions in mind. It is for this reason that this book—in contradistinction to other work on medieval Islamic mapping—does not hold up the work of Idrīsī as the hallmark of the medieval Islamic mapping tradition.

Another well-known Muslim cartographer is the sixteenth-century Ottoman naval admiral Pīrī Re'īs (ca. 875–961/1470–1554). Famous for the earliest extant map of the New World,[3] Pīrī Re'īs and his surprisingly accurate early sixteenth-century map of South America (919/1513) has been the subject of many a controversial study (fig. 3.2).[4] In keeping with the emphasis placed on Western products in the field of the history of cartography—especially maps from the Renaissance onward—scholarship on Pīrī Re'īs's map has focused on its connections with early modern European cartography.[5] Like Idrīsī's work, Pīrī Re'īs's work is unique. While one can argue that Pīrī Re'īs's work has connections with the medieval Islamic mapping tradition, it cannot be used to study the medieval Islamic cartographic worldview.

The striking mimesis of these cartographic traditions has caused the work of Idrīsī and Pīrī Re'īs to be elevated above the rest of the Middle Eastern mapping corpus in contemporary scholarship.[6] Aside from the problems of attribution that abound with these two cartographers, scholarly focus on this *more mimetic* end of the Islamic mapping tradition has occluded an enormous body of maps that were far more popular in the medieval and early modern Islamic world than the work of Idrīsī or Pīrī Re'īs.

The same could be said of the *Book of Curiosities* manuscript, which has received considerable attention since its acquisition by the Bodleian Library of Oxford in 2002. This manuscript contains a medley of hybrid maps, some unique and others, such as the map of Sind, that reflect the influence of the KMMS tradition. The square world map is unusual and merits further investigation (fig. 3.3). The Bodleian's *Book of Curiosities* is, like the maps of the Idrīsī manuscripts, plagued by dating issues. Although the text of the manuscript has been dated to the eleventh century by Emilie Savage-Smith and Yossef Rapoport, the maps reflect a late twelfth-/early thirteenth-century provenance.[7] While unique manuscripts such as the *Book of Curiosities* are intriguing additions to our repertoire of the Islamic cartographic tradition, we must bear in mind that they are not representative of the popular mapping tradition that was widespread in the medieval Islamic world.[8]

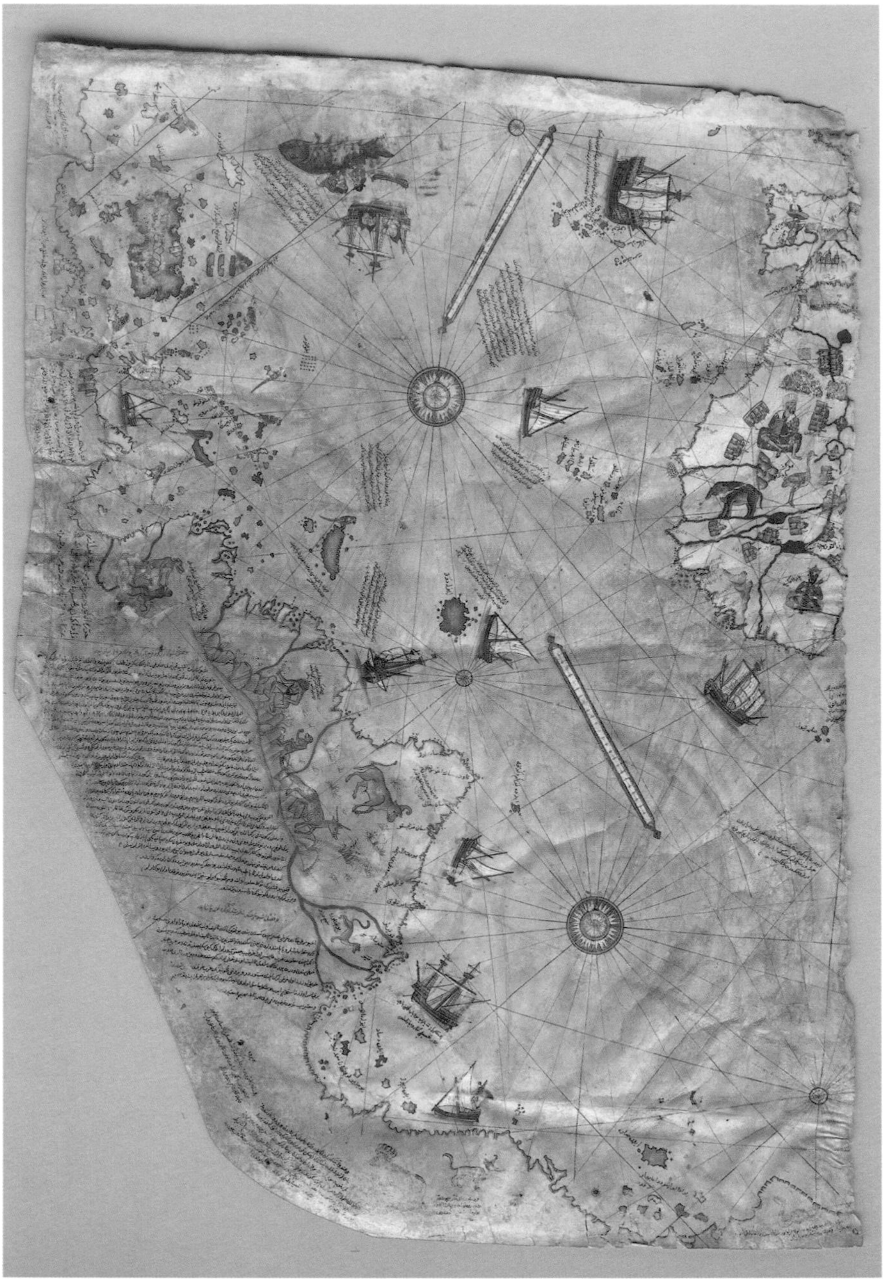

Fig. 3.2. Pīrī Reʾīs world map fragment showing South America. 919/1513. Gouache on parchment. 90 × 63 cm. Courtesy: Topkapı Saray Museum, Istanbul. Revan 1663m.

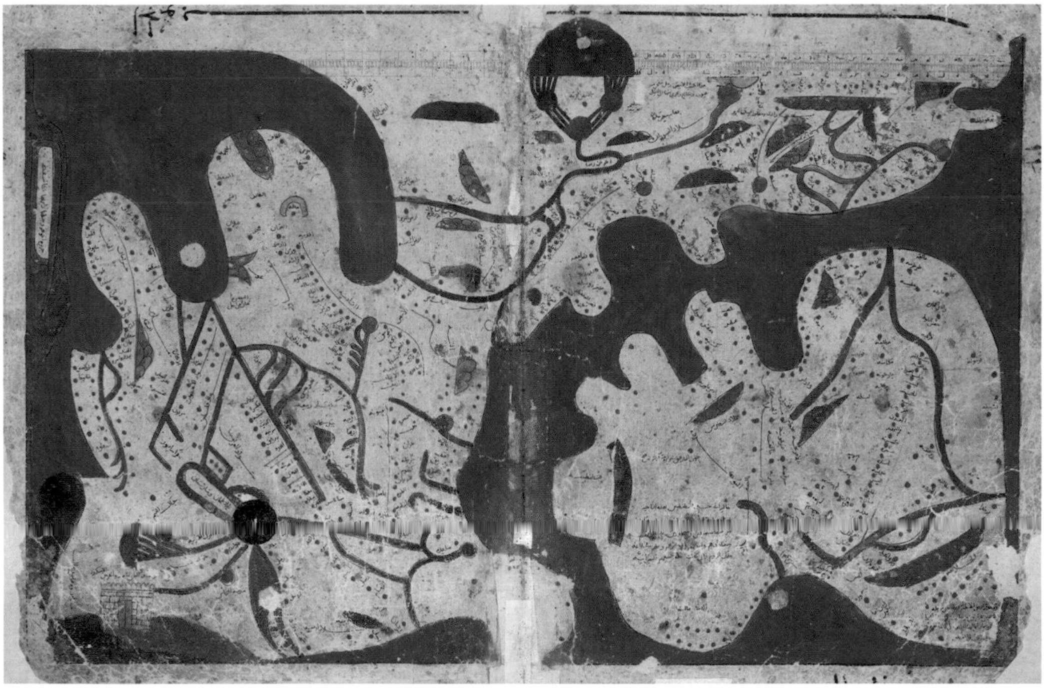

Fig. 3.3. *Book of Curiosities* square map of the world. Circa 12th–13th century. Gouache and ink on paper. 47 × 29 cm. Courtesy: The Bodleian Library, University of Oxford. Ms. Arab. c. 90, fols. 23b–24a.

Islamic Maps Abound

Based on numerous years of research in Oriental manuscript libraries in North America, Europe, and the Middle East, I have determined that maps occur in a wide variety of Islamic texts and contexts. One common location for maps is the so-called *ʿajāʾib* (or "wondrous") literary tradition, which includes descriptions of flora, fauna, architecture, and other wonders of the world, and usually incorporates world and cosmographic maps. Best known of this genre is the work of the thirteenth-century Iranian writer Zakariyāʾ ibn Muḥammad ibn Maḥmūd Abū Yaḥyā al-Qazwīnī (d. 682/1283), whose work *ʿAjāʾib al-makhlūqāt wa-gharāʾib al-mawjūdāt* (The Wonders of Creatures and the Marvels of Creation) focuses on the wonders of the world—real and imaginary. Although Qazwīnī's original manuscript is not illustrated, copies from the late thirteenth century onward—during the lifetime of the author—incorporate illustrations of flora and fauna as well as world maps usually based on the model found in Bīrūnī manuscripts (compare figs. 3.4 and 3.5).[9]

Abū Rayḥān Muḥammad ibn Aḥmad al-Bīrūnī was an eleventh-century polyglot with regional expertise in Iran, Central Asia, and India. He is considered one of the most accomplished scientists of the medieval world. A specialist in mathematics, astronomy, and astrology, Bīrūnī also dabbled in pharmacology, meteorology, and, relevant to our discussion, geography. It is in this context that he wrote in 420/1029 his *Kitāb al-tafhīm li-awā'il ṣinā'at al-tanjīm* (Book of Instruction in the Elements of the Art of Astrology), which is divided into the five subject areas of geometry, arithmetic, astronomy, the astrolabe, and astrology.[10] Under the section of astronomy, Bīrūnī includes a discussion of geography, cosmology, and chronology. It is in the context of this discussion that a most unusual world map appears (fig. 3.5).[11] According to Tibbetts, this map first occurs in a thirteenth-century Arabic copy of Bīrūnī's *Kitāb al-tafhīm* at the British Library (Ms. Or. 8349 copied 635/1238).[12]

This Bīrūnī map shows a single world continent spread like a nibbled pancake on the surface of a voluminous Encircling Ocean. This continent encompasses the lion's share of the Northern Hemisphere and shows Asia, Africa, and Europe without any divisions. The Indian Ocean and the Encircling Ocean are one in the face of a significantly foreshortened Africa. This unusual vision of the world speaks to a locus somewhere in Central Asia or northeastern Iran because it is the central focal point of the image around which the form of the world radiates. What is even more curious is that it is this Bīrūnī-type image of the world that circulates alongside the KMMS vision of the world in copies of encyclopedic medieval Arabic and Persian works in the later Middle Ages.

World maps based on a modified Bīrūnī model appear, for instance, in Yāqūt's thirteenth-century geographical encyclopedia *Kitāb mu'jam al-buldān* (Book Compendium of Lands) (compare figs. 3.4 and 3.5 with fig. 3.7). Although these maps do not at first glance appear to resemble the KMMS mapping tradition (chapter 4), closer examination reveals some parallels, for instance, between the form ascribed to the Mediterranean and the lands arrayed around it. Three major differences stand out between the Bīrūnī and the KMMS world maps models. One is that Africa is foreshortened. This results in the assignment of a much larger area to the Encircling Ocean (Baḥr al-Muḥīṭ) that includes within its girth the Indian Ocean as well. The second is the emphasis on South Asia as implied by the double promontory demarcating Makrān along the coast of Sind and India (Hind) separated by a yawning gulf. The third is the clear indication of the Baltic Sea (Baḥr Warank) as an offshoot of the northernmost section of the Encircling Ocean along with the mention of the "Wiznīk" people or place astride it (see figs. 3.5 and 3.7).[13]

W. Jwaideh and M. Kowalska have both shown that Qazwīnī relied heavily on Yāqūt's *Kitāb mu'jam al-buldān*.[14] Thus, it is no surprise to find that Qazwīnī's *'Ajā'ib*

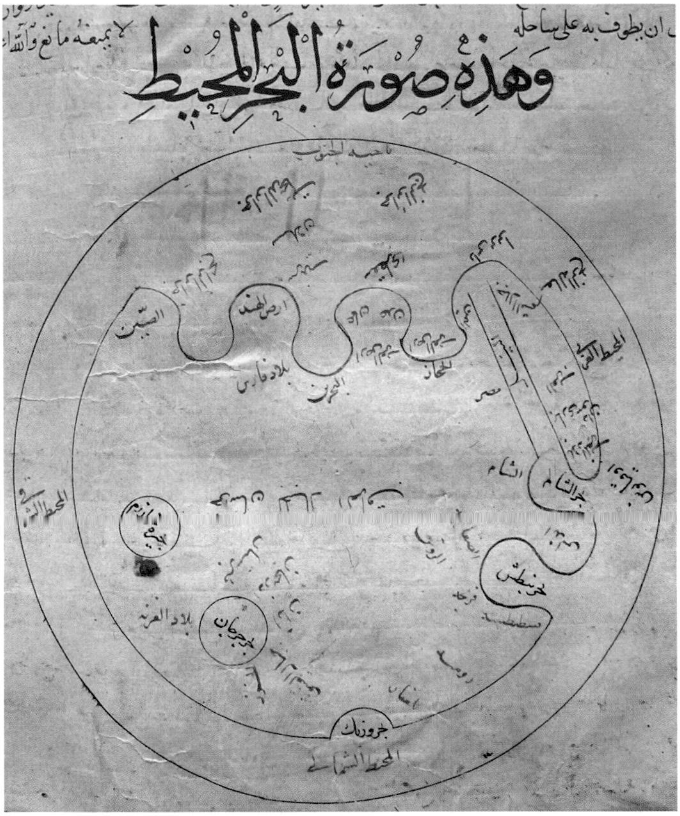

Fig. 3.4. Typical al-Qazwīnī world map based on the al-Bīrūnī model. 14th century. Red, blue, and black ink on paper. Diameter 16.75 cm. Courtesy: The Institute of Oriental Manuscripts, Russian Academy of Sciences, St. Petersburg. Ms. E7, fol. 63b.

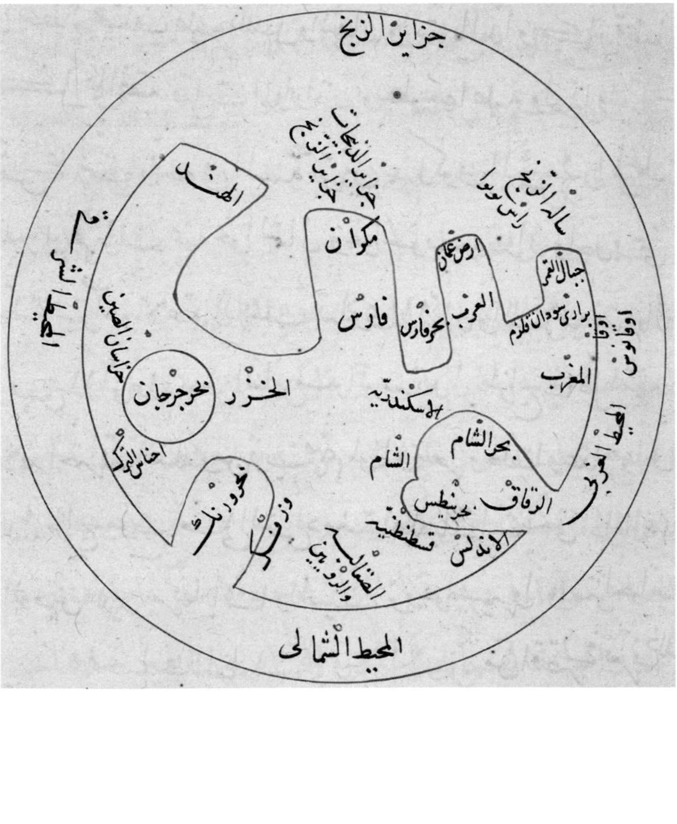

Fig. 3.5. Earliest extant al-Bīrūnī sketch map of the world showing distribution of land and sea. 635/1238. Red and black ink on paper. Diameter 9.5 cm. The British Library Board, London. Ms. Or. 8349, fol. 58a. © The British Library Board.

Fig. 3.6. Incomplete KMMS-type world map in an al-Qazwīnī manuscript. 790/1388. Gouache and black ink on paper. Dimensions unknown. Courtesy: Bibliothèque nationale de France, Paris. Supplément Persan 332a, fol. 58a.

al-makhlūqāt contains a Bīrūnī-type world map. Versions from the fourteenth century onward even contain multiple world maps. In such cases, the Bīrūnī-type world map (shown in fig. 3.4) accompanies a map of the KMMS tradition (shown in fig. 3.6), suggesting the importance of both to the general public. The large number of extant *'ajā'ib* manuscripts containing these world maps indicates that, at least from the fourteenth century onward, both the Bīrūnī and KMMS models had a wide audience. (For details on the KMMS world mapping tradition, refer to chapter 4).

ناحية الجنوب

واختلفوا في سبب ملوحة آما البحر فزعم قوم انه لماطال مكثه وابحت الشمس
عليه بالاحراق صار مرا لما واحتدب الهوا اما الطف من اجزايه فهوبقيه
ما صعنه الارض من الرطوبة فلط وزعم احرون ان ملح البحر وفاع نما
البحر تلد لك صار مرا زعافا وزعم بعضهم ان الماء الاستحالات فطعمه كل

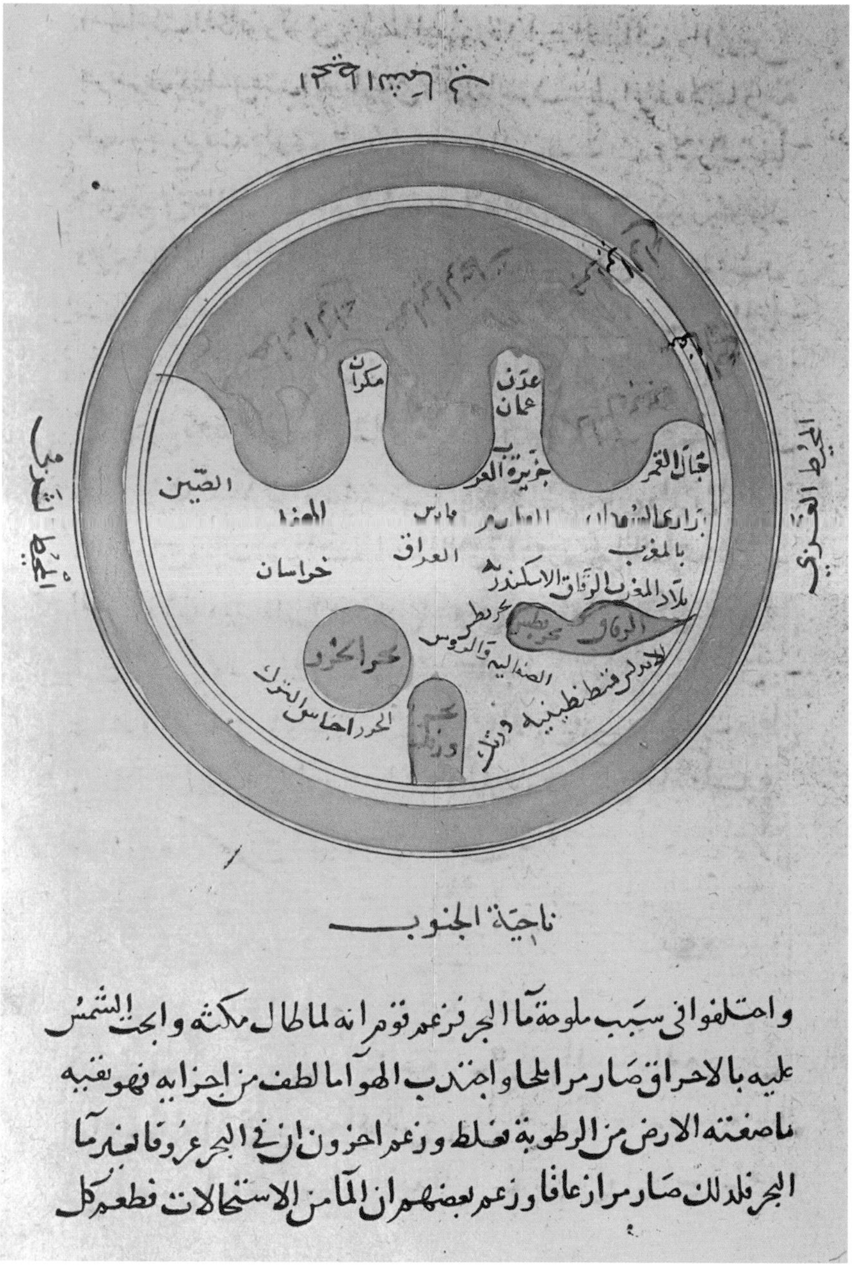

Fig. 3.7. Typical Yāqūt world map variation of al-Bīrūnī model. 827/1424.
Blue gouache and red ink on paper. Diameter 9.5 cm. Courtesy: Topkapı
Saray Museum, Istanbul. Ahmet 2700, fol. 16b. Photo: Karen Pinto.

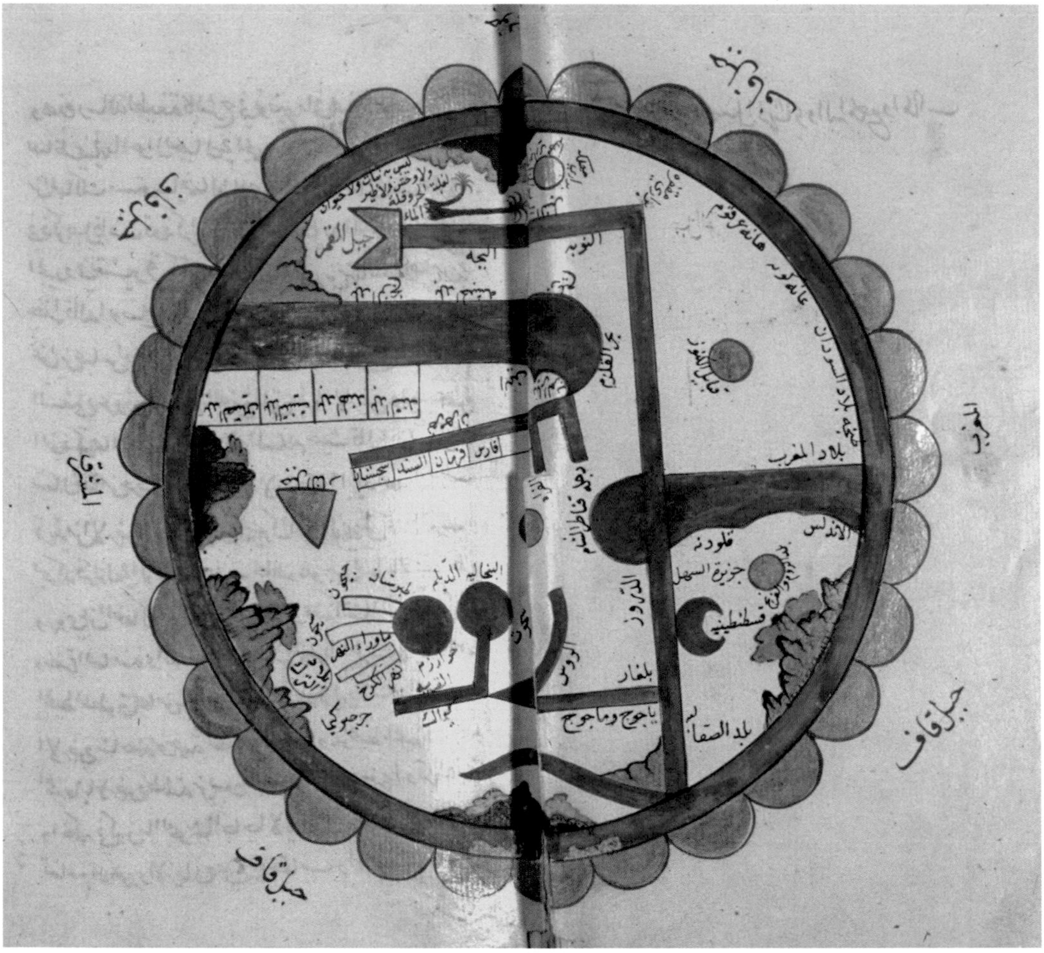

Fig. 3.8. Typical Ibn al-Wardī world map based on KMMS model. Late 17th century. Gouache and ink on paper. Diameter 14.4 cm. Courtesy: Leiden University Libraries. Cod. Or. 158, fols. 3b–4a.

Even more popular than Qazwīnī's *ʿAjāʾib al-makhlūqāt* was Ibn al-Wardī's *Kharīdat al-ʿajāʾib wa farīdat al-gharāʾib* (The Unbored Pearl of Wonders and the Precious Gem of Marvels), a unique pocket-book encyclopedia tradition that incorporated *ʿajāʾib* lore. Ibn al-Wardī (d. 861/1457) manuscripts include at least one world map per copy, usually based on the KMMS model or some variation of it (fig. 3.8),[15] along with other images that could be interpreted as cartographic, such as a Qibla map (way-finding diagram and instrument for pinpointing the direction to Mecca for the purpose of prayer;

see fig. 7.17), and an inset map of Qazwin and other cities. This is a contested series because there is no clear consensus on Ibn al-Wardī being the original author.[16] Regardless of attribution, every copy that I have examined contains at least one world map. Multiple copies of this manuscript, ranging in date from the fifteenth century to the nineteenth century, are found in virtually every Oriental manuscript collection. In Istanbul, particularly at the Sülemaniye library, the copies are so numerous (numbering in the hundreds) that it has been impossible thus far to identify and classify all the world maps that exist in this series. Judging by the plethora of pocket-book size copies that still abound, the *Kharīdat al-ʿajāʾib* was extremely popular in the sixteenth, seventeenth, and even eighteenth centuries. This best seller must have had a significant impact on the popular geographic imagination of the early modern Islamic world.[17] It must be judged as significant that this late medieval Arabic best seller always incorporated, usually within the first four or five folios, a modified KMMS-type world map.

Suggestive of their growing popularity, world maps make their appearance in historical treatises from the thirteenth century onward. Some copies of the famous Islamic historian Abū Jaʿfar Muḥammad ibn Jarīr ibn Yazīd al-Ṭabarī's *Taʾrīkh al-rusul wa-al-mulūk* (History of Prophets and Kings), for instance, include as part of the manuscript's frontispiece a "clime"-type map of the world.[18] These clime maps show the world divided into seven climatic zones from the first clime, which is closest to the equator and the desert, to the seventh clime, which is closest to the lands of the Ṣaqāliba (Slavs) in the north. (For an example of a typical clime map, see fig. 3.9.[19]) The origin of these clime-type maps is unclear. All we know is that they begin to appear in manuscripts after 1200. Tibbetts suggests medieval European mapping influence, citing the diagrams of Petrus Alphonsus from the 1100s and the Macrobian zonal map tradition. (See fig. 6.16 for an example of a Macrobian map.) Tibbetts also acknowledges that the tradition of clime maps could have been started by Bīrūnī under Indic influences and diffused to western Europe via the Muslim world.[20] Some clime maps, such as the one in the British Library's 729/1329 copy (Ms. Or. 3623, fol. 5a) of Qazwīnī's *Āthār al-bilād* (Monuments of the Countries), also referred to as Qazwīnī's *Geography*, contain superimposed outlines of a Bīrūnī-type world map.[21] Manuscripts of the *Muqaddimah* (Prologue) by another well-known historian, Ibn Khaldūn (732–808/1332–1406), sometimes begin with an Idrīsī-type world map. (See fig. 3.1 for an example of this type of map.)[22]

By the fifteenth century, KMMS-type world maps even occur in album collections of calligraphy and miniatures for Timurid rulers, indicating that world maps were sought after for their intrinsic value and not just for the geographical texts accompanying them. Such is the case with the map shown in figure 3.10 prepared circa 1413 for the Timurid prince Iskandar Sultan (d. 1415), who was blinded shortly thereafter for his territorial ambitions. Complete with underlying clime demarcations and

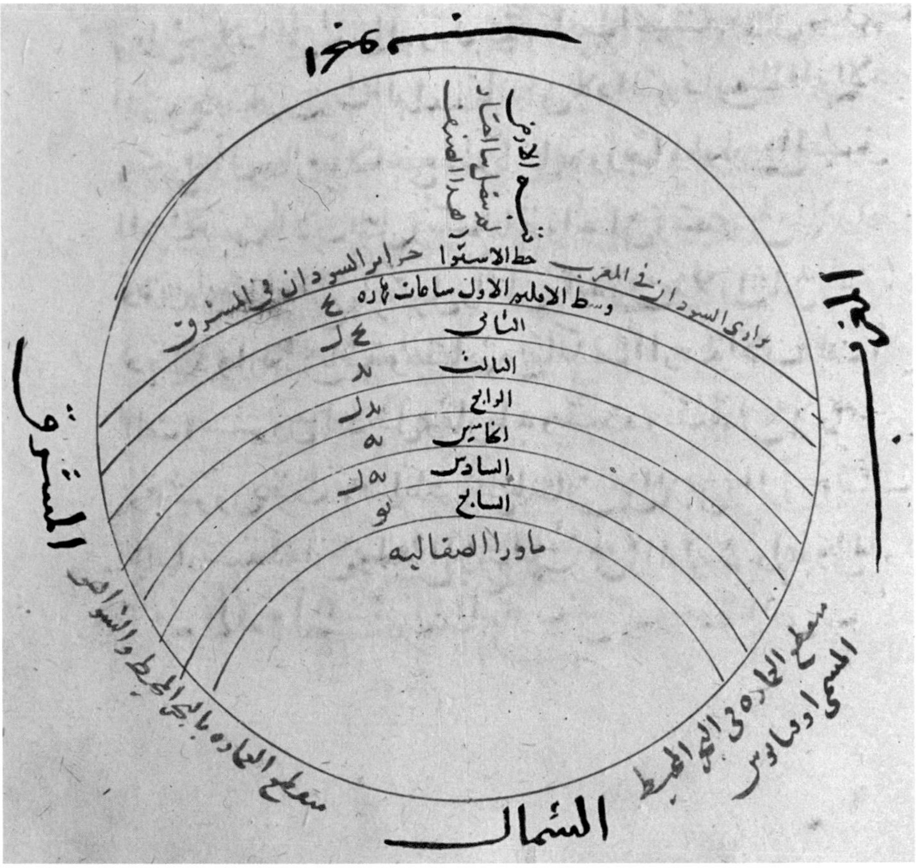

Fig. 3.9. Clime-type map with arched boundary lines from Yāqūt's *Kitāb mu'jam al-buldān* (Compendium of Lands). 827/1424. Red and black ink on paper. Diameter 10.5 cm. Courtesy: Topkapı Saray Museum, Istanbul. Ahmet 2700, fol. 18a. Photo: Karen Pinto.

elements of typical Timurid chinoiserie as communicated by the illumination style of the mountains, this map is based on the KMMS model with one major exception: instead of the customary KMMS world map nomenclature indicating regions and provinces, it lists towns and cities, in particular those of Timurid Iran. Territorial borders are noticeably absent—a testament perhaps to Iskandar Sultan's ambitions. (See chapter 4 for a discussion on KMMS maps.) Instead, we see for the first time on a KMMS map the imposition of a longitudinal grid in the uppermost reaches of the map along with an ever so subtle graticule etched into the paper (invisible in the

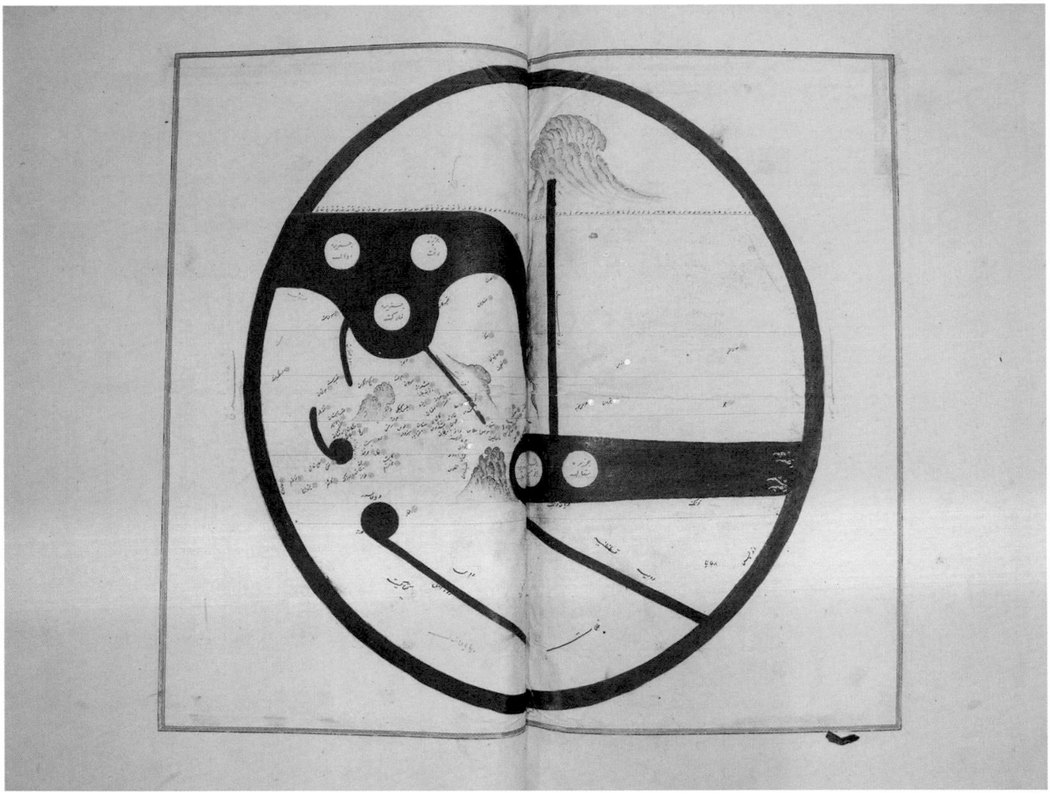

Fig. 3.10. KMMS-type stand-alone world map in a Persian *muraqqaʿ* album made for Iskandar Sultan. This KMMS world map is unusual for its graticule and its emphasis on cities in Iran and Central Asia. Circa 1413. Isfahan. Gouache, ink, and gold on paper. Diameter 32.9 cm. Courtesy: Topkapı Saray Museum, Istanbul. Bağdat 411, fols. 141b–142a.

picture), suggesting that the production of this map was supervised unusually by an astronomer. This reading fits with other material in the manuscript including an abridged section on astronomy, an elaborate double-page elaborately illuminated horoscope, star charts, instructions titled "Knowledge of Astronomy by Means of an Astrolabe," and designs for amulets.[23]

An unusual variant of a KMMS-type world map (fig. 3.11) surfaces as late as the mid-sixteenth century in a gigantic (31.6 m × 79 cm) imperial Ottoman genealogical scroll, referred to as the *Tomar-ı hümayun* (Imperial Scroll). It was started by ʿArifī (Ārif Çelebi), the first court historian (şehnameci) of Süleyman I (r. 1520–1566), continued by Eflātūn, and completed by Selim II's (r. 1566–1574) *şehnameci* (court historian), Seyyid Loḳmān (Luqmān), in 1569.[24] In later iter-

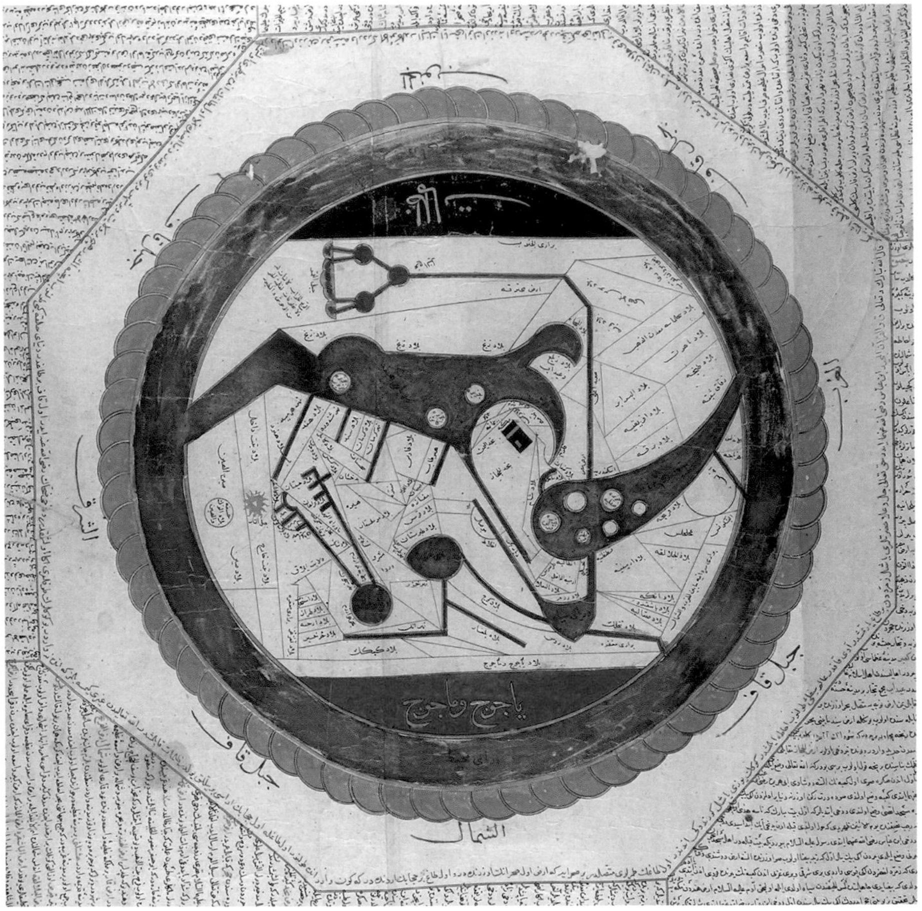

Fig. 3.11. KMMS-type world map from *Tomar-ı Hümayun* (Imperial Scroll) (sometimes referred to as *Zübdetü't-tevārīḫ* [Cream/Quintessence of Histories]). Mid-16th century. Ottoman. Gouache, ink, and gold on paper. Scroll dimensions: 31.6 m × 79 cm. Started by Arifi, continued by Eflatun, and completed by Seyyid Loḳmān/Luqmān in 977/1569. Courtesy: Topkapı Saray Museum, Istanbul. Ahmet 3599.

ations, Loḳmān converts this scroll into codex form.[25] Updated to include the latest ruling sultan, these manuscript renditions are named by Loḳmān *Zübdetü't-tevārīḫ* (Cream/Quintessence of Histories) after the title of the *Tomar*. To distinguish the scroll from the later codex form, the scroll is referred to as *Tomar-ı hümayun*.[26]

Loḳmān chooses not to continue the tradition of incorporating a KMMS-type world map. Instead he excises it from the codex renditions of the *Zübdetü't-tevārīḫ* along with the seven-sphere cosmographical chart that preceded the KMMS map and replaces it with a more mimetic map of the world of the kind found in the *Tārīḫ-i Hind-i Gharbī* (History of the West Indies) series—the earliest version of which was presented to Mehmed III in 1583.[27] It is tempting to see Loḳmān's decision (based perhaps on an imperial request) as a signal that Ottoman imperial flirtation with the classical Islamic tradition was at an end and that a different kind of world map was becoming vogue.[28]

If so, the KMMS tradition in the imperial Ottoman context ends with a bang as the *Tomar-ı hümayun* KMMS world map (fig 3.11) is dramatically painted with green scallops representing the mythical mountains of Jabal Qāf (discussed in chapter 6) encircling the world and its circumambulatory ocean, a bright red for the lands of Yājūj wa Mājūj (Gog and Magog) in the northernmost extremity, and black for the southernmost pole of Ẓulumāt (Darkness). While retaining the underlying traces of the KMMS template, which closely resemble the hybrid Ibn al-Wardī model (fig. 3.8), the interior of the map apportions space with significant variations. For starters, red lines indicating territorial boundaries abound. No longer is the landmass of Africa a sea of white space. Instead it, like the other landmasses, is carved up and apportioned with careful attention to detail. The Mediterranean comes with three extra, albeit unlabeled islands, and the tributaries feeding into the Aral and Caspian Seas are represented comprehensively. Some provinces have been rearranged, new ones added, and a number have been renamed. The central focus of the map is on the Persian Gulf with Mecca just off center and represented usually with a miniature Kaʿba symbol.[29]

From the late twelfth century onward, maps related to the hajj (pilgrimage) make their appearance. These maps are to be found in certificate scrolls confirming the completion of the hajj (fig. 3.12) and in travelogues containing maplike pictures of the holy sites. These too can be read as an indication of the growing demand for visual images of spaces, sacred and otherwise. Eventually, the scope of the images in these pilgrimage scrolls expands into an illustrated hajj manuscript series (fig. 3.13) and a collection of prayer books: *Futūḥ al-ḥaramayn* (Conquests of the Holy Sites), which first appears in the early sixteenth century, and proliferates in a multiplicity of copies thereafter; and the *Dalāʾil al-khayrāt* (Ways of Edification) prayer book that was extremely popular in the eighteenth century, which also includes bird's-eye views of the holy cities of Mecca and Medina. The latter was initiated by a pious Berber scholar and Sufi, Abū ʿAbd Allāh Muḥammad ibn Sulaymān ibn Abī Bakr al-Jazūlī (d. 869/1465), in the fifteenth century and the former by Muḥyī Lārī (d. 933/1526–1527) in the early sixteenth century as a tribute to his Gujerati sultan Muẓaffar. The pocket-book version of these books became so popular that hundreds of copies were produced from the sixteenth to nineteenth centuries.[30]

In tandem with these hajj manuals, a tradition begins of including a glazed Kaʿba tile in mosques adjacent to the *miḥrāb* (prayer niche) showing a schematic maplike

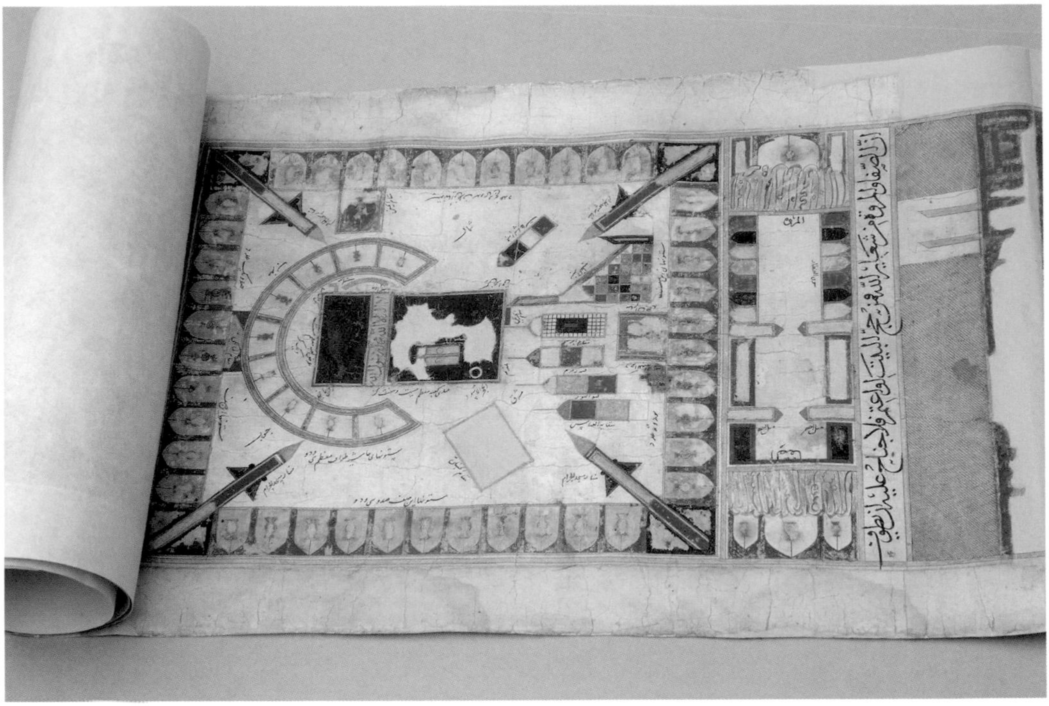

Fig. 3.12. Pilgrimage certificate scroll showing the Ka'ba and the stages of Hajj. 837/1433. Gouache, ink, and gold on paper. Scroll dimensions: 34 × 666 cm. Courtesy: The Museum of Islamic Art, Qatar. MIA MS. 267.

representation of the Ka'ba framed by a rectangle with a series of place-names indicating the direction of prayer (Qibla) from around the world. The tile example shown in figure 3.14 in blue and white luster is typical of seventeenth-century Ottoman ware from Iznik. There are many more examples of these tiles located in mosques and museum libraries around the world.[31]

The inspiration for these Ka'ba tile maps comes from Qibla charts (way-finding diagrams for locating Mecca) that emerged as a result of the Islamic injunction to pray, bury the dead, and sacrifice animals, among other ritual rites, specifically in the sacred direction of Mecca and to perform nonreligious acts, especially those related to bodily functions, deliberately not in that direction. In other words the sacred direction of Mecca governed the lives of Muslims from the earliest years of the rise of Islam in the seventh century, and it is not surprising to find that charts and instruments developed to help Muslims determine the direction to Mecca. What is surprising is that it took

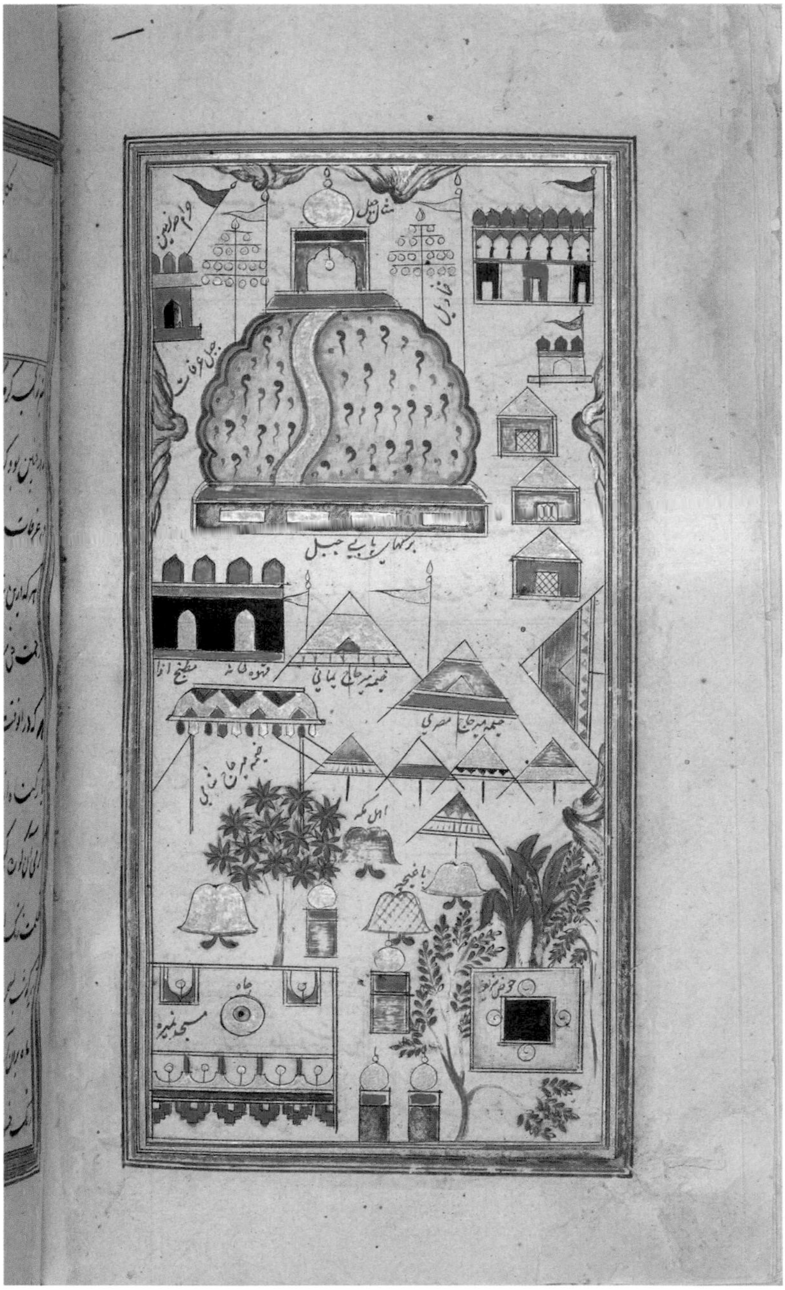

Fig. 3.13. Mount ʿArafāt and the pilgrimage stages around it from *Futūḥ al-ḥaramayn*. 1089/1678. Deccan, India. Gouache, ink, and gold on paper. 15.1 × 7.6 cm. Source: The Metropolitan Museum of Art, New York. Purchase, funds from various donors. Elizabeth S. Ettinghausen Gift, in memory of Richard Ettinghausen, and Louis E. and Theresa S. Seley Purchase Fund for Islamic Art, 2008. 2008.251.

Fig. 3.14. Kaʿba tile with bird's-eye representation of the Great Mosque in Mecca with the Kaʿba slightly off center surrounded by other important structures encased in a keyhole shape. 17th century. Iznik, Turkey. Fritware with underglaze painting. 62.4 × 35.8 × 3.5 cm. Courtesy: The Walters Art Museum, Baltimore. 48.1307.

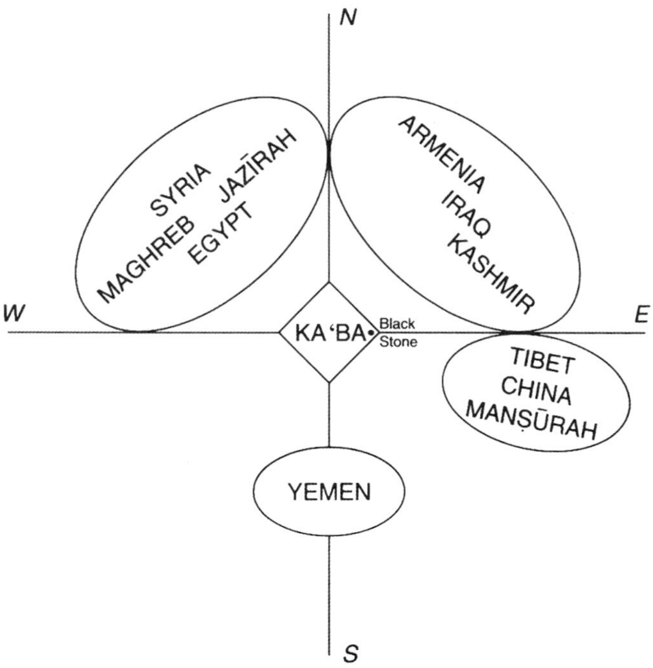

Fig. 3.15. David King's reconstruction of Ibn Khurradādhbih's scheme of sacred geography, in *Kitāb al-masālik wa-l-mamālik*, based on de Goeje's edition, *Liber viarum et regnorum*. Courtesy: David King, professor of the history of science and director of the Institute for the History of Science at the Johann Wolfgang Goethe University in Frankfurt.

so long for these Qibla charts to enter manuscripts and everyday life. The earliest extant Qibla charts herald from the twelfth century, even though the earliest mention of Qibla schemes occurs in Ibn Khurradādhbih's *Kitāb al-masālik wa-al-mamālik* (Book of Routes and Realms), penned in the mid-ninth century but of which the earliest extant copy is from the mid-fourteenth century.

David King created a sketch of this earliest known no longer extant Qibla scheme (fig. 3.15).[32] It is based on the most basic four-sector scheme as related to the four corners of the cube-shaped Ka'ba. A variety of Qibla schemes developed based on a combination of folk and mathematical science. These schemes appear in a multitude of sectors ranging from the simplest of four- and eight-sector schemes to ten-, eleven-, twelve-, and even seventy-two-sector schemes. The number of sectors is the chief method by which different Qibla charts are classified, although it should be noted that variations can be found within copies of the same manuscript series by the same author. Such is the case with Ibn al-Wardī's *Kharīdat al-'ajā'ib* manuscript series in which Qibla charts range from eleven-sector diagrams to seventy-two-sector ones.[33] (See, for instance, the eleven-sector example shown in fig. 3.16 from a late sixteenth-century Ibn al-Wardī manuscript and the seventy-two-sector example also from an Ibn al-Wardī manuscript shown in fig. 7.17.)

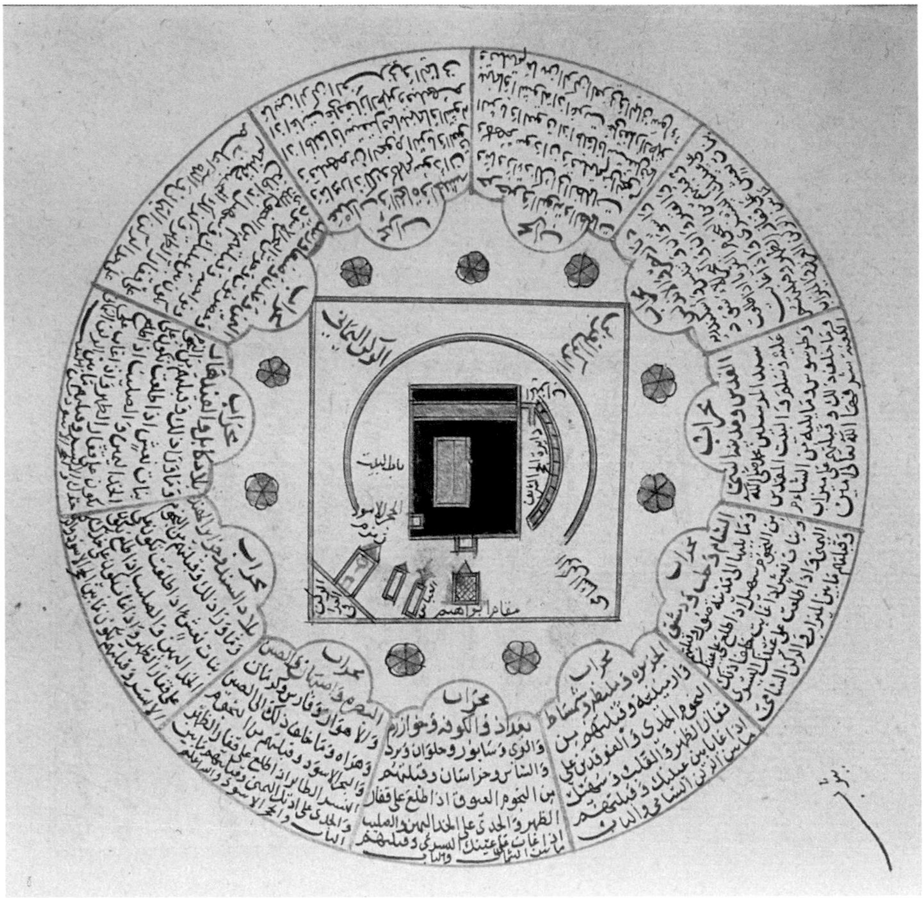

Fig. 3.16. Eleven-sector Qibla scheme of sacred geography from a copy of Ibn al-Wardī's *Kharīdat al-ʿajāʾib wa-farīdat al-gharāʾib* (The Unbored Pearl of Wonders and the Precious Gem of Marvels). 988/1580. Ottoman. Black and red ink and gold on paper. Diameter 15.2 cm. Courtesy: Topkapı Saray Museum, Istanbul. Ahmet 3012, fol. 40a. Photo: Karen Pinto.

If we stretch our definition of what constitutes a map, we can incorporate into this repertoire maplike images that occur in miniature paintings and landscape murals, such as the mid-fourteenth-century image of an angel (possibly Gabriel) presenting the Prophet Muḥammad (seated on the upper left side of the image) with a model of an unidentified city (fig. 3.17). Some scholars have suggested that the model shows the city of Jerusalem because of the focus of *Miʿrājnāma* (Night Journey) manuscripts on Muḥammad's miraculous trip in the middle of the night to Jerusalem from Mecca

Fig. 3.17. Model of a city (possibly Jerusalem) being presented by the angel Gabriel to the Prophet Muhammad while he describes the city to Abū Bakr, Abū Jahl, and members of the Quraysh tribe from the Ilkhanid *Miʿrājnāma*. Circa 1317–1335. Baghdad, Iraq, or Tabriz, Iran. 35 × 25 cm. Gouache, ink, and gold on paper. Courtesy: Topkapı Saray Museum, Istanbul. Hazine 2154, fol. 107a.

with the angel Gabriel. Others suggest that it could be a miniature of Damascus, the oldest continuously occupied city in the Islamic world and the location of many sacred sites.[34] Showing two mosques and a series of rivers, the model is vague enough to represent almost any city. Perhaps it seeks to present a vision of the ideal city in map form that any city can take. Be that as it may, the fact that maplike models start to appear in miniature paintings suggests that the penchant for mapping was growing in the medieval Middle East.[35]

Source of Islamic Maps

What is the source of this rich, variegated, and widespread medieval Islamic propensity to map? Some scholars cite the silver globe (*Ṣūrat al-Ma'mūniyya*) that the Abbasid caliph Ma'mūn (r. 197–218/813–833) is said to have commissioned from the scientists who worked in his Bayt al-ḥikma (House of Knowledge).[36] There are multiple problems associated with the ascription of all Islamic cartography to this Ma'mūnid silver globe, foremost of which is that it is no longer extant and we cannot definitively determine what it looked like. The only descriptions that we have of this map are a few scattered and vague passages, such as the one, cited by Abū al-Ḥasan ʿAlī ibn al-Ḥusayn ʿAlī al-Masʿūdī (d. 345/956) in his *Kitāb al-tanbīh wa-al-ishrāf* (Book of Instruction and Supervision): "I have seen these climates represented in another book, in different colors . . . in the *Ṣūrat al-Ma'mūniyya* that al-Ma'mūn ordered to be constructed by a group of contemporary scholars to represent the world with its spheres, stars, land, and seas, the inhabited and uninhabited regions, settlements of peoples, cities, etc."[37]

Masʿūdī's description suggests a flat two-dimensional model of the KMMS type with multiple Encircling Oceans incorporating stars along the outer edge.[38] Some scholars have read the use of the word *aflāk* (spheres) in Masʿūdī's description to indicate that it was a globe. If it were a globe, then Masʿūdī's description of it suggests a complicated celestial map superimposed on a world map. Fuat Sezgin goes so far as to suggest that the world map in a fourteenth-century ʿUmarī manuscript should be considered a copy of the long lost Ma'mūnid globe (fig. 3.18). Aside from the unexplained five-century hiatus, if Masʿūdī's description is to be read as describing a globe, then it must have been an armillary sphere and a very sophisticated one at that. The earliest extant examples of armillary spheres date from the mid-fourteenth century and none of them contain detailed maps of the world. It is more likely that the cartographer who made the ʿUmarī world map based it on a model of a fourteenth-century globe or that he was influenced by Idrīsī's work.[39]

The other problem with Sezgin's thesis is his out-of-date view of the role played by the Bayt al-ḥikma. Revisionist work by Dmitri Gutas on Ma'mūn's Bayt al-ḥikma suggests that it was not a center for new scholarship or scientific invention, or even

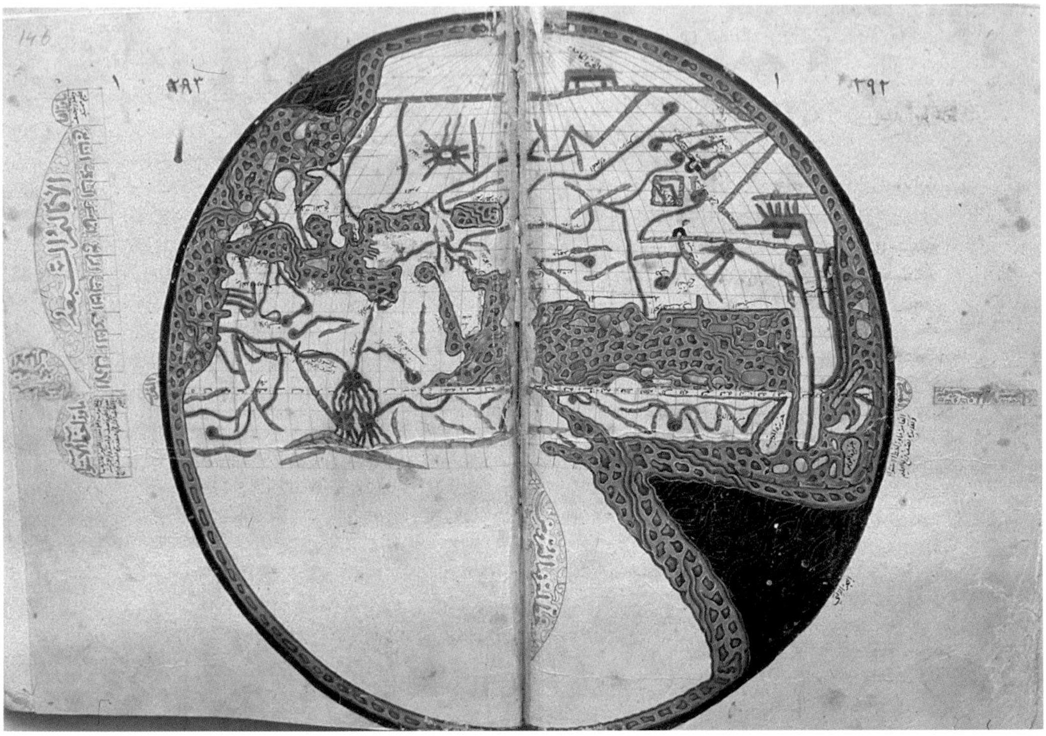

Fig. 3.18. Controversial al-ʿUmarī world map with longitudinal and latitudinal graticule. Circa 14th century. Diameter 26.5 cm. Gouache and ink on paper. Courtesy: Topkapı Saray Museum, Istanbul. Ahmet 2797, fols. 292v–293r. Photo: Karen Pinto.

a project that was exclusively the brainchild of Maʾmūn. It was set up either by Maʾmūn's great-grandfather, the Abbasid caliph Manṣūr (r. 136–158/754–775), or by his son the caliph Mahdī (r. 158–169/775–785) as part of Abbasid political and intellectual accommodation of Sassanian tradition in order to appeal to the Iranians who had supported their revolution. Gutas argues that the Bayt al-ḥikma should be seen either as an administrative clearinghouse for the Abbasid regime or as a continuation of the Sassanid tradition of large royal libraries with vast book collections open to the elite scholarly public for perusal and copying with the facility of translations available on demand. The primary emphasis was on the translation of Middle Persian and Avestan texts into Arabic, not Greek texts as previously thought.[40]

Instead of a version of maps that look like the fourteenth-century ʿUmarī maps, five centuries after the original, few in number, and not representative of the bulk of the Islamic mapping tradition, it is more likely that Masʿūdī's was referring to a world map from an early version of the KMMS mapping tradition. This interpretation is reinforced both by the logic of the corpus of extant Islamic maps of which the KMMS tradition represents by far the vast majority and by what Masʿūdī says about 150 pages later in another reference in his *Kitāb al-tanbīh wa-al-ishrāf* (Book of Notification and Review) to the so-called *Jughrāfiyā* book that he saw, which Sezgin argues is identical to *Kitāb ṣūrat al-arḍ* (another title commonly used by manuscripts in the KMMS tradition), "In the book known as *Jughrāfiyā* the philosopher describes the earth, its sites, mountains, seas, islands, rivers, water sources, inhabited sites, well-preserved places; the number of sites in his day amounted to 4,530, and he has listed them individually, one after the other in each clima. In that book he described the colors of the mountains of the known world accordingly in red, yellow, green, and so on. There are more than two hundred mountains. He provided information on their size, their deposits, and what they contained by way of metals and precious stones."[41] This is exactly what the KMMS series looks like.[42] Embedded within detailed descriptions of the world and its twenty regions—which could be described as clima—are colorful maps showing seas, rivers, other water sources, islands, mountains, and sites (cities and towns), along with historic places. They are listed one after another and the presence of each regional map is announced with a header. The mountains and deserts in the KMMS maps are indicated in a variety of colors including red and yellow; and the seas "are drawn in different colors," exactly as Masʿūdī states, in "various sizes and shapes,"[43] and as Masʿūdī notes, they are consistently depicted in the KMMS maps of the world as five: namely, the Encircling Ocean, the Mediterranean Sea, the Persian Gulf/Indian Ocean, the Caspian and Aral Seas. Even Sezgin agrees that Masʿūdī's descriptions imply that the maps were accompanied by a descriptive text.[44] This is how the KMMS manuscripts are laid out: extensive descriptions accompany each map.

This reading is further confirmed by the other key reference to the Maʾmūnid *Jughrāfiyā* that comes from the work of the twelfth-century Andalusi scholar Muḥammad ibn Abī Bakr Zuhrī whose *Kitāb al-jaʿrāfiyā* says, "Their objective is the depiction of the earth, even if it does not correspond to reality. Because the earth is spherical but the *jaʿrāfiyā* [presumably a reference to a foldable map] is simple [uncomplicated] and its form expands the way astrolabes and collections of eclipses unfold. In it to the viewer of the earth is [visible] all its parts, its regions, its borders, its climes, its seas, its rivers, its mountains, its populous [parts], and its wastelands, and where all the cities are located in its east and in its west, and the viewer can see its wondrous [ʿajāʾib] places and what constitutes all the parts of its famous wonders and the depicted [pictured] buildings from its lands of time immemorial [or its lands of antiquity]."[45] The operational terms here are "depiction" and "simple," as the KMMS maps are, and "does not correspond to reality," as the KMMS maps do not because they depict the coastlines with smooth lines and use geometric shapes for the landmasses and islands. In spite of the lack of overt mimesis, as Zuhrī notes, one can, thanks in part to the labels, make out the regions, borders, climatic zones, seas, rivers, mountains, deserts, inhabited areas, and the locations of cities in relation to each other, whether in the east or the west because the corners of all the maps are labeled with directions. This is exactly how the KMMS maps are set up.[46]

Prior to the *Ṣūrat al-Ma'mūniyya*, we have a few scattered references to maps in the early chronicles from the period of the Umayyad caliphate (41–132/661–750). From the end of the first-century *hijra* (ca. 83/702), we have a report of Ḥajjāj ibn Yūsuf (d. 95/714), the Umayyad governor of the eastern part of the empire, commissioning maps of the region of Daylam (south of the Caspian Sea) for military and *jizya* (poll tax) purposes. He specifically requested a depiction of the region's plains, mountains, steep roads/mountain passes, and forests, and showed the finished product to the local people. Similarly, in preparation for an attack on the city of Bukhara, it is reported that Ḥajjāj requested a plan of the city.[47] References to requests for maps for military purposes are highly unusual in medieval Islamic history. It is not until the time of Meḥmed II (mid-fifteenth century) and the early Ottoman rulers that we learn of similar requests for maps for military purposes.[48] Unfortunately, none of Ḥajjāj's requests are extant and there are no detailed descriptions of these maps. It is possible that some of the later regional maps of the KMMS series, such as that of the province of Daylam, may be related to these early Ḥajjāj requests to chart Iranian terrain, but in the absence of extant examples and detailed descriptions, there is no way to know for certain.

Do these textual references to singleton maps suggest that we can trace the beginnings of the Islamic mapping tradition even earlier to the Umayyad period instead of the Abbasid? In addition to the Ḥajjāj references, the earliest extant star fresco in the Islamic context comes from Quṣayr ʿAmra, the desert bath/playhouse of the Umayyad prince and future caliph Walīd II (706–744; r. 125–126/743–744) (fig. 3.19). Tucked away in the sand and sun of the Syro-Arabian desert that separates Jordan from Syria and Iraq, this fresco is well known among historians of science since it was first studied in 1932 and has been placed on par in importance with the Farnese Atlas.[49] Much ink and detailed work has gone into determining the constellations depicted on this rare astronomical cupola fresco to discern how this chart matches up with other star charts of the ancient and medieval periods. The best and most up-to-date work on the subject is the joint article "The Fresco of the Cupola of Qusayr ʿAmra" by J. P. Brunet, R. Nadal, and Cl. Vibert-Guigue. The authors argue that the inspiration of some of the constellations was from Ptolemy's catalogue, albeit with some significant differences and a lot of mistakes in the positioning of the constellations because the painter was trying to accommodate the skylights of the bathhouse.

In addition to this star chart, the bathhouse contains other intriguing images that have mapping-related themes, such as the "Cosmokrator"-type representation of an

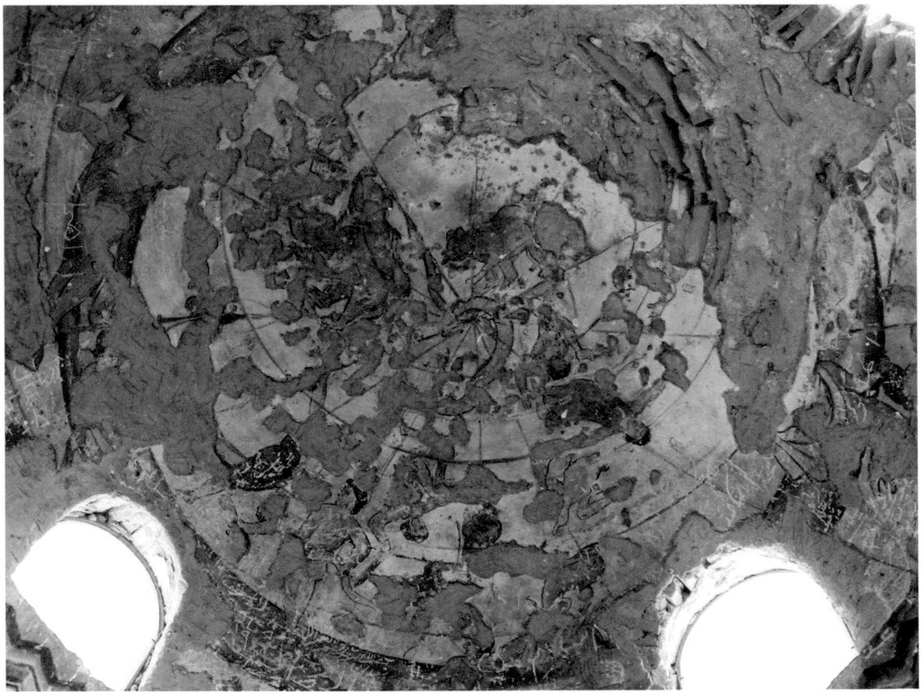

Fig. 3.19. Cupola star fresco. Quṣayr ʿAmra, Jordan. Circa early eighth century. Dark- and light-brown and yellow fresco painting applied to mortar. Photo: Karen Pinto.

enthroned figure directly opposite the entrance that has been identified by scholars as the throne apse because it contains an image of an enthroned figure. The inscription above designates the person in the throne as an *amīr* (prince) and is one of the reasons why, since the early twentieth-century discovery of the bathhouse, the patron was presumed (correctly as we now know) to have been Walīd II. The image is seriously damaged in places, but it is still clear enough to show an enthroned figure being fanned by two servants holding fly-whisks (*flabella*), with birds ringing the arch above and fish in the sea below. The latter image of water, fishes, and a boat (which was stripped out by Musil and Meilich and is now at the Pergamon Museum in Berlin) has been referred to as a "Nilotic" scene reflective of Coptic Egyptian influence.[50] In *Arab Painting*, Ettinghausen suggests that this image of the enthroned ruler could be interpreted as a Cosmokrator motif inspired by Sassanian motifs: "Might not the central and most important painting have expressed the idea that the caliph was ruling the earth, which was thought to be bounded by the ocean, designated by sea monsters, and beneath the all-covering sky, symbolized by birds."[51] The soffits of the arches in the main entrance halls are decorated by female figures with large outlined eyes holding up a cloth with fruits that have been interpreted as symbolizing Gaea, the earth goddess of the ancient world.[52]

Last but not least, there is the intriguing globular image (fig. 3.20) with maplike markings that a cherub is handing to a reclining androgynous figure located on the spandrel above and to the right of the door as one enters the site. This can be interpreted as a mid-eighth-century Umayyad globe fresco of either the landscape immediately surrounding Quṣayr ʿAmra, the region of the Dead Sea, the Red Sea, and the Mediterranean, or possibly the earliest depiction of the moon, which is a striking celestial object in the dry flat desert and is easily observed for mapping purposes.[53] (For further proof of reading this fresco as a possible map, see the parallels between fig. 3.20 and fig. 3.17.)

The Umayyad caliphs may have been interested in the stars, the moon, and other cosmographic images, but it is the Abbasid caliph Manṣūr (r. 136–169/754–775; considered the founder of the dynasty because the first caliph, Saffāḥ, was sickly and died within five years of taking power) who seems to have been the earliest Arab caliph to take an active interest in mapping for the purposes of planning and control. Balādhurī (third/ninth century) reports that in order to resolve a dispute a delegation from Basra, comprising Manṣūr Sawwār ibn ʿAbdallah al-Tamīmī (later known as al-ʿAnazī), Abū Dāwūd ibn Abū Hind, and Saʿīd ibn Abū ʿArūba (also known as Bihrān), showed the caliph Manṣūr an image (ṣūra) of the swamps of Baṭīḥa near Basra in order to make the case that the caliph should call off the construction of the Subaiṭīyah canal because it would make the waters of Basra salty.[54] The swamps near Basra were a volatile area in this period, known for slave revolts that plagued later Abbasids (the Zanj rebellion of 255–270/868–883 in particular). In addition, Yaʿqūbī (d. 284/897) reports that a plan was drawn up in 141/758 for Manṣūr prior to the building of the famous round city of Baghdad.[55] This is an unusual citation since the active use of architectural drawings for construction is not heard of until the time of the Ottomans.[56] Wendell uses Yaʿqūbī's description of Baghdad to argue that the city was deliberately laid out according to the plan of an awe-inspiring cosmic *imago mundi*.[57]

Two Egyptian chroniclers mention maps in the context of caliphs. The ninth-century historian Abū ʿUmar Muḥammad al-Kindī (d. 350/961) mentions a large world map produced on a scroll (darj) for the fifth Abbasid caliph, Hārūn al-Rashīd (r. 170–193/786–809), famed in the West for his role in the *Thousand and One Nights*, and that the caliph found the province of Asyūṭ in Upper Egypt to be the most interesting depiction on this map.[58] Four centuries later, Maqrīzī mentions that a magnificent map on fine blue *tustarī* silk with gold lettering was prepared for the Fatimid caliph Muʿizz (r. 341–365/953–975). On this was pictured "parts of the earth with

A

Fig. 3.20. *A*, Cherub bringing globe to melancholic prince or princess. *B*, Northwestern spandrel adjacent to entrance with close-up inset of globe. Quṣayr ʿAmra, Jordan. Circa mid-eighth century. Blue, dark- and light-brown, and yellow fresco painting applied to mortar. Diameter of globe approx. 28cm. Photo: Karen Pinto.

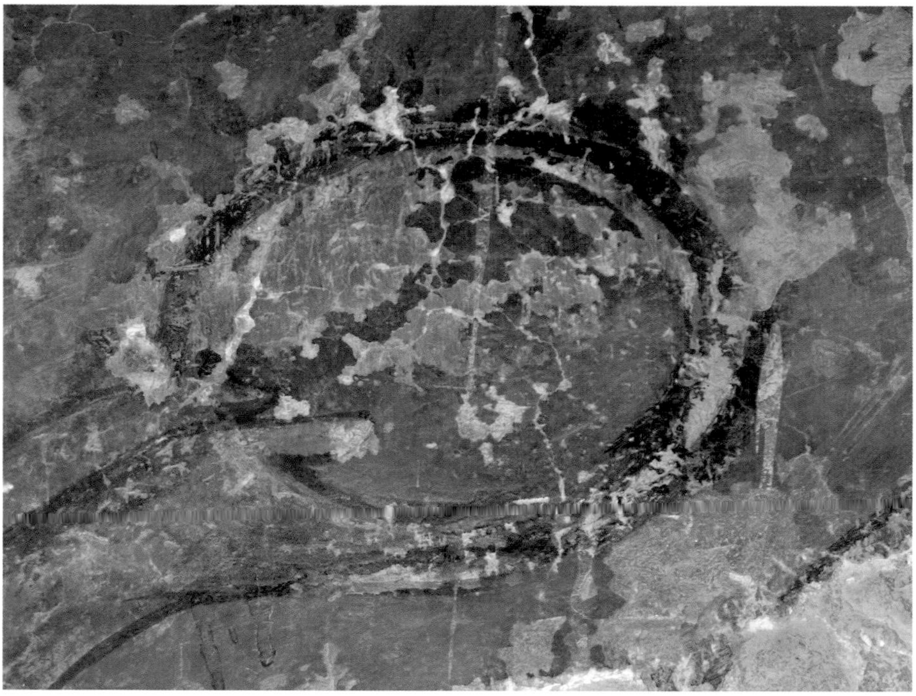

B

Fig. 3.20. (continued)

all the cities and mountains, seas and rivers." Mu'izz was so entranced with this map that he had it entombed with him in his mausoleum in Cairo—and thus we no longer have access to it.[59] Reinaud mentions another silk map made for a Buyid prince in the tenth century that showed the different countries of the earth and its mountains, seas, rivers, towns, and routes in different colors.[60] Ibn al-Nadīm's tenth-century *Fihrist* (List/Catalogue) mentions a cloth map made by Qurrah ibn Qamīṭā on unbleached Dubayqī cloth with waxed dyes.[61] Regrettably none of these maps are extant. Given that we have a number of references mentioning the existence of maps on cloth in the tenth century, we can presume that these were popular, at least among members of the ruling class. Cloth maps would have been easier to roll and carry on trips unlike maps in manuscripts. Descriptions of the elaborate use of color to mark places, mountains, rivers, and seas on these maps conjures up images akin to the KMMS mapping series.

Prior to the KMMS series, the earliest extant source containing maps is an early eleventh-century copy (428/1037) of Abū Ja'far Muḥammad ibn Mūsā al-Khwārazmī's (d. ca. 332/847) *Kitāb ṣūrat al-arḍ* (Picture of the Earth).[62] Composed mostly of a

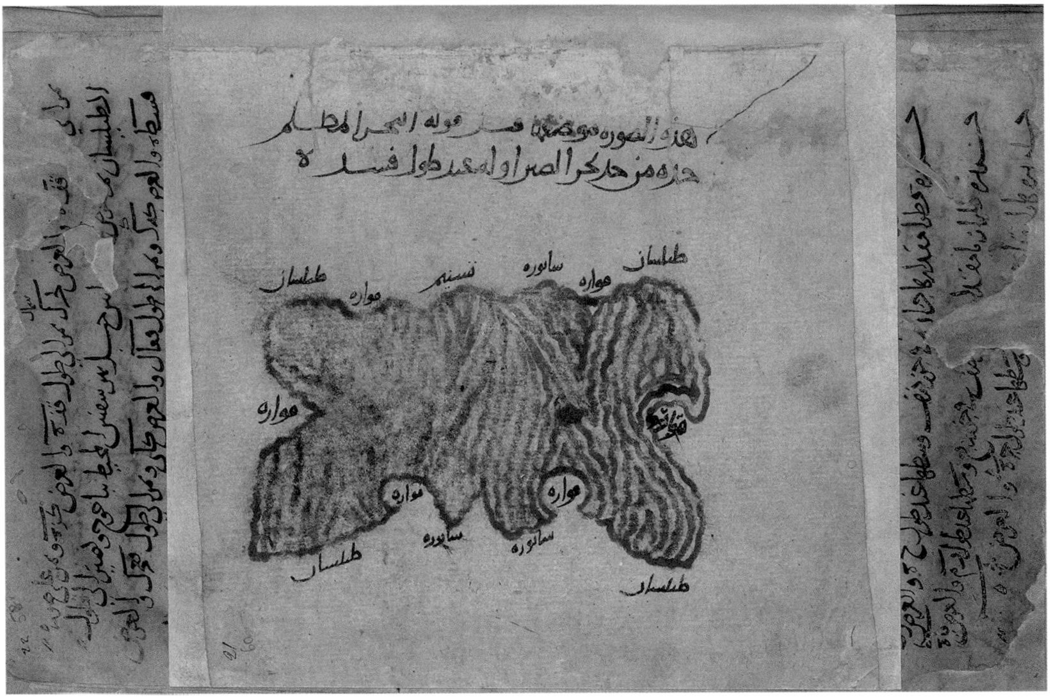

Fig. 3.21. Sketch of anonymous sea bound into al-Khwārazmī's *Kitāb ṣūrat al-arḍ*. This could be a depiction of the Indian Ocean or the Encircling Ocean or even of one of the inland seas, such as the Caspian, surrounded on all sides by land. 428/1037. 30.5 × 20.5 cm. Blue gouache and black ink on paper. Courtesy: Bibliothèque nationale et universitaire, Strasbourg. Ms. 4247, fol. 21a.

series of *zīj* tables (i.e., tables containing longitudinal and latitudinal coordinates), it includes four maps.[63] One of these maps has been vaguely identified as Sri Lanka on the basis of the label "Jazīrat al-Jawāhir" (Island of Rubies) and the inclusion of a line for the equator. Of the other two, one is a map of the Sea of Azov (linked by the Strait of Kerch to the Black Sea) and the other is a map of the Nile. The fourth map on a loose sheet bound into the manuscript is of an ocean, which has yet to be definitively identified (fig. 3.21). Some suggest that it is an image of the Indian Ocean. It is also possible that it is an image of the Black Sea or even of the Encircling Ocean surrounded by the earth. The labels on the landmass surrounding the sea are vague and repetitive and include words such as *ṭaylasan* (meaning "cape" or "handkerchief"), *sābūra* (meaning "horn," possibly "horned line" in this context) or *shābūra* (meaning "fog" and "mist"),[64] *tasnīm* (with *mā'*, refers to "waters of paradise"), and a thus far mysterious, undecipherable word, *miqawāra* (meaning "gouge" or "hollow").[65] This map is bound into

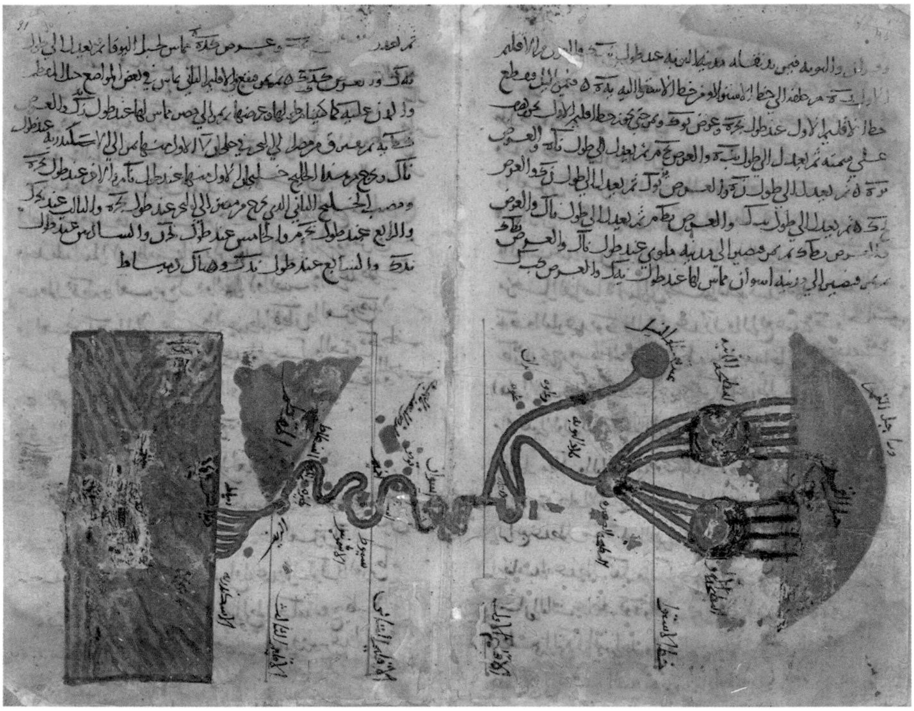

Fig. 3.22. Earliest extant map of the Nile in al-Khwārazmī's *Kitāb ṣūrat al-arḍ*. 428/1037. 33.5 × 41 cm. Blue, green, and brown gouache and red and black ink on paper. Courtesy: Bibliothèque nationale et universitaire, Strasbourg. Ms. 4247, fols. 30b–31a.

the manuscript between folios 20 and 22 on a much smaller piece of paper and lacks any specific place-names. For this reason it cannot be definitively identified nor can we be sure that it was intended as a map for this manuscript.[66] Of all the maps in this manuscript, the map of Egypt showing the Nile is the only one directly related to the KMMS series (fig. 3.22). This is where the parallels end. A map of the Sea of Azov does not occur in any of the KMMS maps or in any other collection of Islamic maps.

Before moving on to a detailed discussion of the KMMS tradition and its maps, we should discuss one more unique and crucial early map that comes into the Islamic world courtesy of Turkic migrations. I speak here of the intriguing world map tucked away in the earliest Turkish-Arabic dictionary, Maḥmūd ibn al-Ḥusayn ibn Muḥammad al-Kāshgarī's *Dīwān lughāt al-Turk* (Compendium of Turkic Dialects) (fig. 3.23). The original manuscript was compiled sometime between 464/1072 and 469/1077 but the only extant copy dates from 664/1266.[67] Thus we cannot be sure whether the original manuscript also contained a map nor can we

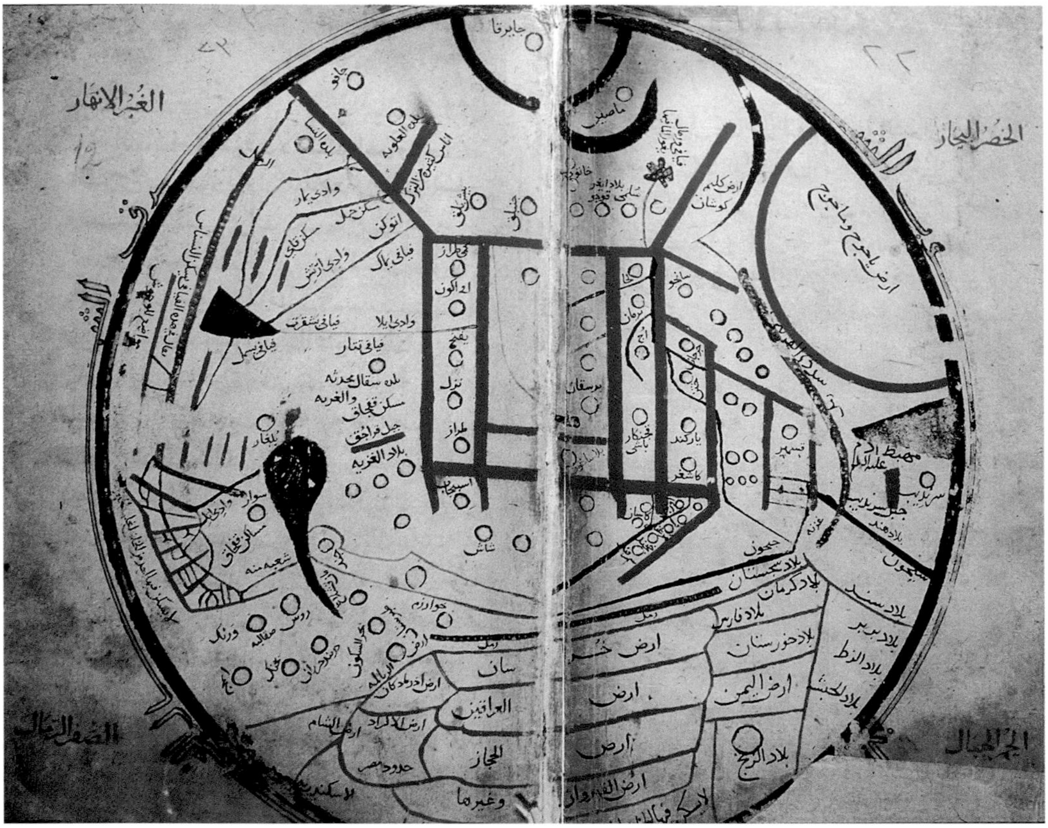

Fig. 3.23. World map showing Turkic tribes in Central Asia from al-Kāshgarī's *Dīwān lughāt al-Turk* (Compendium of Turkic Dialects). 664/1266. 27.5 × 11.5 cm. Dark-green and yellow gouache and red and black ink on paper. Millet Genel Kütüphanesi, Istanbul. Ms. Ali Emiri 4189, fols. 22b–23a.

be sure of its form, although the layout of this copy hints at answers.[68] Although this map is oriented eastward toward China, centers on Central Asia, and focuses on the location of Turkic tribes, in its illustration style it betrays Islamic cartographic influences. Red lines demarcating boundaries, dark-green copper (now black because of oxidation) for the seas, and slate gray for the rivers, encased in an encircling band symbolizing the Baḥr al-Muḥīṭ (Encircling Ocean), with a keyhole form for the Caspian Sea are all common iconographic tropes also used on the KMMS world maps. The grid of lands in the Islamic world laid out at the bottom of the map is evocative of Bīrūnī-type maps that lay out the lands in a grid-like structure in the lower half of the map. (Compare the bottom portion of fig. 3.23 with figs. 3.4, 3.5, 3.7, and especially 3.9.) It is, as the reigning expert on the Kāshgarī map,

Andreas Kaplony, suggests, akin to a modern "dialect atlas" speaking the "same visual language" as Arabo-Islamic cartographers, and it is for this reason that we need to take this map into consideration when reviewing the Islamic cartographic tradition. Given the Kāshgarī map's close visual connection to Islamic models and the fact that the earliest extant Islamic manuscript maps herald from the eleventh century, after the Turkic entry into the Islamic theater, it leaves us wondering if it was the Turks who brought a world-envisioning mapping tradition to the Islamic world from their Icarian vantage point atop the highest peaks in the world.

The KMMS Mapping Tradition

Most of the KMMS maps occur in the context of geographical treatises devoted to an explication of the world in general and the lands of the Muslim world in particular. These "map manuscripts" generally carry the title of *Kitāb al-masālik wa-al-mamālik*, although they are sometimes named *Ṣūrat al-arḍ* (Picture of the Earth) or *Ṣuwar al-aqālīm* (Pictures of the Climes/Climates). These manuscripts emanate from an early tradition of creating lists of pilgrim and post stages that were compiled for administrative purposes.[69] They read like armchair travelogues of the Muslim world with one author copying prolifically from another.[70]

Beginning with a brief description of the world and theories about it—such as the inhabited versus the uninhabited parts, the reasons why people are darker in the south than in the north, and so on—these geographies methodically discuss details about the Muslim world, its cities, people, roads, topography, and so forth. Sometimes the descriptions are interspersed with tales of personal adventures, discussions with local inhabitants, debates with sailors as to the exact shape of the earth and the number of seas, and so on. They have a rigid format that seldom varies, with a territorial sequence as follows: first the whole world; then the Arabian Peninsula, the Persian Gulf, the Maghrib (North Africa and Andalusia), Egypt, Syria, the Mediterranean, and upper and lower Iraq; and finally twelve maps devoted to the Iranian provinces, beginning with Khuzistan and ending in Khurasan, including maps of Sind and Transoxiana. The maps, which usually number twenty-one—one world map and twenty regional maps[71]—follow the same format as the text.[72]

Not all these geographical manuscripts contain maps, however. Only those referred to generally as part of the Balkhī-Iṣṭakhrī tradition, also referred to as the "Classical School" of geographers. For this reason the cartographically illustrated manuscripts of this genre, which I refer to as the KMMS mapping series, are sometimes also referred to as the "Atlas of Islam."[73] A great deal of mystery surrounds the origins and the architects of this manuscript-bound cartographic tradition. This is primarily because *not a single manuscript survives in the hand of the original authors*.[74] In fact, the earliest extant manuscript of this tradition dates from the late eleventh century, almost a century after the death of the last reported author. As a result, it is not clear who initiated the tradition of accompanying geographical texts with maps. Scholars of the eighteenth, nineteenth, and twentieth centuries held that Abū Zayd Aḥmad ibn Sahl al-Balkhī (hereafter Balkhī) (d. 322/934), who—as his *nisba* (patronym) suggests—came from Balkh in Central Asia, initiated the series and that his work and maps were later elaborated on by Abū Isḥāq Ibrāhīm ibn Muḥammad al-Fārisī al-Karkhī al-Iṣṭakhrī (hereafter Iṣṭakhrī) (fl. early tenth century) from Istakhr in the province of Fars. Iṣṭakhrī's work was, in turn, elaborated on by Abū 'l-Qāsim ʿAlī al-Naṣībī Ibn Ḥawqal (hereafter Ibn Ḥawqal) (fl. second half of tenth century), who came from upper Iraq (the region known as the Jazira). Finally, Shams al-Dīn Abū ʿAbd Allāh Muḥammad ibn Aḥmad ibn Abī Bakr al-Bannāʾ al-Shāmī al-Muqaddasī (hereafter Muqaddasī) (d. ca. 1000) from Jerusalem (al-Quds) is considered the last innovator in the series.[75]

The problem is that with the exception of Balkhī virtually no biographical information exists on the other authors. Because of this, we have to rely on scraps of information scattered here and there in the geographical texts themselves for information about the authors. Furthermore, in all the forty-three titles that the tenth-century bibliographic compiler Ibn al-Nadīm credits to Balkhī, not one even vaguely resembles the title of a geographical treatise.[76] According to the biographers, Balkhī was most famous as a philosopher and for his *tafāsīr* (commentaries on the Qurʾān)—in particular one known as *Naẓm al-qurʾān*, which was praised by many judges. He is not, however, known in the biographical record for his geographical treatises. Yet stories of how Balkhī sired the Islamic mapping tradition abound and endure. It is for this reason that the genre is generally referred to as the "Balkhī school of mapping."[77] This attribution of a whole school of mapping to a shadowy, mythical father is misleading. In order to discontinue the misnomer, I have opted instead for the use of a new acronym: KMMS. It represents a deliberate departure from the previous nomenclature and symbolizes a reinvestigation of the tradition that makes a break with past treatment.

In addition, many of the surviving KMMS copies contain either incomplete colophons or no colophons at all. The geographical texts are sometimes so mixed up in the surviving manuscripts with a variety of titles that it is often difficult to disentangle them. The numerous incomplete and anonymous manuscripts, sometimes abridged, along with versions translated into Persian, cloud the matter further.[78]

For the purposes of this book, the question of exactly who authored the first carto-geographical manuscript and precisely what it looked like is moot, as I date the maps according to when they were made, not to the lifetime of an original author because none of the extant manuscripts date to the time of the original authors.[79] What is relevant for the purposes of this book is that with these geographical manuscripts, we get some of the earliest cartographic images of the world in an Islamic context. Since all images are socially constructed, these iconic *carto*-ideographs contain valuable information about the milieus in which they were produced. They are a rich source of historical data that can be used as alternate gateways into the past. They can tell us about the time period in which they were copied and lead to greater knowledge of the period in which they were originally conceived. It is through this alternate cartographic window into the histori*ographic* imagination that I seek to peer in this book. Since the extant examples stretch in time from the eleventh century to the nineteenth century and range from the heart of the Middle East to its peripheries, they can provide us with a broad range of historical insights across time and space.

The earliest extant KMMS manuscript is by Ibn Ḥawqal and is housed at the Topkapı Saray museum library (Ahmet 3346). It is firmly dated to 479/1086 by its colophon. It is the world map version of this manuscript that proliferates in a more embellished form via the Ibn al-Wardī manuscript copies from the fifteenth century onward. (See fig 4.7 for an example of the world map from the earliest extant KMMS manuscript and compare to figs. 3.8 and 7.4, examples of Ibn al-Wardī world maps.) The striking mimesis of the maps in this manuscript stands in stark contrast to the later KMMS map copies, which over the centuries abandon any pretense of mimesis entirely.[80] Moving through the KMMS set, we travel through a series of more and more stylized maps that shift further into the realm of objets d'art and away from direct empirical inquiry. By the nineteenth century, the KMMS maps become so stylized that, were it not for the earlier examples, they would be hard to recognize as maps. (Compare fig. 4.7 with fig. 4.8 in the next chapter.)

With the foundational basis of the KMMS corpus established, I turn to a more detailed look at the nature, layout, and variants of the world maps found in these manuscripts. The next chapter will provide readers with an overview of the KMMS world maps that are essential for understanding the analyses that unfold in later chapters.

4

KMMS World Maps Primer

Medieval Islamic maps present a millennia-old conception of the world that we may never fully understand. Much of their appeal lies in their conceptual opacity. They are beautiful puzzles that hint at a hidden, larger picture. In this chapter, we will look more closely at the pieces of these puzzles and discern how they fit together. This is the primer where we begin to understand the visual language of the KMMS images in preparation for the iconographic element of the tripartite approach that I pursue in this book.

Muslim cartographers of medieval times did not seek to show the earth's precise shape. They chose instead to create stylized images that can be best described as mnemonic ideational maps presenting a world both as they saw it and as they wished it to be understood. Transported through memory and easily redrawn, these images serve some thousand years later as reminders of bygone worldly imaginings. Figure 4.1 is an example of the typical KMMS *ṣūrat al-arḍ* (picture of the world).

We might well look on this image, with its peculiar shapes and bold strangeness of line, its gold leaf and exquisite floral

Fig. 4.1. Timurid KMMS world map from al-Iṣṭakhrī's *Kitāb al-masālik wa-al-mamālik* (Book of Routes and Realms). 827/1424. Ultramarine gouache and black ink against a foliated gold background on paper. Diameter 24 cm. Courtesy: Topkapı Saray Museum, Istanbul. Bağdat 334, fol. 2b. Photo: Karen Pinto.

decoration, and ask, how can this be a map? Our modern vantage point and our familiarity with precise, mimetic representations of the world make it challenging at first to reconcile this image as a map of the world.[1]

The initial inscrutability of medieval Islamic maps hinders the application of the traditional iconographic exercise of the art historian. In his 1939 classic, *Studies in Iconology: Humanistic Themes in the Art of the Renaissance*, Erwin Panofsky, the doyen of the classical model of iconographic and iconological analysis in the discipline of art history, explains the process thus:

> When an acquaintance greets me on the street by removing his hat, what I see from a *formal* point of view is nothing but the change of certain details within a configuration forming part of the general pattern of colour, lines and volumes which constitute my world of vision. When I identify, as I automatically do,

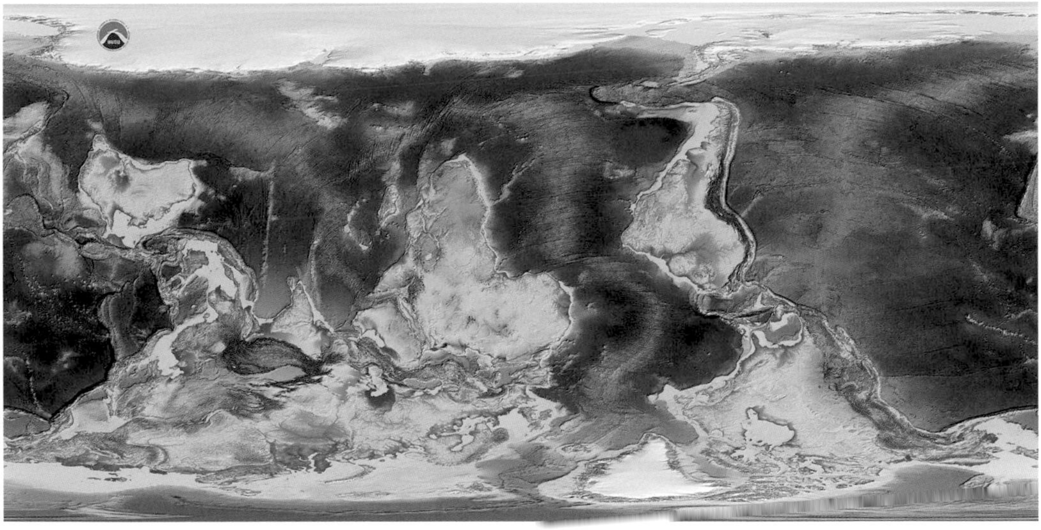

Fig. 4.2. Relief map of the world with south on top. Source: C. Amante and B. W. Eakins, 2009. *ETOPO1 1 Arc-Minute Global Relief Model: Procedures, Data Sources and Analysis*. NOAA Technical Memorandum NESDIS NGDC-24. National Geophysical Data Center, NOAA. doi:10.7289/V5C8276M [September 15, 2014].

> this configuration as an *object* (gentleman), and the change of detail as an *event* (hat-removing), I have already overstepped the limits of purely *formal* perception and enter a first sphere of *subject matter* or *meaning*. The meaning thus perceived is of an elementary and easily understandable nature, and we shall call it the *factual meaning*; it is apprehended by simply identifying certain visible forms with certain objects known to me from practical experience, and by identifying the change in their relations with certain actions or events.[2]

Panofsky is presuming from the outset that the hat can be identified as a hat. What if the object cannot immediately be identified? How do we proceed to the next stage of factual meaning?[3]

To show how these maps are a representation of the real world we need to first make them transparent through the lens of mimesis. It is only after this step that we can move on to other modes of analysis such as iconological, linguistic, structural, and postphenomenological, which will be pursued in subsequent chapters of this book.

Our first step in decoding the world map above is to take a typical late twentieth-century map of the world and turn it with south facing up—that is, with Africa on top instead of Europe (fig. 4.2). Our present custom of orienting maps with north on top is a practice begun by Renaissance mapmakers that indicates the rising importance of northern Europe from the late fifteenth and sixteenth centuries onward. Medieval European cartographers oriented their maps with east on top toward the *Oriens* (Orient), hence the source of the meaning of the verb "to orient."[4] (See chapter 6 for more detail on the subject of medieval European T-O maps.) Muslims chose to orient their world maps southward. We are still not sure why. One hypothesis is that since these maps were produced somewhere in the heartlands of the Abbasid caliphate, north of the holy cities of Mecca and Medina, they would naturally point south toward the Arabian Peninsula just as medieval European maps pointed east because of Jerusalem.

Our next step is to cut off the Americas (fig. 4.3). Since these maps were created prior to the discovery of America, one needs to work with a map that does not show the New World. Elimination of this landmass from our worldview prompts a merging of the Atlantic and Pacific Oceans into an all-encompassing Encircling Ocean, which was regularly featured in premodern maps of the world.

Finally, we need to skew the projection (fig. 4.4). This causes Africa to loom large instead of Europe.[5] If we can envision Australia merging with the tip of Africa, causing it to appear overextended,[6] a most surprising metamorphosis of the modern map takes place, such that the Islamic map actually starts to resemble it. Suddenly our deep-seated need for mimesis and accuracy as the necessary ingredients of cartographic knowledge and visual practice are satisfied, and we are now more comfortable accepting the quaint objects with their strange drawings and funny squiggles as maps.[7]

From the overextension of Africa, we can surmise that Muslim ships must have hugged the coasts of East Africa and that they did not travel past the Horn because Africa is foreshortened and the distance between East Africa and India as shown on the maps is much shorter than we know it to be today. This fits with received knowledge that Muslim sailors did not venture beyond Sofāla in present-day Mozambique and Madagascar, and that because of ship design, they were compelled to stay close to the coast to service the need for fresh water.[8]

Thus we begin to grasp how the Muslims arrived at their conception of the world, and we can see how it corresponds in some "mimetic" fashion to the map of the world as we know it today. Taking into consideration that Muslim scholars first developed this form of the world map sometime in the vicinity of the tenth century, when they had only rudimentary tools and a pre-Islamic cartographic heritage to draw on, supplemented by travels and sophisticated mathematical calculations, we can develop an admiration for the unique cartographic achievements of the medieval Muslim mapmakers. Those achievements were not superseded until the early Renaissance.

The classical ideograph is made up of a double-edged circle in a square or rectangular frame (fig. 4.5).[9] Placed within this circle is the image of a pre-Columbian

Fig. 4.3. Relief map of the world with south on top and the American continent removed. Source: C. Amante and B. W. Eakins, 2009. *ETOPO1 1 Arc-Minute Global Relief Model: Procedures, Data Sources and Analysis.* NOAA Technical Memorandum NESDIS NGDC-24. National Geophysical Data Center, NOAA. doi:10.7289/V5C8276M [September 15, 2014].

Fig. 4.4. Relief map of the world with south on top skewed to show how a modern world map can resemble the typical KMMS map of the world. Source: Karen Pinto. Image based on the NOAA map in figure 4.3.

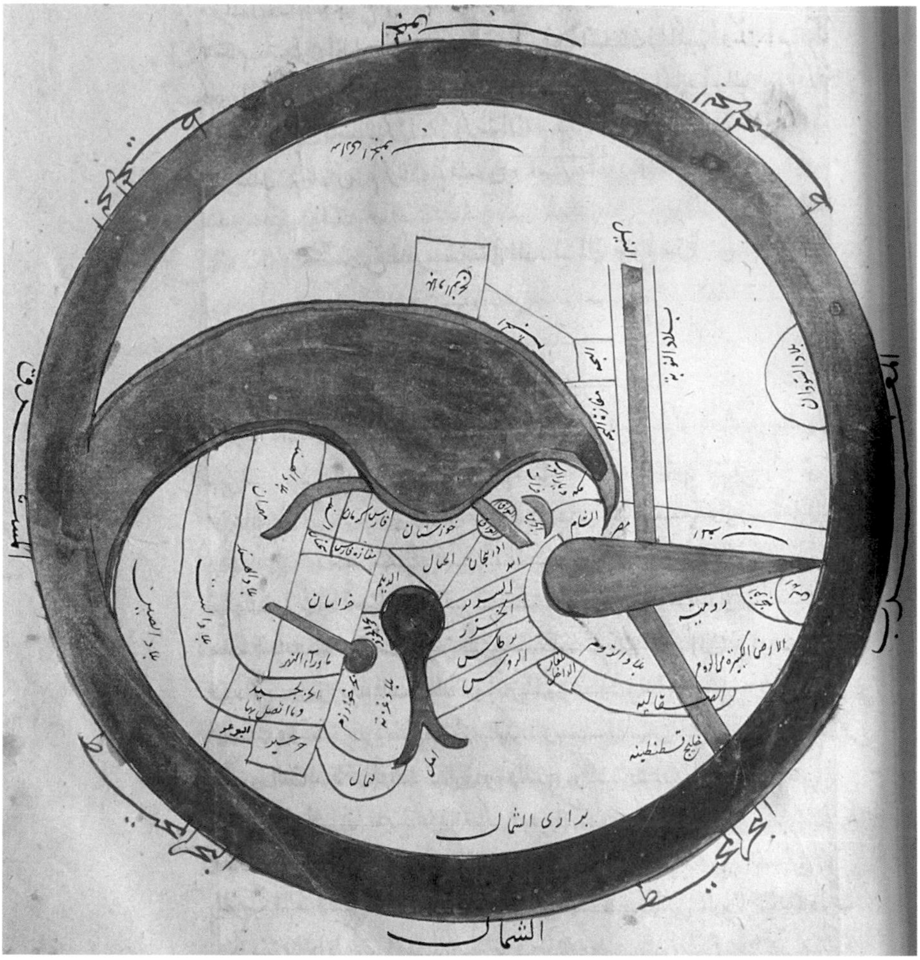

Fig. 4.5. KMMS World map from Ottoman Cluster. 878/1473. Ottoman. Light-blue gouache and red and black ink on paper. Diameter 19.5 cm. Courtesy: Sülemaniye Camii Kütüphanesi, Istanbul. Aya Sofya 2971a, fol. 3a. Photo: Karen Pinto.

world, punctuated by seas and rivers. The circle is surrounded by the names of the eight directions (north, south, east, west, southeast, southwest, northeast, and northwest), evenly spaced around the perimeter of the circle, either in red, gold, or black ink.

One is struck by the large, hooked shape that sweeps into the center of the image from the left, menacing a form with outspread arms. Both shapes emerge from a blue encircling band and are accompanied by two smaller blue shapes resembling keyholes. When we realize that the blue is being used as

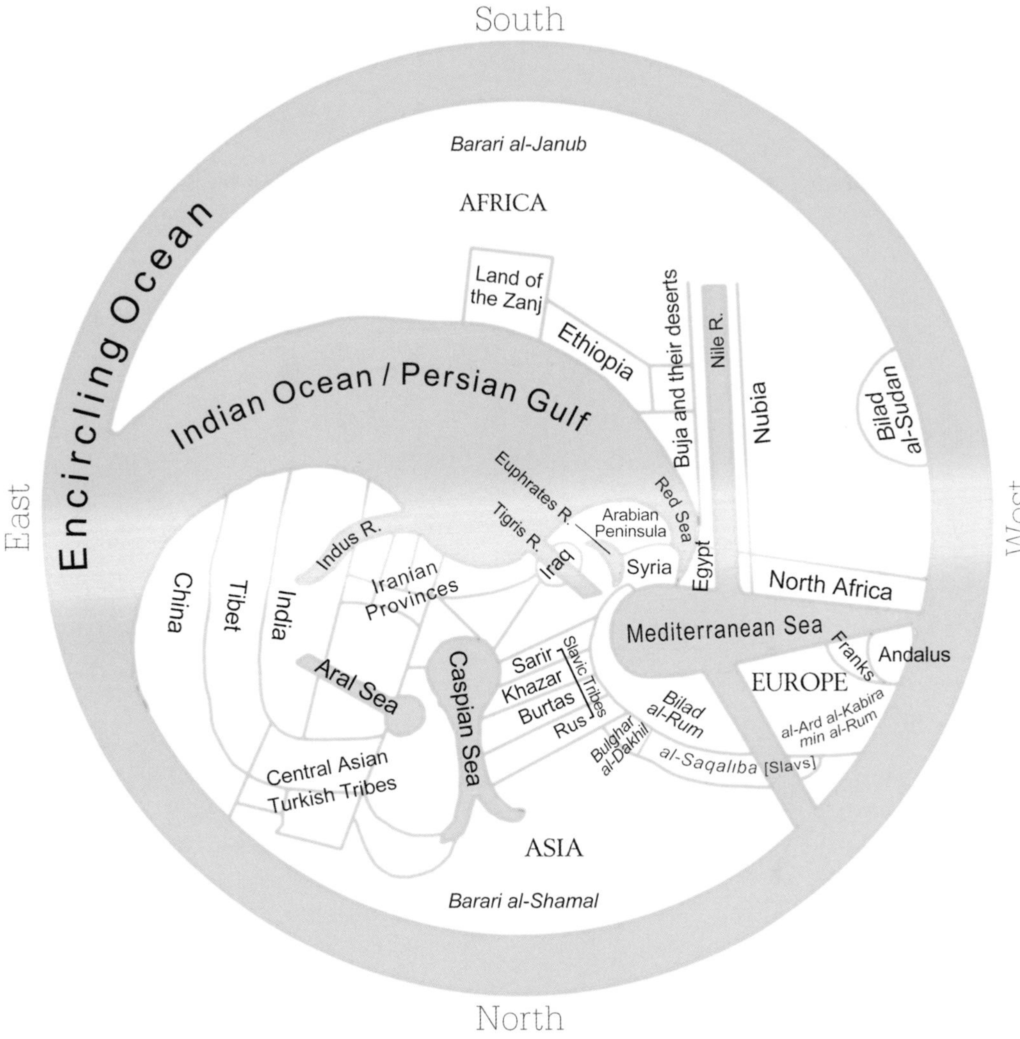

Fig. 4.6. Translation of KMMS world map based on Ottoman Cluster map (fig. 4.5). Source: Karen Pinto with assistance from Damien Bovlomov, photo editor of *Imago Mundi*.

the familiar cartographer's metaphor for water, and that all forms colored blue are seas, then the white, "paper-colored" crescent-tailed shape emerges as the landmass of the world, curled fetally inside the encircling waters.

Within this neat, aesthetically well-packaged ideograph are all the features standard to the classical medieval Islamic vision of the world. We see the Encircling

Ocean (Baḥr al-Muḥīṭ) that rings the world along with four other seas, seven rivers, and the three major landmasses of Africa, Asia, and Europe (listed here in order of their size on the map). The key to comprehending the medieval Muslim conception of the world is being able to assimilate the basic shapes of the landmasses and the seas, along with the map's southerly inversion (fig. 4.6).

Comparing the image with the manipulated modern world map shown earlier (fig. 4.4), we realize that the sharp, crescent-shaped landmass is in fact the continent of Africa. Once we make this association, we recognize that the double-headed, bulging form in the lower left-hand corner corresponds to the continent of Asia. The bulge connecting Africa to Asia is the Arabian Peninsula, and the tiny triangle marooned in the lower right-hand sector of the image is none other than Europe. Africa loomed large in the medieval Islamic imagination whereas, comparatively, Europe didn't count for much.

From the middle of the eastern edge of the circle, the Indian Ocean/Persian Gulf sweeps into the center of the image as a menacing hook pointing at the teardrop-shaped Mediterranean Sea in the lower right-hand quadrant of the image. The world maps indicate two other seas—the inland seas of the Caspian and the Aral—as two small, usually identical, keyhole shapes that puncture the Asian landmass and separate it from its barren northern extremities. They are not as dominant in the image as the Encircling Ocean, the Persian Gulf/Indian Ocean, and the Mediterranean Sea.

Looking beyond the most prominent forms of the image, we notice the importance of certain rivers. The Nile and the Bosphorus in the western (right-hand) side of the image, stretch out on opposite sides of the Mediterranean. Portrayed in a more subdued fashion are the Tigris and Euphrates Rivers at one end of Asia, located between Iraq and the Arabian Peninsula; and on the other end, the Indus River is marked between Kirmān, Sind, and India. Out of the Aral Sea emerges the famous Oxus River (Nahr Jayḥūn/Āmū Daryā) that marks the start of Transoxiana/Central Asia while the double-forked Volga River (Nahr Itil) is shown flowing into the northern end of the Caspian Sea.

Demarcations of countries, provinces, and people and their locales are shown variously in red, gold, or black ink. At first glance the boundaries on each map might appear the same; in fact they are distinctly different: some are straight lines, others curved and undulating. The northern and southern poles are always indicated as wastelands (barārī). Each territory is inscribed with its name, and the result is a curious assortment of toponyms and ethnonyms that can be explored in depth.

Counterintuitively, the earliest extant maps in the KMMS tradition are more mimetic than the later ones. Such is the case of the world map dating to 1086 CE (AH 470) shown in figure 4.7. In this map, the Mediterranean is bigger than in the fifteenth-century map and almost equals the size of the Indian Ocean. The Nile is no longer just a vertical shaft reaching up into Africa from the Mediterranean; it

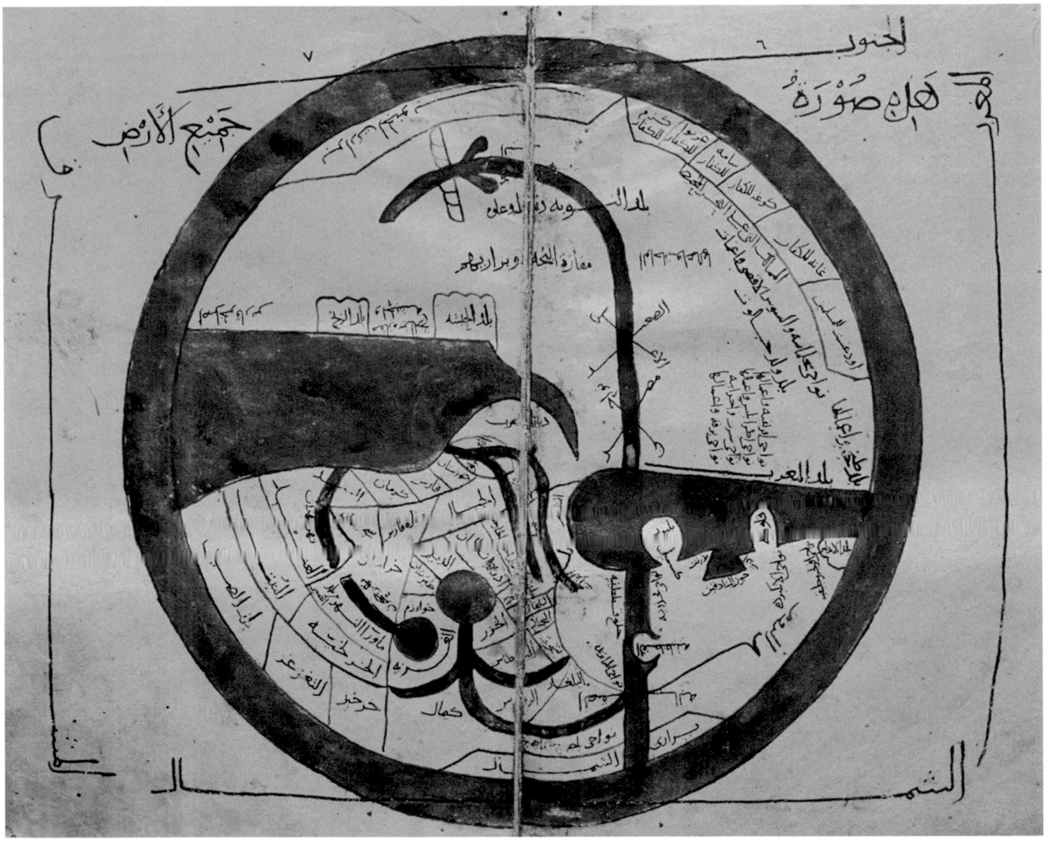

Fig. 4.7. Earliest extant KMMS world map from Ibn Ḥawqal's *Kitāb ṣūrat al-arḍ* (Book of the Image of the World). 470/1086. Iraq(?). Dark-forest-green and dark-blue and red and black ink on paper. Diameter 36.5 cm. Courtesy: Topkapı Saray Museum, Istanbul. Ahmet 3346, fols. 3b–4a. Photo: Karen Pinto.

arches through the upper reaches of Africa and even the cataracts are marked. The detail of places in Africa is greater than later versions, and the southeastern flank of Africa marks sites such as the West African kingdoms of Kanem and Ghana. In contrast, the Asian landmass is densely marked with names while the now triple-forked Volga River (Nahr Itil) is shown as stretching all the way from the Caspian to the Black Sea.[10] The Indus River seems almost to meet up with the Oxus. Most surprising is the greater mimesis of the European landmass. The Peloponnese peninsula of Greece juts out as a circular form, as does the bulge marking the Golden Horn of Constantinople on the Bosphorus. Further along the European coast, we see that the Calabrian peninsula—clearly identified on the map as Qalawriya—also juts out in a distinctively elongated shape.

This map manuscript challenges the traditional narrative of the history of cartography, which is customarily viewed as a logical progression of less mimesis to more. The Islamic mapping tradition went from a state of greater mimesis to lesser, and this curious anomaly defies easy explanation. We know that later copies of the maps in this particular manuscript morph into either elaborately illuminated objects of art with eschatological messages or simpler schematic versions as indicated by the fifteenth-century model (see fig. 4.5). The mid-nineteenth-century rendition of the KMMS Islamic world map shown in figure 4.8 represents the extreme end of this tradition in which the map styles have shifted away from mimesis and fully into the realm of objet d'art.[11]

Gone are the dramatic hook of the Indian Ocean and the bulbous shape of the Mediterranean. Instead they are elongated rectangles with rounded ends. The Nile is diminished in size and length, but the symbols for the cataracts have ballooned into large rectangles with white pillars resembling a forbidding gate. The origin of the Nile in the so-called Mountains of the Moon (*Montes Lunae* of ancient Greek lore)—a reference to the Rwensori range that separates modern-day Uganda and the Congo and is the source of the White Nile—dominates the southernmost sector of the African landmass. The symbolic marker for the site is depicted as a large saffron-colored object with a lighter orange zigzag line in the middle, which together resemble a pair of open lips. The Indus and the Oxus Rivers are two parallel blue slashes and the Oxus is no longer connected to the Aral Sea. All that is left of the Aral Sea is an oval ringed in blue. The Tigris and Euphrates appear to reach up to the Volga, but one cannot be sure because the artist who painted this map forgot to fill in the parallel scallops indicating these rivers with blue. The Volga now reaches from the Caspian Sea all the way to the Mediterranean Sea along the pathway of the Bosphorus. As a result, the Bosphorus no longer cuts through the northern end of the Encircling Ocean, and Europe is now fused with the Asian landmass instead of being an isolated triangle. This rendering reflects the growth in importance of Europe in the nineteenth century. The primacy of the Encircling Ocean is reinforced by a series of additional encircling bands. Surrounding the landmasses is a yellow band of the same color as the zigzag line in the marker for the Mountains of the Moon. Next is an olive-green band with a sinusoidal pattern and hatching representing the familiar Encircling Ocean. Enclosing this is a narrow yellow band followed by a broader orange band decorated with hash marks and painted in the same bright-saffron pigment that is used for the outer lips of the symbol marking the Mountains of the Moon. All these bands are followed by one final slate-blue band embellished with a delicate foliated scroll pattern with leaves. This is the widest band and is perhaps an artistic allusion to the cosmos encircling the world as we find in other Asian mapping traditions.

Why do the Muslim-produced maps move from greater mimesis to less? Is it related to the decline of rigorous intellectual inquiry and the closing of the gates of

Fig. 4.8. Nineteenth-century rendition of a KMMS world map from India. Mid-19th century. Mughal. Dark-blue, forest green, orange, and yellow gouache and red and black ink on paper. Diameter 19 cm. The British Library Board, London. Or. Add. 23542, fol. 60a. © The British Library Board.

ijtihād (Islamic legal term denoting "independent reasoning") after the eleventh century as some scholars argue? Or is it related to the rise of Sufism and a growing romantic perspective of the world such as that which flourished in the Timurid era in Iran? Or is it simply that certain members of the upper class developed a taste for elaborately illustrated carto-geographical manuscripts without the slightest interest in whether they mimicked the real world? If the latter is the case, then the question is why were they not interested in mimetic maps? Is this related to the injunction against representation in Islam?[12] These are examples of the many fascinating questions that arise out of a close study of Islamic maps.

Between the nineteenth-century copies and the earliest extant eleventh-century predecessors there exist a host of renditions with their own distinctive hallmarks. Some are plain yet distinctive, such as the mid-fifteenth-century map (see fig. 4.5) produced in Constantinople following the conquest by Meḥmed II. Other versions, such as the one shown in figure 4.9, are lavishly illuminated in lapis lazuli and gold leaf. Created for presentation to a sultan, this map along with other maps in this Aqqoyunlu[13] manuscript will be examined at length in chapters 10, 11, and 12.

While the mid-fifteenth-century Aqqoyunlu variant represents the height of elegant illumination, the late thirteenth-century map produced in Cairo during the period of the Mamluks, and depicted in figure 4.10, is no less intriguing. Everything is enclosed in thick red lines and the seas have a detailed scalloped pattern resembling waves. The contrast between land and sea is so dramatic that the landmass appears almost to be floating on the seas. The key site markers for this map are images of castles and fortresses. The illustrator of this map has even gone so far as to mark the famous chain that the Byzantines used to guard the mouth of the Golden Horn on the Bosphorus.

Once again the shapes and relative sizes of the seas, rivers, and landmasses vary, confirming Bryson's dictum that essential copies cannot exist.[14] Each map manuscript copy varies from time to place, and these variations, as I will show in later chapters, can be explained by situating each map manuscript in the time period in which it was illustrated.

In the twelfth and thirteenth centuries, a new recension appears in which the Mediterranean and the Indian Ocean/Persian Gulf are depicted as containing three islands each: present-day Sicily (Ṣiqilliya), Crete (Iqrīṭish), and Cyprus (Qubrus) in the Mediterranean; and the islands of Khārg (Khārak), Baḥrain (Awāl/Uwāl), and Lāft (Qishm) in the Persian Gulf/Indian Ocean.[15] The selection of this specific set of islands out of the myriad options suggests that these had special significance. One of the earliest extant examples occurs in the best-known medieval Islamic KMMS world map. I refer here to the distinctive and intriguing KMMS world map from Leiden that opens this book (see fig. 1.1).

This map from a manuscript housed in Leiden caught the attention of map scholars early and was the first Islamic map to be printed in color in the first full-chapter discussion of Islamic cartography in Leo Bagrow's 1964 edition of the *History of Cartography*.[16] Especially striking are the six red islands that dominate the space of the Mediterranean Sea and Indian Ocean/Persian Gulf, and the dark-reddish-brown

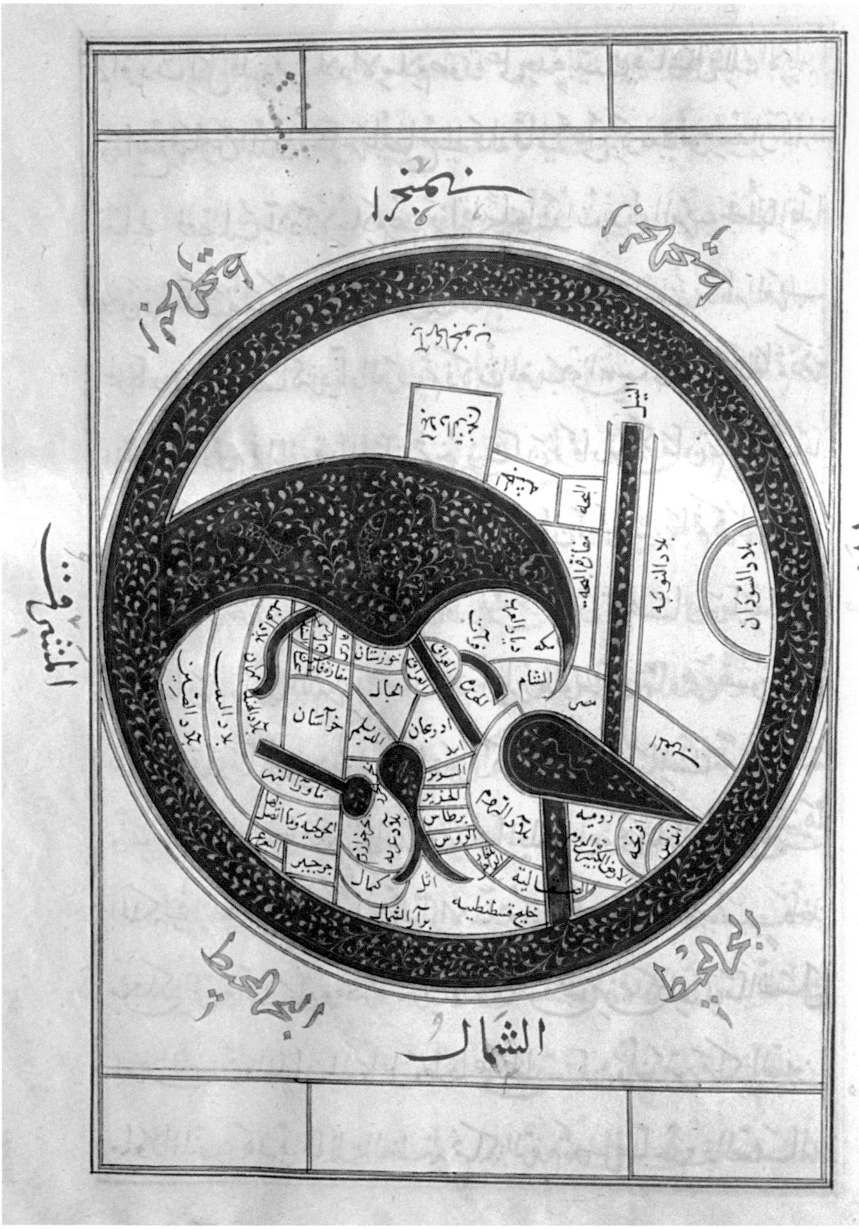

Fig. 4.9. Aqqoyunlu KMMS world map from al-Iṣṭakhrī's *Kitāb al-masālik wa-al-mamālik* (Book of Routes and Realms). Circa mid to late 15th century. Lapis lazuli (ultramarine) gouache with gold and black ink on paper. Diameter 26.4 cm. Courtesy: Topkapı Saray Museum, Istanbul. Ahmet 2830, fol. 4a. Photo: Karen Pinto. (For a close-up, see fig. 12.1.)

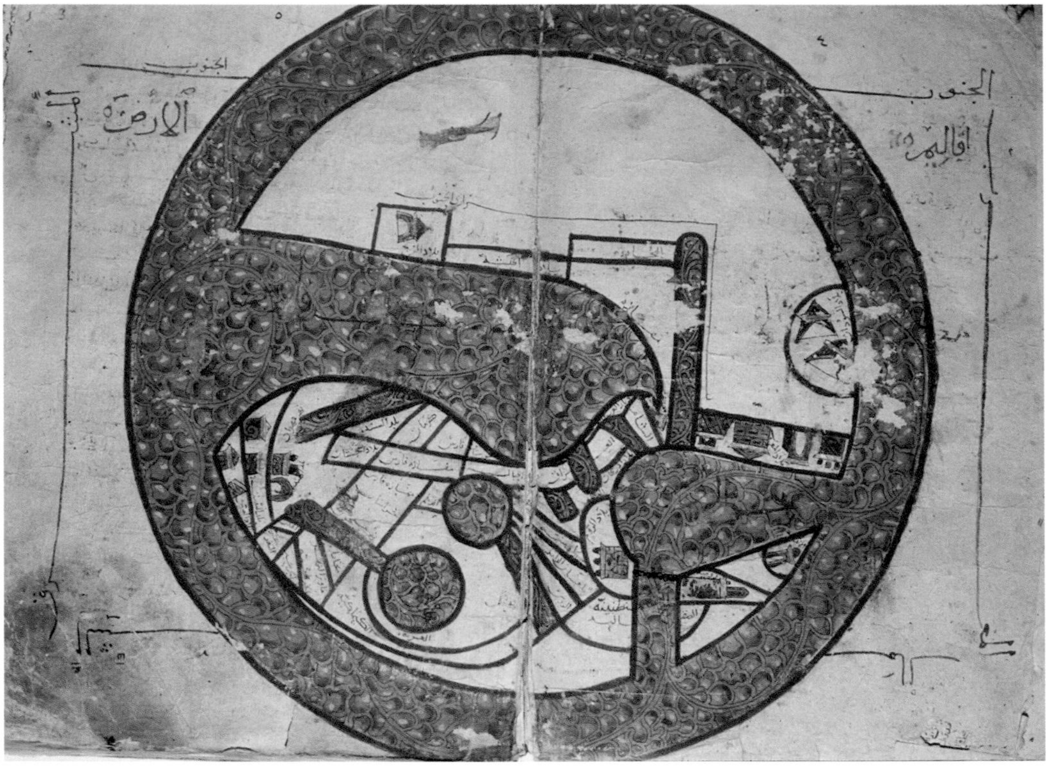

Fig. 4.10. Early Mamluk KMMS world map. 684/1285. Shades of blue, green, red, and brown gouache along with red and black ink on paper. Diameter 35.8 cm. Courtesy: Topkapı Saray Museum, Istanbul. Ahmet 3348, fols. 2b–3a. Photo: Karen Pinto.

gate-like symbol marking Abyssinia. This is a lavish, larger-than-normal-sized map[17] with elaborate calligraphy indicating north, south, east, and west along the outer edges of the double folio.

The KMMS world map in the late thirteenth-century manuscript housed at the Bodleian (fig. 4.11) follows on the Leiden world map model (fig. 1.1). The resemblance between these two maps is so close as to give the impression that the Bodleian world map is based on the Leiden manuscript.[18] As with the Leiden world map, the bright-red islands dominate the image, and the landmasses have similar dramatic flourishes. The execution, however, seems to have been hasty, as suggested by some incomplete patches. It seems that the illustrator forgot to paint the Aral and Caspian Seas with the mauve pigment that is used to set the other waterways apart from the landmasses.

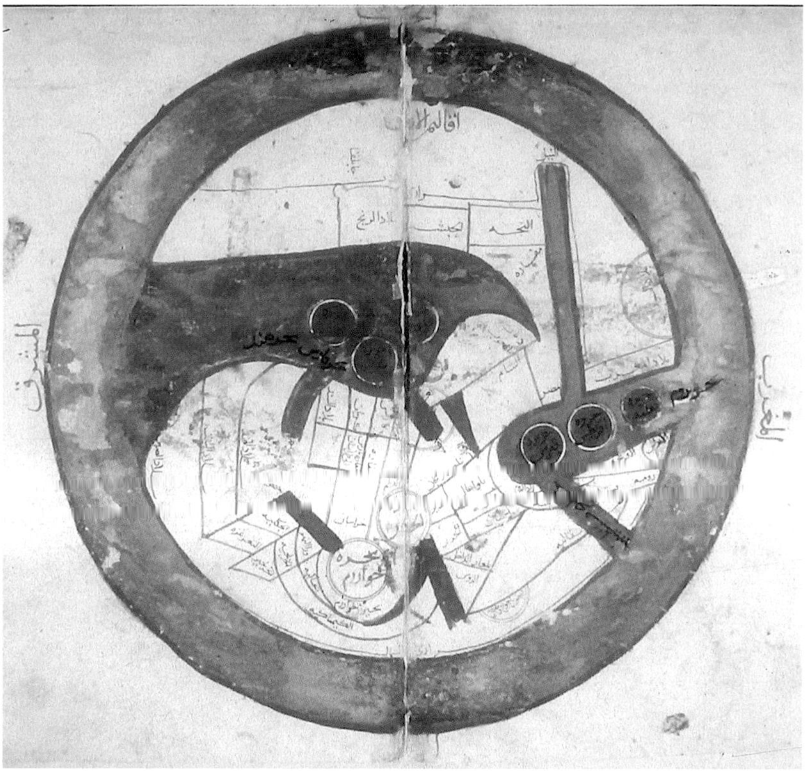

Fig. 4.11. Commoner's version of a KMMS world map. This version represents a break from the elaborately illustrated renditions. 696/1297. Unusual pale-lilac-colored gouache along with red and black ink on paper. Diameter 29 cm. Courtesy: The Bodleian Library, University of Oxford. Ms. Ouseley 373, fols. 3b–4a.

The patchiness of the paintwork and the smudged red lines suggest that this was a hurriedly completed copy that was not intended for high-level/elite presentation as was the Leiden manuscript.[19]

Curiously it is the maps in the Persian manuscripts that follow the stemma of the maps in the Arabic Leiden-Bodleian model. Compare the Timurid world map (fig. 4.1) and the late Timurid/early Safavid example (fig. 4.12) with the Leiden-Bodleian models (figs. 1.1. and 4.11). The three islands in the Mediterranean and Persian Gulf/Indian Ocean are the hallmarks of this stemma. How did the Leiden-Bodleian model generate copies in the Iranian world? There are two possibilities. Either the Leiden-Bodleian model reached Iran and triggered a new stemma of copies. Or the model was developed first in Iran when Arabic was the lingua franca and before the onset of the Persian vernacular from the late tenth century onward.

In the Persian maps shown in figures 4.13 and 4.14 we see that often there is no distinction made between the rivers and the seas, while in the earlier maps sharp distinctions are drawn through the use of color and pattern. For instance, the seas

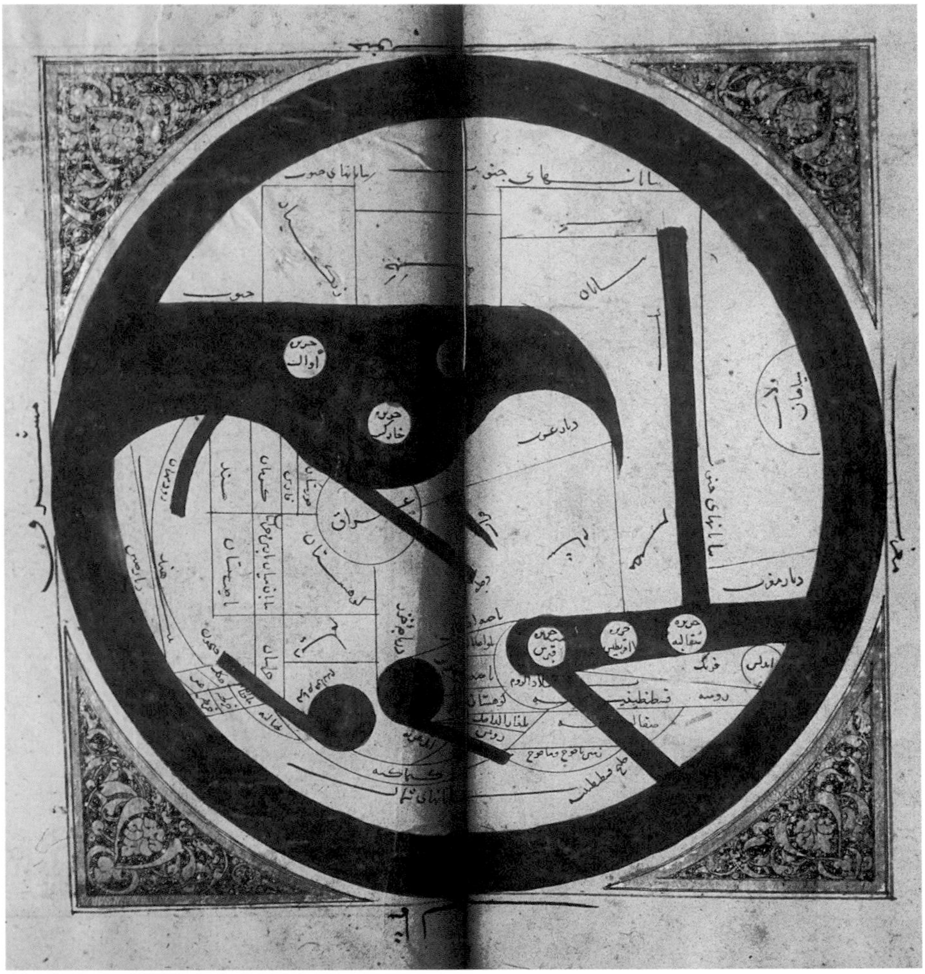

Fig. 4.12. Another Timurid KMMS world map. 15th century. Ultramarine gouache and black ink against a foliated gold and red background on paper. Diameter 19.65 cm. Courtesy: Sülemaniye Camii Kütüphanesi, Istanbul. Aya Sofya 3156, fols. 2b–3a. Photo: Karen Pinto.

and rivers of the earliest extant map of the world (see fig. 4.7) are distinguished by green and blue respectively. Sometimes the seas of the Persian maps are elaborately decorated with wavelike forms. One of the best examples of elaborate "sea-decoration" comes from the Safavid map manuscript shown in figure 4.14. The slate-gray seas are elaborately decorated in a pattern of delicate silver waves.

In addition to circular world maps, three KMMS manuscripts also contain an oblong world map variation. The hall-

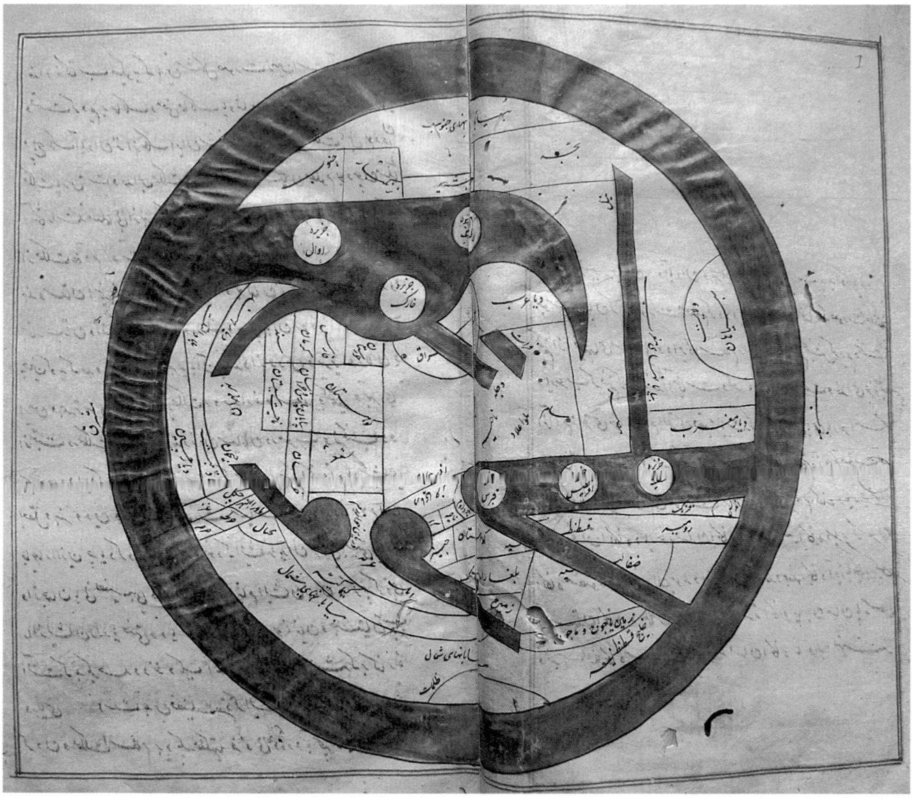

Fig. 4.13. Safavid KMMS world map. Another simple KMMS manuscript probably intended for everyday use. 1165/1751. Plain blue gouache and ink on paper. Diameter 19.2 cm. Courtesy: The Institute of Oriental Manuscripts, Russian Academy of Sciences, St. Petersburg. C-610, fols. 6b–7a.

mark of this oblong version is its emphasis on the Indian Ocean, which is filled with more than a dozen unnamed islands. These oblong maps do not show the provinces, nor do they mark the lands of China and Tibet in the Far East. Instead they indicate the climes using horizontal lines, and in this context the equator is marked. Other than Mecca (marked with a gold square), no other places are indicated. The map from the Topkapı Saray collection in Istanbul, shown in figure 4.15, distinguishes the Indian Ocean and a section of the Atlantic with a dark-green pigment whereas the Mediterranean and the remainder of the Encircling Ocean are illustrated in blue with a wavelike design. Since oblong maps of this form are rare, this book focuses on analyzing the widespread standard circular KMMS world map.[20]

Fig. 4.14. Another Safavid KMMS world map. 1075/1664. Gouache and ink on paper. Diameter 28.4 cm. Courtesy: Topkapı Saray Museum, Istanbul. Revan 1646, fols. 1b–2a. Photo: Karen Pinto.

The modern carto-ideograph of the world as we know it is so firmly planted in our subconscious that it is difficult to see beyond it. We can, however, comprehend that other forms existed, developing slowly over time, and that the earlier maps are as expressive of the societies that produced them as the ones that we know today. Just as we need our familiar modern map to interpret satellite images, similarly we need to bring it into play to conceptualize the perspective of earlier maps.

How did the medieval Muslim cartographers arrive at their image of the world? Why the emphasis on symmetry and geometric forms? Why the elaborate ornamentation? Why are the coastlines perfectly even and, more often than not, mirror images of each other? Why do certain forms dominate the template? What lies behind the seemingly repetitive harmony of forms? How do these combine to generate a dy-

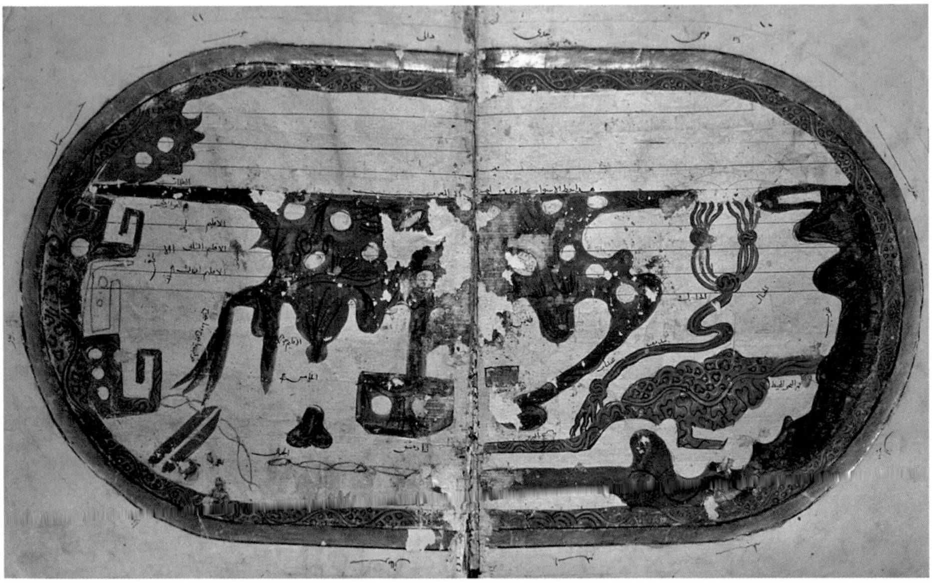

Fig. 4.15. Oblong KMMS world map from a manuscript titled *Ashkāl al-arḍ* (Shapes of the Earth). Circa 14th century. Dark-forest-green, blue, and gold gouache with black and red ink on paper. 49 × 27 cm. Courtesy: Topkapı Saray Museum, Istanbul. Ahmet 3347, fols. 5b–6a. Photo: Karen Pinto.

namic image that incorporates a sense of movement? These are some of the questions that come to mind when one encounters classical medieval Islamic *mappamundi* and lead to the analysis of one unifying feature, the Encircling Ocean, in the next chapter.[21]

Now that we have in hand a basic understanding of the typical KMMS world map and can recognize the constituent forms, we can proceed to the tripartite analysis. In the next chapter, I begin an in-depth iconographic analysis of a particular recurring feature—the Baḥr al-Muḥīṭ (Encircling Ocean). In the course of this analysis, I will show that the origin of medieval Islamic maps is not rooted exclusively in the Greco-Roman tradition, as has been previously presumed. This alternate view has been gaining ground in recent years, and it is my intention here to firmly establish this within the narrative of the history of Islamic cartography.[22]

5

Iconography of the Encircling Ocean

[S]he looked into the water and saw that it was made up of a thousand thousand thousand and one different currents, each one a different colour, weaving in and out of one another like a liquid tapestry of breathtaking complexity; and Iff explained that these were the Streams of Story, that each coloured strand represented and contained a single tale. Different parts of the Ocean contained different sorts of stories, and as all the stories that had ever been told and many that were still in the process of being invented could be found here, the Ocean of the Streams of Story was in fact the biggest library in the universe. And because the stories were held here in fluid form, they retained the ability to change, to become new versions of themselves, to join up with other stories and so become yet other stories; so that unlike a library of books, the Ocean of the Streams of Story was much more than a storeroom of yarns. It was not dead, but alive.[1]

SALMAN RUSHDIE, *Haroun and the Sea of Stories* (pp. 71–72)

Metaforms influence visual practice and thought and emerge in all premodern maps of the world, including Islamic mappamundi, as manifestations of a geographical discourse that transcends time and place. Among the basic forms that make up the KMMS image of the world we find numerous metaforms: the girding shape of the Baḥr al-Muḥīṭ (Encircling Ocean) and the square of the page give rise to the circle within the square; the twin forms of the Caspian and Aral Seas; the reciprocal forms of the Persian Gulf/Indian Ocean and the Mediterranean; the balance of three islands in the Mediterranean matched with three islands in the Persian Gulf/Indian Ocean; ring- and arch-shaped place markers. This assortment of forms make up a whole that is both unique in the combination of its individual signifiers and universal in the overarching form of the double-ring encircling band that can be identified as a logo for an imago mundi (image of the world).

The Encircling Ocean is a crucial metaform that conceals behind its circular simplicity the quintessential stamp of every premodern imago mundi.[2] This chapter is the first of three that collectively demonstrate the applicability of iconographic analysis to medieval Islamic maps. I have chosen the Encircling Ocean as the focus of this analysis for two reasons. First, the form is present in all medieval Islamic world maps, without exception. Second, these maps show a high degree of idiosyncrasy in their terrestrial representations, but in their portrayal of the Encircling Ocean, we find a view shared with many other world-mapping traditions. Thus we have in this form a handle that enables us to grasp the place of these maps within a wider cultural context.[3]

The iconographic roots of the Encircling Ocean plunge deep in time. I begin this chapter with a brief glance at prehistoric examples of the form and proceed to a broad survey of non-European encircling forms. This textual and visual tour of oceanic conceptions will serve as the cornerstone of a foundation for future iconographic exploration of the more unique aspects of medieval Islamic maps.[4] In subsequent chapters I will delve more deeply into Greco-Roman and medieval European connections to the encircling ocean image.

Prehistoric Worlds[5]

Here I will provide just a few examples of numerous extant images that can be read as cosmographic maps.[6] The Neolithic Tulaylāt Ghassūl star fresco from Jordan, dated to the mid-fourth millennium BCE, and the Gold Disk from Moordorf, circa 1500 BCE, found near Aurich, Germany, have both been interpreted by Unger as cosmographical maps of the world surrounded by not one but two encircling seas (figs. 5.1 and 5.2).[7]

Of interest is the recent discovery of the site of Göbekli Tepe in southeastern Anatolia, which may be the earliest extant Neolithic site, dating back 11,000 years. Earlier than Jericho, the site predates settled civilization, the pyramids, Stonehenge (another site with a curious double-ringed circle layout), and even the earliest

Fig. 5.1. "The Star Fresco from Tulaylāt Ghassūl, Jordan." Possible cosmological map with the world at the center surrounded by multiple encircling oceans. 5th–4th millennium BCE. Red, black, and yellow painted wall plaster. Diameter 1.8 m. Source: From George Kish in *HC 1* (plate).

Fig. 5.2. Gold Disk, Moordorf, near Aurich, Germany. Interpreted as a cosmological map with a central continent surrounded by multiple encircling oceans. 1500–1300 BCE. Purified gold mixed with a small percentage of silver and lead. Diameter 14.5 cm. Source: Niedersächsisches Landesmuseum, Hanover/Creative Commons.

Fig. 5.3. Concentric circle structures with T-shaped pillars at Göbekli Tepe, near Şanlıurfa, Turkey. Earliest known religious structures with double-walled concentric circles enclosing T-shaped pillars with bas-relief of animals and people. 10th–8th millennium BCE. Carved stone pillars. Average diameter 300 m. Source: Creative Commons.

evidence for agriculture. It is now recognized as containing the oldest known human-made religious structures, probably constructed by hunter-gatherers. The site consists of a series of double-walled concentric circles with two T-shaped pillars at the center of each set of concentric rings. They bear an uncanny resemblance to early medieval mappamundi forms. (See chapter 6 for a discussion of T-O maps and the like.) Scholars, including the chief archaeologist of the site, have interpreted these structures as symbolic of humans stretching their arms out to the cosmos symbolized by the double-ringed circles (fig. 5.3).[8]

Egyptian Pharaonic Worlds

Ancient Egyptian concepts of the world were dominated by the Nile. Even the sky was seen as a vast body of water composed of the celestial river counterpart to the Nile that encircled the earth:[9]

> Along a ledge that ran like a cornice all around the sides and ends of the box, at a level a little above that of the crest of the mountain-chain, they supposed a great celestial river to flow continually round and round the earth; from near the eastern peak via the south to near the western peak the celestial river flowed between open flat banks, but from near the western peak via the north near the eastern peak its course lay within a deep valley filled with dense shadows. . . . They imagined the sun, moon, and planets to be gods, carried along on separate boats which sailed at different rates round the earth on the waters of the celestial river.[10]

Every Egyptian creation story begins with the idea that before anything, there was only an omnipresent "primordial abyss of waters" personified by the God Nu (later referred to as Nun), the father of all gods.[11] Resonances of this are to be found in the Qurʾān (e.g., 21:30 and 25:54), which also subscribes to a water-based cosmogony, and ḥadīth (accounts about what the Prophet Muhammad said and did). Kisāʾī's late twelfth-century *Qiṣaṣ al-anbiyāʾ*, based in a well-established tradition of tales of the prophets, tells the story of the great serpent that God created to surround the Canopy and that greeted the Prophet Muḥammad warmly during his Night Journey ascent into the heavens.[12] And, in his *Sunan*, Tirmidhī reports the Prophet Muḥammad discussing the skies as a "preserved canopy of the firmament whose surge is restrained."[13] This tale bears an uncanny resemblance to the fearful dragon-snake of ancient Egyptian cosmology that "daily endangered the Cosmic Order by attacking the boat of the sun at dawn and at sunset."[14]

The sky was personified by the Goddess Nut (meaning "the watery one"), who was often depicted as stretching over her male counterpart, the earth god, Geb. From dynastic Egypt there is a cosmographical image etched onto the cover of a stone sarcophagus of Wereshnefer, from Saqqara, dated to the beginning of the Ptolemaic period (the Thirtieth Dynasty, ca. 350 BCE), depicting the land of Egypt with its neighbors sheltered by the overarching body of the Goddess Nut (fig. 5.4). Encasing the lands of Egypt is a narrow encircling band encasing a depiction of the world. This double-encircling band can be interpreted as an Encircling Ocean stamp, signifying it as one of the earliest imagines mundi.[15]

Mesopotamian Worlds

From Mesopotamia, we have the most famous ancient map of the world, known generically in history of cartography circles simply as the Babylonian world map. Found on the lower half of the obverse side of a tiny (122 × 82 mm) unbaked Babylonian clay tablet of circa 600 BCE, it is the earliest clearly identified map of the world (fig. 5.5). One of the most famous artifacts of the British Museum collection, it was adopted in the 1930s by Leo Bagrow as the logo of *Imago Mundi*—today the oldest and most respected journal for the history of cartography.

Fig. 5.4. Cosmographical map of the land of Egypt with the sky personified by the Goddess Nut. South is on top. The double encircling band indicates the Encircling Ocean. Early Ptolemaic period, 380–300 BCE. Saqqara, Egypt. Gray granodiorite sarcophagus. Diameter of map 72 cm. The Metropolitan Museum of Art, New York. Gift of Edward S. Harkness, 1914 (14.71a, b). Image copyright © The Metropolitan Museum of Art. Image source: Art Resource.

Unlike other ancient and prehistoric images, which involve—as we shall see—a lot of subjective interpretation, the text surrounding the image in the clay tablet, along with the sites marked on it, sets it out clearly as a map of the Mesopotamian world with cosmographic overtones. Without expert assistance, the features of the world map are at first hard to discern (see the decoded template in fig. 5.6).

The map, oriented with north on top, places Babylon at the center, encircled by a variety of important neighbors such as Assyria, Urarṭu, the land of Ḫabbān, the city of Deri, and so on, indicated by small circular icons. From the north, a parallel line descends southward. Experts have identified this line with the Euphrates River.[16] The parallel line ends in an area referred to as the land of "Bīt-Yakīn" (Sea-Country), which terminates in a broad band ambiguously marked on the map as "channel," on one side, and "swamp," on the other. This band in combination with its hook-like end is the earliest visual representation of the Persian Gulf.[17] The channel then finds its way to the "Encircling Ocean / Bitter River," which rings the image. Beyond this lie a

Fig. 5.5. Babylonian world map. Earliest identified map of the world. Oriented with north on top, it shows key city-states of the period with Babylon at the center. Circa 700–500 BCE. 12.2 × 8.2 cm. British Museum, Inv. 92687. © The Trustees of the British Museum / Art Resource, NY.

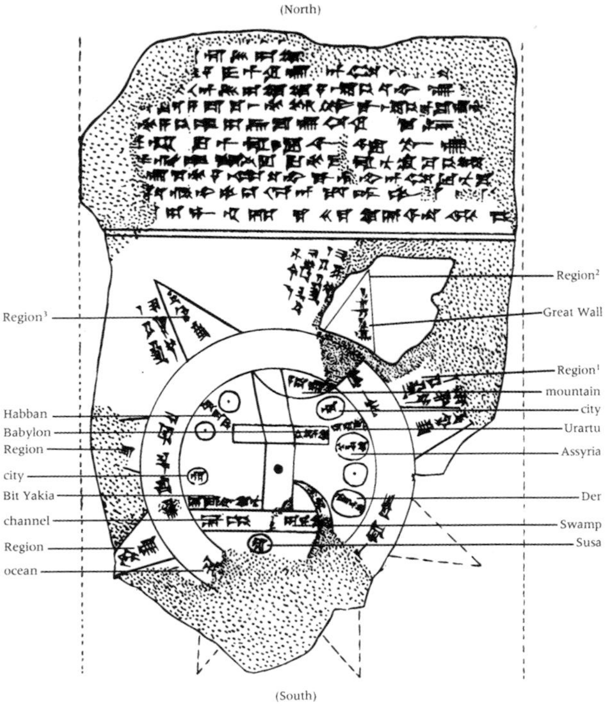

Fig. 5.6. English identification of features of the Babylonian world map. Drawn by R. Campbell Thompson and amended by C. B. F. Walker. Courtesy: Catherine Delano-Smith and The Imago Mundi Limited.

series of seven triangular spokes, which are a reference to mythical lands/islands that lie between the encircling "Bitter River" and the "Cosmic Ocean" beyond.[18]

In the context of the Islamic maps, the feature that catches our eye is the double-ringed circle composed, as the enclosing text informs us, of *Marratu*—"the Ocean" (also referred to as the "Bitter River").[19] The way in which the seven triangular islands mediate between the earthly encircling ocean and the heavenly encircling one also has parallels in Islamic geo-cosmography: namely, with the Jazā'ir al-Khālidāt (the "Heavenly Islands," taken as a reference to the Canary Islands), which lie at the edge of the world in the Encircling Ocean.[20] The emphasis in the accompanying cuneiform text is on these seven islands and their attributes.[21] Just like the Muslim Encircling Ocean, the Babylonian one is filled with terrifying animals and monsters.[22] According to the Babylonian cosmogony *Enûma Elish*, both the earthly and the cosmic oceans emerged out of the primeval ocean and the battle to overthrow the old world order of Apsû and Tiamat:

> When on high the heaven had not been named,
> Firm ground below had not been called by name,
> Naught but primordial Apsû, their begetter,
> (And) Mummu-Tiamat, she who bore them all,
> Their waters commingling as a single body;
> No reed hut had been matted, no marsh land had appeared,
> When no gods whatever had been brought into being,
> Uncalled by name, their destinies undetermined—
> Then it was that the gods were formed within them.[23]

It was the discovery of this incredible little clay tablet that opened up unlimited possibilities of interpretation of premodern images, such as those embarked on by Eckard Unger in the 1930s and Catherine Delano-Smith in the 1990s.[24] As Delano-Smith says in her reexamination of this map, "[For a change,] the historian of cartography is spared no doubt as to the interpretation of the schematic representation by the labeling of the various lines and circles."[25] It is its inherent clarity of meaning and purpose that makes the Babylonian clay tablet an excellent basis for interpretations of other ancient and prehistoric objects bearing similar encircling motifs as a marker for the imago mundi.[26]

Iranian Worlds

Here we have a larger set of extant works with dates of origin spanning from prehistory to the Islamic conquest of Iran in the mid-seventh century. The best and most intriguing prehistoric examples come from Susa in Khuzistan (southern Iran,

Fig. 5.7. Bowl with geometric pattern from Susa showing the Mazdean cosmological concept of complementary seas: the Vurukasha (Wide-Shored or Wide-Formed Pure) and the Pūitika (Impure). Circa 3100–3000 BCE. Susa, Iran. Courtesy: Louvre, Paris / Bridgeman.

bordering Iraq). Dating as far back as the middle of the fourth millennium, they are coterminus with the Tulaylāt Ghassūl star fresco and the Gold Disk from Moordorf. The intricate anthropomorphic and geometric designs on these Susa artifacts have intrigued archaeologists and art historians alike since the turn of the twentieth century.[27]

What is striking about this early Susa ware is the numerous circular, semicircular, rectangular, and hexagonal motifs mixed with stylized animal and bird forms.[28] Phyllis Ackerman, in a seminal piece, "Symbol and Myth in Prehistoric Ceramic Ornament," sees cosmographical maps in these images. Ackerman interprets these designs of two semicircular halves joined by a square channel, such as the one shown in the bowl in figure 5.7, as a representation of the Mazdean cosmological concept of complementary seas: Vurukasha (Wide-Shored or Wide-Formed Pure) and Pūitika (Impure). Water flowed from the impure sea to the pure one and back again in a cosmic

Fig. 5.8. Templates of Susian bowls showing double-sea encircling ocean motifs. Created by Phyllis Ackerman. Courtesy: Mazda Publishers, Inc.

cycle encircling the earth—above and below.[29] Ackerman classifies the range of these double-sea encircling ocean motifs in template form as shown in figure 5.8.

Given the importance of Iranian influence in the development of medieval Islamic culture, especially at the beginnings of Islamic cartographical inquiry, it is customary to infer that the development of the Islamic world map was influenced by pre-Islamic Iranian concepts of the world. The influence of the Zoroastrian *kishwar* (also referred to as the *iqlīm*) system on Muslim geography with Iraq and Iran at the center is widely accepted in scholarly circles (fig. 5.9). In this system, which Muslim geographers adopted from Sassanid legacy, the world was divided into seven equal circles, each representing a region, with the fourth circle containing Iraq, Fars, Khurasan, and other Iranian provinces at the center.[30]

There is one image that emerges from the Achaemenid material record that fits the encircling motif perfectly and supports the idea that the double ring morphs into a signifier for power and dominion over the world (fig. 5.10). This is the classic and ever-present symbol for Ahura Mazdā—the chief God of the Mazdean (Zoroastrian) pantheon with whom the prophet Zarathustra communicated.

The symbol has traditionally been interpreted as a bird figure around a symbol of the sun.[31] In this book I propose that the inner ring of the symbol can also be read as an imago mundi symbol. This deceptively simple sign is ubiquitous in Achaemenid architecture, occurring as a stamp above a variety of scenes such as royal hunts, battles, and investiture ceremonies, as well as arches. This interpretation is

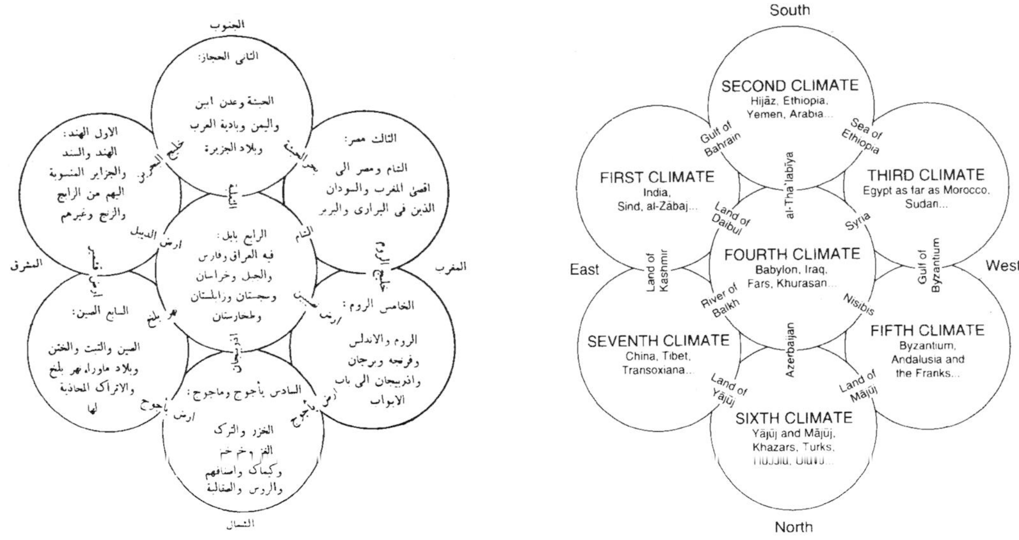

Fig. 5.9. Iranian *kishwar* system as represented in al-Bīrūnī's *Kitāb taḥdīd nihāyāt al-amākin*. 14.5 × 13 cm. Source: Ahmed Zeki Velidi Togan, ed., *Biruni's Picture of the World*, Memoirs of the Archaeological Survey of India, no. 53 (Delhi, 1941), 61. Courtesy: Archaeological Survey of India, New Delhi. Translation by Gerald Tibbetts.

reinforced by the use of the Ahura Mazdā symbol in the Behistun Inscription of Darius I (522–486 BCE) as he boasts to the enslaved people lined up in front of him the many regions under his control from Persia, Elam, Babylonia, Assyria, Arabia, Egypt, Lydia, Greece, Media, Armenia, to Bactria, Sogdia, Gandara, and Scythia (fig. 5.11). Darius credits the Zoroastrian God Ahura Mazdā for granting him this empire. It is not surprising to find that the inscription is sanctified with the Ahura Mazdā symbol reinforcing the cosmos-ruler connection through control of the world, which is literally spelled out below the image and further reinforced by the many people shown as enslaved subjects of the Achaemenids.

With the rise of the Arsacid Parthians,[32] we have, on the one hand, Hellenistic influences and, on the other, the onset of what some have termed a vigorous Neo-Iranian movement, which reasserted Achaemenid ideas syncretized with those of their (Arsacid) northeastern Iranian origins.[33] Like their Achaemenid predecessors, the Parthians promoted monumental art in the form of rock carvings and giant-sized statues, and they serve as the bulk of the extant record for the period.[34] Of note is the Parthian penchant for building "round" cities—Ḥaṭrā, Dārāb (also known as Dārābjird) and Firuzābād. These circular Parthian cities provide one possible answer to the

Fig. 5.10. Ahura Mazdā symbol. The symbol of the chief Zoroastrian god shows a double-ringed circle surrounded by a pair of wings. Circa 550–330 BCE. Persepolis, Iran. Bas-relief. Courtesy: Reza G. Ebrahimi. Photo: Reza G. Ebrahimi.

enduring mystery of why the caliph Manṣūr selected a circular design for the Abbasid capital, Baghdad, that he founded in AH 145/762 CE.[35]

We find further support for this interpretation of the double ring as a symbol of the world in another stone relief, this time of the investiture of Ardashīr I (ca. 226–242 CE), founder of the Sassanid Empire (fig. 5.12). This image shows Ahura Mazdā bestowing the "ring of kingship" upon Ardashīr. We know from extant examples of medieval Persian cosmographies that these ancient rock carvings in Iran were valued as wonders of the world. As discussed in chapter 3, some of these cosmographies contained renditions of KMMS maps.[36] The Ahura Mazdā image fits with an Islamic tradition of conceiving the landmasses of the world as a bird. The consensus among Orientalist scholars, Kramers, Ahmad, Tibbetts, and others, is that this world-as-bird concept is Iranian or Oriental in origin, and definitely pre-Islamic.[37] Like Ibn Ḥawqal, I believe that if the KMMS world is viewed from the right angle, it is possible

Fig. 5.11. Achaemenid inscription of Darius I at Behistun with the Ahura Mazdā symbol amid an inscription naming the regions that the Achaemenids control above figures representing enslaved peoples. The inscription includes a life-size bas relief of the Achaemenid ruler Darius I. 522–486 BCE. Mount Behistun, Kermanshah, Iran. Carved on limestone cliff. 15 × 25 × 100 m. Source: Creative Commons.

to spot the bird in the map.[38] (Readers can attempt this exercise using a sampling of the KMMS maps in chapter 4.)

With the Sassanians (224–642 CE), we move into a period of abundant material remains. Although there are no known extant Sassanian maps, there is no question that the circle, specifically the double-ringed circle, was a favorite form of Sassanian artists.[39] There are an enormous number of finely crafted round silver Sassanian plates in museum collections around the world (fig. 5.13). These generally consist of figures and scenes involving a Sassanian king surrounded by a double-edged circle, which is in turn surrounded by a constellation of repeat forms that resemble the prehistoric and ancient forms discussed earlier.[40]

The handsome gold bowl with an image of King Khusraw II carved in rock crystal at the center, surrounded by a ring of red bead studs, and a galaxy of red, green, and white glass inlay, presents a resplendent example of this classical Sassanian model (fig. 5.14). The bowl, which is housed at the Bibliothèque nationale de France and dated to the sixth century CE, is also referred to as the "Cup of Solomon." There is much debate over how the cup came to be in France but its Iranian/Central Asian provenance is not disputed. There is a myth surrounding this cup that claims it had "world-showing" powers and belonged over the ages only to the most powerful of leaders—such as Alexander the Great, the Sassanid Kay Khusraw, Charlemagne, and eventually the Ottoman sultan Süleymān the Magnificent.[41]

Fig. 5.12. Rock relief of the investiture ceremony of the first Sassanid king, Ardashīr I, being handed the "ring of power" by Ahura Mazdā. The inscription contains the oldest attested reference to the use of the term "Iran." 226–242. Naqsh-i Rustam, Iran. Carved stone relief. Source: Creative Commons. Photo: Ginolerhino.

In the absence of extant examples of Sassanian maps, we do know that Mazdean Zoroastrianism, the state religion in Iran from the Achaemenids until the arrival of Islam, built an elaborate cosmographical system that conceived of the earth as encircled by an ocean called the Vurukasha Sea. The world surrounded by the Vurukasha Sea was thought of as divided into seven climes—*haft kishwar* (fig. 5.15). At the center was a gigantic mountain, Hara, through whose 360 apertures the sun moved to create 360 days and nights. The whole world was surrounded by a series of high mountains,[42] which were encased by a stone sky above and water below.

Water is a crucial element in Zoroastrian cosmography, revered for its purity. All water was thought to have a common source, circulating from below the earth to the peak of Mount Hara from where it poured into the Vurukasha Sea. This water then circulated through the *haft kishwar* before going below the earth and then back up to Mount Hara again for purification.[43] A hymn to the river goddess Arədvī Sūrā Anāhitā tells us about the Vurukasha Sea, its attributes, and the purifying quality of water:

Fig. 5.13. Sassanian silver bowl with bust of ruler surrounded by a double-ringed circle amid a constellation of circles. Late 3rd–4th century. Iran. Cast and hammered silver. 7.6 × 23.7 cm. Courtesy: Cincinnati Art Museum / Bridgeman. Gift of Mr. and Mrs. Warner L. Atkins, 1955.71.

> Arədvī Sūrā Anāhitā . . . immense, far-famed, who is as great in her immensity as all these waters which flow forth upon the earth; who, mighty, flows forth from Mount Hukairya upon the sea Vurukasha—all the edges of the sea Vurukasha are turbulent, all the middle is turbulent, as Arədvī Sūrā Anāhitā pours forth upon them. . . . The outflow of that one sea will pour forth over all the seven regions. She pours down her waters summer and winter alike. She purifies the waters, she purifies the seed of males, the womb of females, the milk of females.[44]

In his seminal book on the Greco-Arabic translation movement in Baghdad during the initial years of the Abbasid caliphate, Dmitri Gutas argues that "Zoroastrian imperial ideology and political astrology fused together [to form] the cornerstone of al-Manṣūr's Abbasid dynastic ideology." Following the Abbasid revolution (747–750 CE) and the transfer of the seat of the caliphate to Iraq and the new capital of Baghdad (founded by the caliph Manṣūr in 762), the Abbasids found themselves in a completely different milieu. One that was Persianized and not dominated by the

Fig. 5.14. Cup of Solomon or Khusraw. Gold, rock crystal, red bead studs, red, green, and white glass. Courtesy: Bibliothèque nationale de France, Paris, CA-MEE 379. Photo credit: Erich Lessing/Art Resource, NY.

influence of Byzantine culture as had been the case of the Syria-and-Damascus-based Umayyad caliphate (661–750) that the Abbasids had overthrown. In order to keep the support of their new and powerful Persian constituency—on whose backs the Abbasids had come to power—the caliph Manṣūr (considered the founder of the Abbasid state and builder of Baghdad) promoted the thinking that "the ʿAbbāsid dynasty, in addition to being the descendants of the Prophet and hence satisfying the demands of both Sunnī and Shīʿī Muslims, was at the same time the successor of the ancient imperial dynasties in ʿIrāq and Iran, from the Babylonians through the Sasanians, their immediate predecessors. In this way they were able to incorporate Sasanian culture, which was still the dominant culture of the large masses of the population east of ʿIrāq, into mainstream ʿAbbāsid culture." Manṣūr, the architect of this policy, did this by initiating a major translation movement of traditional Zoroastrian material into Arabic so that it could be harnessed by the caliphate as a tool of propaganda to reach the Arabized Persians. Since all the Zoroastrian sciences derive from its canon, the Avesta, the drive was to make this available in Arabic, and it was in this vein that Manṣūr and his successors revived the Sassanian tradition of "houses

Fig. 5.15. Ancient Iranian picture of the world. Reconstructed from texts by Mary Boyce (namely, *Textual Sources for the Study of Zoroastrianism*, 17).

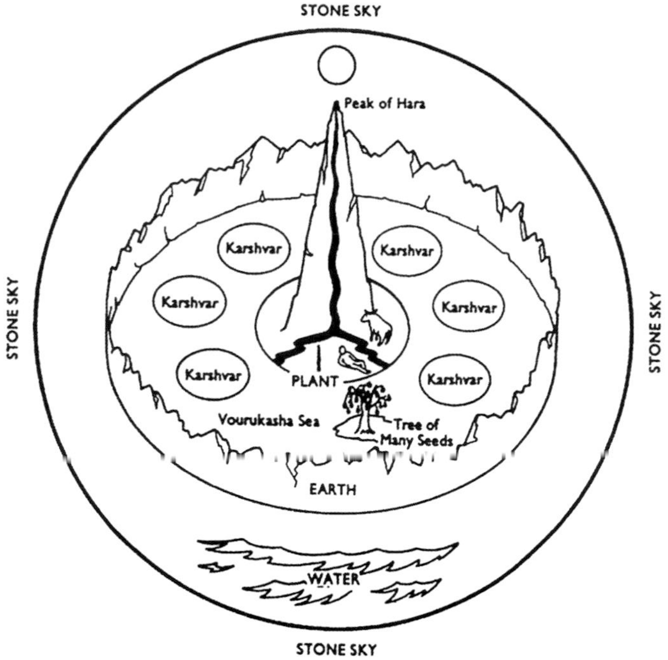

of wisdom" (buyūt al-ḥikma)—royal libraries—where pre-Islamic Sassanian and Persian historical lore was deposited. It is from this tradition of *buyūt al-ḥikma* that the movement of Abbasid learning and sciences was born. Scholars customarily credit the caliph Ma'mūn, Manṣūr's great-grandson, as the great patron of learning through the *Bayt al-ḥikma* (House of Learning) when, according to Gutas, the process dates back to the roots of the Abbasid dynasty.[45]

Other scholars have argued for the influence of pre-Islamic Iranian ideology on Islamic cartography on the basis of the following: the hallmark of the ancient Iranian *haft kishwar/iqlīm* system, the centrality of the provinces of Iran, and the fact that the earliest known father of this tradition came from the region of Iṣṭakhr. Until now the possible influence of Susian pottery, Zoroastrian cosmography, the Ahura Mazdā symbol, and the Parthian and Sassanian penchant for round cities and double-ringed royal plates have been left out of the analysis.[46]

Semitic Connections

Classical Jewish writing focuses on God rather than the cosmos. As the Rabbi Louis Jacobs notes: "Jews did not develop in any period of their history a special cosmology of their own. They adopted or accepted the cosmologies of the various civilizations in which they lived, but utilized these for the religious purposes with which they were primarily concerned."[47] Jacobs argues that if one were to study Jewish cosmology, then one should think of it in the plural as a syncretic tradition with many interpolations.

The Hebrew Bible is one of the primary influences, but the Bible is notoriously terse and ambiguous when it comes to geographical and cosmological material. The same passages of Genesis as discussed in the medieval European context (see chapter 6) affect Jewish cosmology. Jewish eschatology refers to a subterranean source of water, like the Vurukasha Sea of ancient Zoroastrian cosmology, that contains the notion of a single spring as the source of all water on the earth and around it.[48] Using reconstructions dating back to 1909,[49] the contemporary graphic designer Michael Paukner has created an elegant rendition of the "ancient Hebrew conception of the universe," that shows the "waters above the firmament," with its storehouse for snow, hail, and wind and its connection to God and the Heaven of Heavens through the Gate of Heaven. Below the earth and the sea are the "storehouses and fountains of the Great Deep" (fig. 5.16). In keeping with Psalm 104, the waters encircle the pillars like a garment, standing above the mountains.[50] Ginzburg tells us that the "Great Sea that encompasses the earth" is given to weeping and misery on account of its separation from the upper and lower waters[51]—a notion parallel to that of the Muslim Encircling Ocean, as we shall see in chapter 7.

It has been suggested that the Temple of the Holy of Holies in Jerusalem was constructed as a microcosm of the cosmos, with the outer house corresponding to the earth and the laver to the sea.[52] From Cosmas Indicopleustes (sixth century CE), who is notorious for his "flat earth" beliefs, we get a description of a table in Moses's first Tabernacle: "On this table again he ordered to be daily placed twelve loaves of shewbread. . . . He commanded also to be wreathed all around the rim of the table a waved-molding, to represent a multitude of waters, that is, the ocean; and further, in the circuit of the waved work, a crown to be set of the circumference of the palm of the hand, to represent the land beyond the ocean and encircling it, where in the east lies Paradise, and where also the extremities of the heaven are bound to the extremities of the earth."[53] While discussing the forms of the world, Yāqūt mentions one of a table similar to the construction attributed to Moses.[54]

The Book of Jubilees (composed during the second century BCE), describes the wondrous journey of Enoch as he walked with God on a tour to the ends of the earth. All scholarly attempts to create a map based on this text include an encircling river because it is a standard feature of Enoch's account.[55] Philip Alexander argues that the world map in the Jubilees tradition should be considered the earliest extant imago mundi, and Scott, while skeptical of Alexander's assertion, suggests that "although one can quibble about whether the *Genesis Apocryphon* was actually the earliest witness to this *imago mundi*, nevertheless a case can be made that the *Jubilees 8–9* tradi-

Fig. 5.16. Ancient Hebrew conception of the universe. Courtesy: Michael Paukner.

tion was preserved in apocalyptically oriented Christian circles from the time of the New Testament and for centuries thereafter, and that it may have influenced some of the medieval *mappae mundi*."[56] Based on the Hebrew Bible, the series of images regarding the creation of the world at the outset of the 1493 *Nuremberg Chronicle*—one of the earliest incunabula—reinforces the view that even as late as the end of the fifteenth-century illustrators saw and used the double-edged ring as the baseline symbol to signify the imago mundi (fig. 5.17).

From the Babylonian Talmud comes a number of myths, such as the one about Leviathan—the monster fish of the great sea—created on the fifth day as the plaything of God.[57] We can draw parallels between the monstrous fish that inhabited the

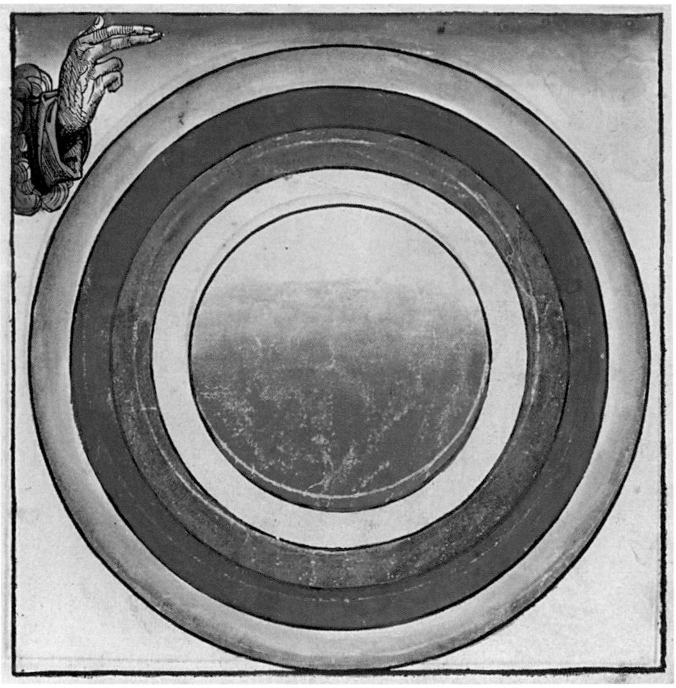

Fig. 5.17. Creation of the world from the *Nuremberg Chronicle.* Image shows the world at the time of creation surrounded by multiple encircling oceans. Comes from one of the earliest incunabula of an illustrated biblical paraphrase and world history. 1493. Nuremberg, Germany. Source: Creative Commons.

Muslim Encircling Ocean and the Talmudic one. Through a reading of the opening passages in Genesis, it would seem reasonable to conclude that water is considered in Jewish cosmology to have a hostile character because of its role as a chaotic elemental presence preceding the creation of the world. The Torah tells us, for instance, that God sealed up the ocean (tehom) so that it wouldn't inundate the world.[58]

This is yet another parallel with the Islamic interpretation of the Encircling Ocean. (See chapter 7 for details on gigantic fish and the negative connotations of the ocean.) The question of *isrā'īliyyāt*—narratives, fables, and historical material borrowed from Jewish sources—is a matter of considerable debate among Islamic scholars, present and medieval. Some say that the Prophet Muḥammad himself gave Muslims permission to borrow from the Jewish tradition.[59]

Indic Worlds

A confirmed influence on both the Iranian and Muslim mentalité is the neighboring Indic tradition. Muslims acquired their system of numerals—which even today are referred to as "arabic numerals"—from the Indian astrologers. We also know that some Indian astrological works, such as the *Siddhānta*,[60] were translated into Arabic, but

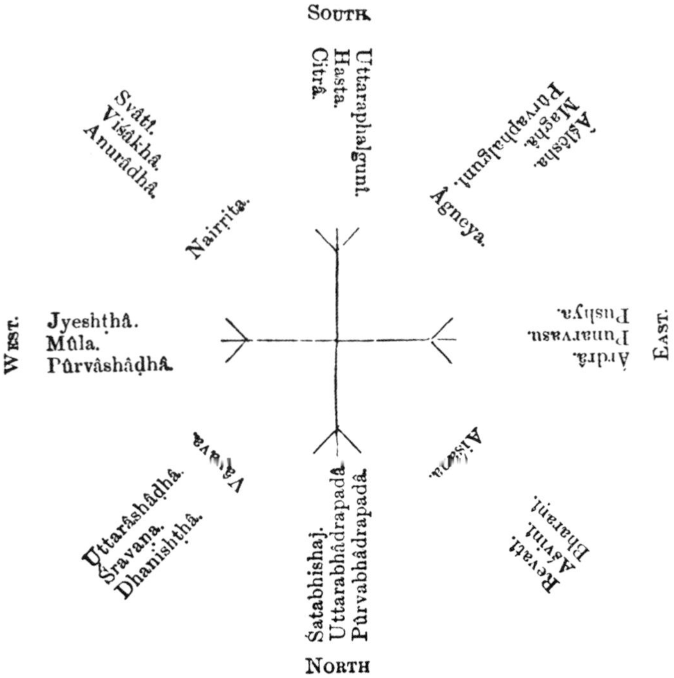

Fig. 5.18. Al-Bīrūnī's sketch of the Purāṇic concept of the earth in the shape of a tortoise resting on the universal sea. Sketch translation by Edward Sachau of al-Bīrūnī's *Kitāb fī taḥqīq mā lil-Hind min maqūla* (Book on the Verification of What Is Said about India). Source: Sachau, *Alberuni's India*, 297.

the majority, according to Dimitri Gutas, "passed into Arabic mainly through Persian (Pahlavī) intermediaries during the Abbasid period."[61] Among scholars of Islamic astronomy, it is widely acknowledged that the *qubbat al-arḍ* (cupola of the earth)—the site of the designated prime meridian from which Muslim astronomers calculated the latitude and longitude of places for their zīj tables—was referred to as Arīn, a derivation from the Indian name of Ujjain.[62] From Ibn Ḥawqal we learn that he consulted an "Indian map located at al-Qawāriyān."[63] And the eleventh-century polyglot Bīrūnī, considered one of the most accomplished scientists of the medieval Islamic period, displays in his book on India, *Kitāb fī taḥqīq mā lil-Hind min maqūla* (Book of the Verification of What Is Said about India), a deep understanding of Hindu cosmographies of the world and the universe, which he contrasts with those of ancient Greek scholars.[64] This indicates Indic cosmogonies were part of the knowledge base of Muslim geographers.[65]

The Purāṇas, the Rāmāyaṇa, and the Mahābhārata all tell of the importance of the Universal Sea, which figures in numerous cosmic stories, such as the "Churning of the Milky Ocean" and the extraction of the elixir of immortality (*Amrita*) from the Universal Sea. Prior to this, the god Vishṇu had to descend to the bottom of the Universal

Fig. 5.19. Purāṇic vision of the world in its cosmic setting showing Krishna and his consort on a tortoise-shaped earth resting on the Universal Sea. From a folio of the *Bhagavata Purana* (Ancient Stories of the Lord) included in *Scenes from the Story of Narakasura*. 1775–1800. Nepal. Gouache and ink on paper. 34.3 × 51.8 cm. Courtesy: Los Angeles County Museum (LACMA). Gift of the Michael J. Connell Foundation, M.72.3.1.

Sea to rescue the freshly created earth.[66] As in the case of the Nile for the ancient Egyptians, the Ganges is thought of as the celestial river that descends from the sky to save the waterless earth because Agastya, offspring of the gods Mitra and Varuṇa, one day unwittingly swallowed the entire ocean.[67] In Hindu mythology the Highest Being is in the form of water, and it is on this primordial water, the Milky Ocean, that Vishṇu sleeps, and it is from his body that new universes are constantly being born and old ones extinguished. One way of representing this vision of the primeval ocean is via a tortoise-shaped earth resting on the oceans, which may have been an attempt to visually translate the Purāṇic concept of Vishṇu lying on the surface of the Universal Sea. In his famous book on India, Bīrūnī discusses and sketches this Purāṇic vision of the world; a copy is shown in figure 5.18. Figure 5.19 shows an eighteenth-century Mughal rendition of Krishna and his consort on a tortoise-shaped earth.[68]

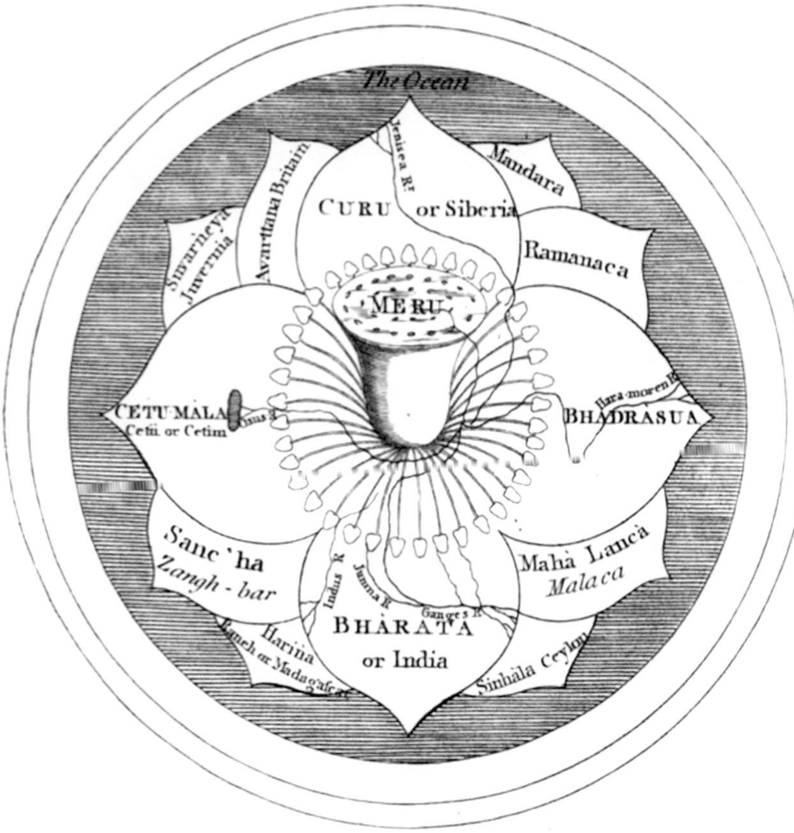

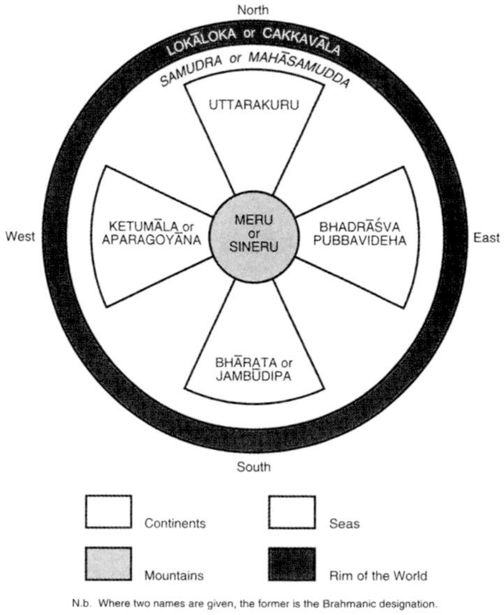

Fig. 5.20. Interpretation of Indian cosmography as a Lotus flower surrounded by an Encircling Ocean with Mount Meru at the center. Drawing by Francis Wilford. 1808. Source: Wilford, "Sacred Isles," 376.

Fig. 5.21. Early Brahmanic Hindu and Buddhist conception of the Catur-Dvīpa Vasumatī (Four-Continent Earth). Idealized view of the ancient Hindu conception of the earth. 1992. Drawing by Joseph Schwartzberg. Source: *HC 2.1*, 336, fig. 16.1. Courtesy: Joseph Schwartzberg and the History of Cartography Project.

Fig. 5.22. Young man constructing a petal-shaped Catur-Dvīpa Vasumatī mandala with colored chalk in an open field behind his house. 1998. Madras, India. Courtesy: Richard Bulliet. Photo: Lucianne Bulliet.

Another way of representing this Purāṇic vision is via the early Brahmanic Hindu conception of the Catur-Dvīpa Vasumatī (Four-Continent Earth), in which the world is conceived as a lotus-shaped form arranged on the surface of an encircling ocean. This is illustrated in two different reconstructions: one by Francis Wilford, from textual descriptions of a lotus-shaped world floating on the ocean that encircles it,[69] and the other by Joseph Schwartzberg, which is a less stylized reconstruction of the same (figs. 5.20 and 5.21).[70] These colorful Catur-Dvīpa Vasumatī mandalas are still being produced in the streets of India today (fig. 5.22),[71] as they were in the eleventh century when the scholar Bīrūnī spent time in India exploring and researching at length the various kinds of *dvīpa*—which he explains as the Indian word for island.[72]

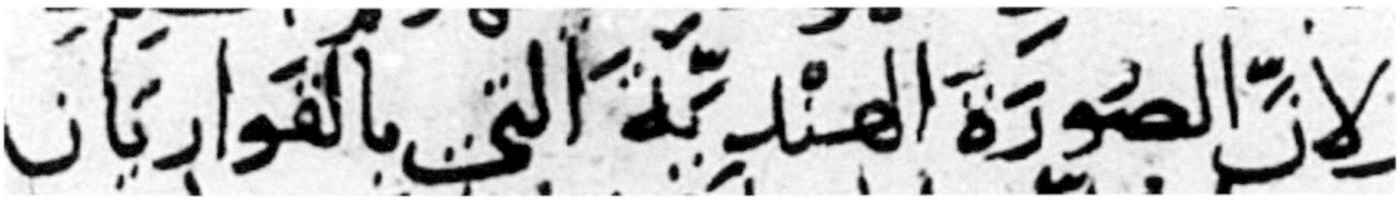

Fig. 5.23. "Qawārīyān" from Ibn Ḥawqal's *Kitab ṣūrat al-arḍ* (Book of the Picture of the Earth). This clip says, "Because the Indian map that is in Qawārīyān . . ." Scholars are still trying to determine what this mysterious "Qawārīyān" Indian map looked like. Courtesy: Topkapı Saray Museum, Istanbul. Ahmet 3346, fol. 1b.

Given the wealth of cosmographical accounts in the Indian tradition, it is not surprising to find that the earliest extant map image from the subcontinent is a stone bas-relief from the Sagarām Sonī temple, Mount Girnar, Saurashtra, Gujarat, dated to somewhere between the eighth and fourteenth centuries CE, depicting the Nandīśvaradvīpa, with the eighth-continent of the Jain cosmos on the outside and the Jambūdvīpa ("Rose-Apple Island"; i.e., the world) at the center. Separating these two are a series of concentric circles representing the other worlds and seas between them.[73]

In what continues to be one of the most intriguing and debated lines in medieval Islamic geographic sources, Ibn Ḥawqal states at the outset of his *Kitāb ṣūrat al-arḍ* (fig. 5.23): "I do not have in mind the seven climates that are used to typically divide the earth, because the Indian map that is in Qawārīyān, even if accurate, is very confusing." He goes on to say, "For all of the earth surrounded by the impenetrable Ocean I have created a map which is consistent partly with the map of Qawārīyān, but which in several respects is different."[74] Scholars are still trying to determine what this mysterious "Qawārīyān" Indian map looked like. Is it possible that the map that Ibn Ḥawqal was referring to resembled examples of the Jambūdvīpa or the Mandala-type images discussed below in this section? Bīrūnī's exhaustive eleventh-century study on India, appropriately labeled *Book of the Verification of What Is Said about India* (*Kitāb fī taḥqīq mā lil-Hind min maqūla*), confirms that Muslim scholars were aware of Hindu Purāṇic cosmologies and Catur-Dvīpa constructions of the world.[75]

The Jain view of the cosmos borrows from the Purāṇic one, and there are many examples of Jain images of the world with multiple encircling oceans.[76] From the late fifteenth century, a gouache on cloth image, now housed at the Victoria and Albert Museum in London, presents a Jain cosmographical image of the World of Man (Manuṣyaloka) (fig. 5.24) with a central continent surrounded by a series of seven encircling seas interspersed with six other circular continents.[77] In this image the upper and lower parts of the body constitute various levels of heaven and hell.[78] In the center, at the point of the midriff, is located the earth (Jambūdvīpa), distinctively encircled by an ocean adorned with a delicate array of fish.[79] This is a recurrent image in renditions of the Jain cosmos. So intrigued was the eleventh-century Islamic scholar

Fig. 5.24. *A*, Jain image of cosmic man with the Jambūdvīpa ("Rose-Apple Island"; i.e., the world) as his midriff. *B*, Close-up shows encircling ocean with fish. 18th–19th century. India. Lacquer painting on wooden tablet. Source: Berthon and Robinson, *The Shape of the World*, 46.

A

Fig. 5.24. (continued)

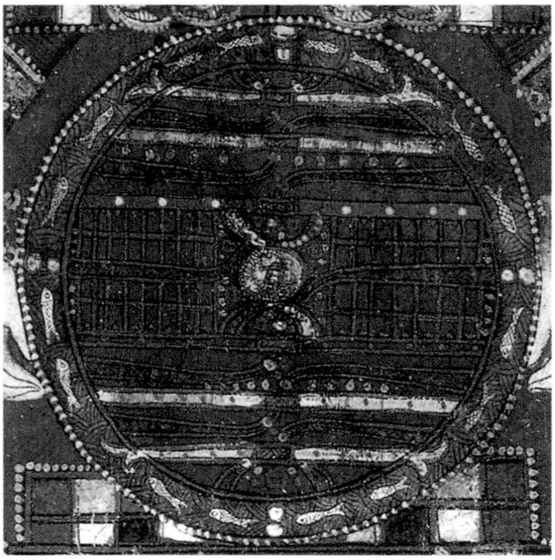

B

Bīrūnī by this Purāṇic vision of the world in the form of a man that he created tables to express it (see fig. 5.25).[80]

Hindu and Jain concepts of circumambulating seas and oceans inform Buddhist cosmographies. The Buddhist Mandala, which literally means "essence, circle" in Sanskrit ("Chilkor" in Tibetan carries the meaning of "center" and "periphery"), incorporates highly abstract microcosmic representations of the universe and its contents, including the terrestrial plane.[81] The terrestrial world is composed of four large continents of which the Jambūdvīpa, where humans reside, is the southernmost. The axis of the Mandala world is the mythical Mount Meru (sometimes referred to as Sumeru). These are surrounded by a series of mountains and encircling oceans that separate mortals from the higher planes of enlightenment. Bīrūnī understood well the importance of Mount Meru in the Hindu tradition as shown by the fact that he devotes an entire chapter of his book to discussing it.[82]

Tibetan Buddhism also employs a four-continent model of the world, including the Jambūdvīpa, with Mount Meru surrounded by a series of concentric oceans and mountains. In its depiction of a central square world with a round encircling ocean, the Tibetan mandala tradition fuses the Indic tradition with the Sino one, as discussed in the next section. The elaborate silk tapestry dating to the period of the Yuan dynasty (1271–1368) presents this vision in a clear and elegant form (fig. 5.26). Mandala construction continues today in Tibet, Burma, and India. In an unusual twist to the Buddhist cosmological formula, a nineteenth-century Burmese rendition depicts the world as a series of detached continents floating on a cosmic encircling ocean (fig. 5.27).

The Number of the Earths.	Âditya-Purâṇa.		Vishṇu-Purâṇa.	Vâyu-Purâṇa.		Vernacular Names.
	What Members of the Sun they Represent.	Their Names.		Their Names.	Their Epithets.	
I.	The navel.	Tâla.	Atala.	Âbhâstala.	Kṛishṇa-bhûmi, the dark earth.	Amśu (?)
II.	The thighs.	Sutâla.	Vitala.	Ilâ (?)	S'ukla-bhûmi, the bright earth.	Ambaratâla.
III.	The knees.	Pâtâla.	Nitala.	Nitala.	Rakta-bhûmi, the red earth.	S'arkara (?) (Sakkaru).
IV.	Under the knees.	Âśâla (?)	Gabhastimat.	Gabhastala.	Pîta-bhûmi, the yellow earth.	Gabhastimat.
V.	The calves.	Viśâla (?)	Mahâkhya (?)	Mahâtala.	Pâshâṇa-bhûmi, the earth of marble.	Mahâtala.
VI.	The ankles.	Mṛittâla.	Sutala.	Sutala.	S'ilâtala, the earth of brick.	Sutâla.
VII.	The feet.	Rasâtala.	Jâgara (?)	Pâtâla.	Suvarṇa-varṇa, the gold-coloured earth.	Rasâtala.

The Number of the Heavens.	What members of the Sun they represent according to the *Âditya-Purâṇa*.	Their Names according to the *Âditya*, *Vâyu* and *Vishṇu Purâṇas*.
I.	The stomach.	Bhûrloka.
II.	The breast.	Bhuvarloka.
III.	The mouth.	Svarloka.
IV.	The eyebrow.	Maharloka.
V.	The forehead.	Janaloka.
VI.	Above the forehead.	Tapoloka.
VII.	The skull.	Satyaloka.

Figs. 5.25. Tables from Edward Sachau's *Alberuni's India* (1888) translation of al-Bīrūnī's *Kitāb fī taḥqīq mā lil-Hind min maqūla* (Book on the Verification of What Is Said about India) showing al-Bīrūnī's tabular explanation of the Purāṇic image of the world in the shape of a man. Source: Sachau, *Alberuni's India*, 230 and 232.

Fig. 5.26. Cosmological mandala with Mount Meru. 1271–1368 (Yuan Dynasty). Cosmic diagram of Indian imagery introduced into China with the advent of Esoteric Buddhism. Shows Mount Meru surrounded by multiple encircling oceans. China. Silk tapestry. 83.8 × 83.8 cm. Source: The Metropolitan Museum of Art, New York. Purchase. Fletcher Fund and Joseph E. Hotung and Michael and Danielle Rosenberg Gifts, 1989. 1989.140.

In addition to the specialist work of Bīrūnī on India, mentioned earlier, it is worth noting that early Islamic literature, including the Qur'ān, ḥadīth, and Qiṣaṣ al-anbiyā' (Tales of the Prophets) mention the existence of seven earths resembling the Indic traditions discussed above.[83] Some of the geographers do too. Citing the Prophet and the Qur'ān, Ibn al-Faqīh al-Hamdhānī notes the existence of seven heavens and a complementary set of seven earths in the introduction to his *Kitāb al-buldān* (Book of Countries).[84] The full tradition is cited in Tirmidhī's *Sunan* as follows:

> Abū Hurayra and others said: "While the Prophet was sitting with his Companions [the *ṣaḥāba*], a cloud appeared, so the Prophet asked them: 'Do you know what this is?' They said: 'Allāh and His Messenger know.' He said: 'These are the clouds that are sent by Allāh to soak the earth of people who are not grateful to Him and do not worship Him.' Next he said: 'Do you know what is above you?' They said: 'Allāh and His Messenger know.' He said: 'Verily it is the canopy of the firmament whose wave surge is restrained [by God].' . . . 'Between you and it is five hundred years [of distance].' . . . 'Do you know what is above that?' He said: 'Indeed, above it are two Heavens and between both of

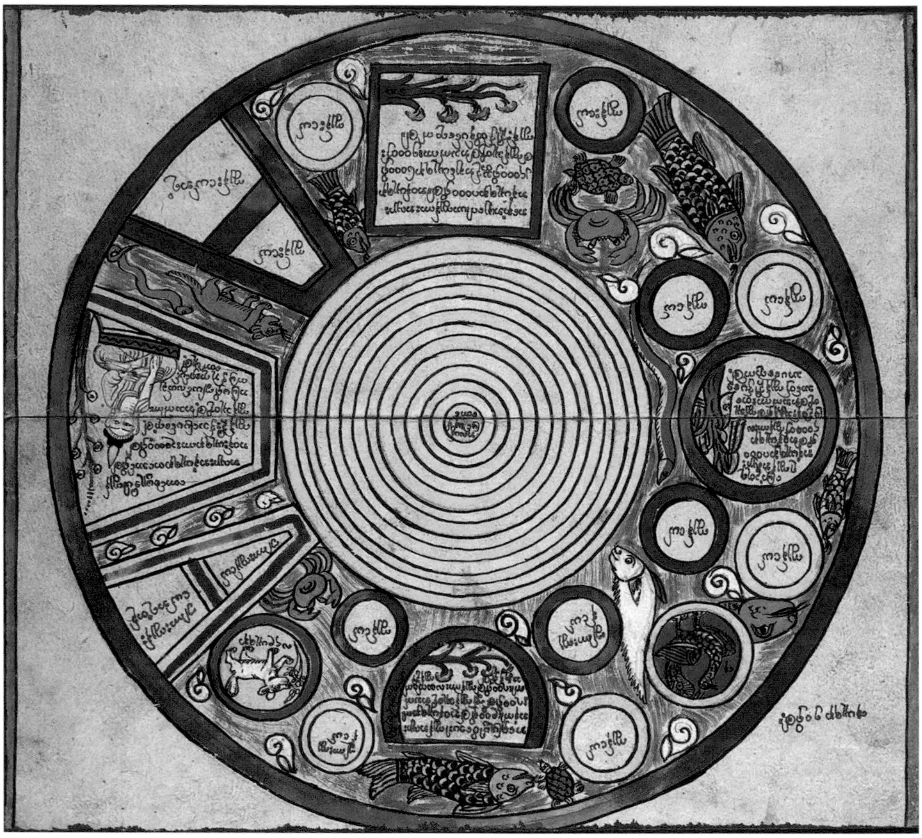

Fig. 5.27. Srid pai khor lo (Wheel of Life). Burmese Buddhist map of cosmos showing the world as a system of continents and ocean floating on an encircling ocean. 20th century. Painting on cloth. Courtesy: Library of Congress, Washington, DC. Asian Division, 109.1.

> them there is [another] distance of five hundred years.' Thus he [the Prophet] enumerated seven Heavens. 'What is between each of the two Heavens is what is between the heavens and the earth.' . . . After this he said: 'Do you know what is below you?' . . . It is the earth.' . . . Then he said: 'Below it is another earth and between them is a distance of five hundred years.' He continued until he had enumerated seven earths: 'Between every two earths is a space of five hundred years.' After this he said: 'By the One in whose hand is the soul of Muḥammad: If you were to send down [a man] with a rope to the lowest earth, he would descend upon Allāh.'"[85]

Thus there is a space of five hundred years between each of the seven heavens and each of the seven earths, and they meet as indicated by the example of the rope. This ḥadīth suggests that the canopy of the heavens, much like the Indic, Semitic, Iranian, and Egyptian traditions, is a gigantic encircling ocean whose waves are held back only by the mercy of God. On the basis of the Prophet Muhammad's statement that "what is between each of the two Heavens is what is between the heavens and the earth," we can infer multiple encircling oceans separating the seven heavens from the seven earths. A curious aspect of the mention of the seven earths in early Islamic eschatology is that it often occurs in connection with punishment for the usurpation of land. "He who took a stretch of earth falsely will have to wear seven earths around his neck on the day of Resurrection" is an oft-repeated ḥadīth.[86]

In his *Qiṣaṣ al-anbiyā'*, Kisā'ī names the seven different earths that lie between the seven seas as: Ratakā, Halda, 'Arfa, Harbā, Maltā (or Maltād), Sijjīn, and 'Ajība. He describes the inhabitants as evil, doomed to suffer through various damnations.[87] Even the eminent historian Ṭabarī discusses the seven earths in his opening volume on history "from Creation until the Flood" in the context of multiple oceans that separate one earth from the other. Distinctly reminiscent of the Indic vision of the world with multiple encircling oceans discussed above, Ṭabarī tells us: "They are seven earths that are flat and islands. Between each two earths, there is an ocean. All that is surrounded by the (surrounding) ocean, and the *haykal* is behind the ocean."[88] Chapter 7 is devoted to visual and textual examples of multiple encircling oceans in the Islamic tradition.

Sino Connections

At first glance the Chinese tradition presents a distinct break from the examples cited thus far.[89] From the Han era onward, the Chinese developed a nonary approach to the world. They saw, described, and discussed the earth and all aspects therein, from palaces to agricultural land use, as a square divided by a three-by-three grid.[90] The heavens, on the other hand, were seen as spherical. Referred to as the *gaitian* (covering heaven) theory, it can be traced back to the early Han arithmetical classic of the Zhou gnomon *Zhou Bi Suan Jing* (ca. 200 BCE):[91] "Turning a carpenter's square creates a circle; combining carpenter's squares forms a square. Squareness belongs to the earth; circularity to heaven. Heaven is round; earth is square."[92]

Without delving too deeply into the debate that surrounds the distinctively different approach of the Chinese to the shape of the world,[93] it is worth noting that their break from the traditional circular form for the earth did not preclude the existence of an all-encompassing sea.[94] We learn, for instance, that: "rain falling on the earth flowed down to form the rim-ocean, the Great Trench; the earth itself was square."[95]

From the Sung period (late tenth century to mid-fourteenth century CE), we have a nautical reference to the "Encircling Ocean" cautioning sailors: "To the east (of Hainan Island) are the 'Thousand-Li Sand-Banks' and the 'Myriad-Li Rocks,' and

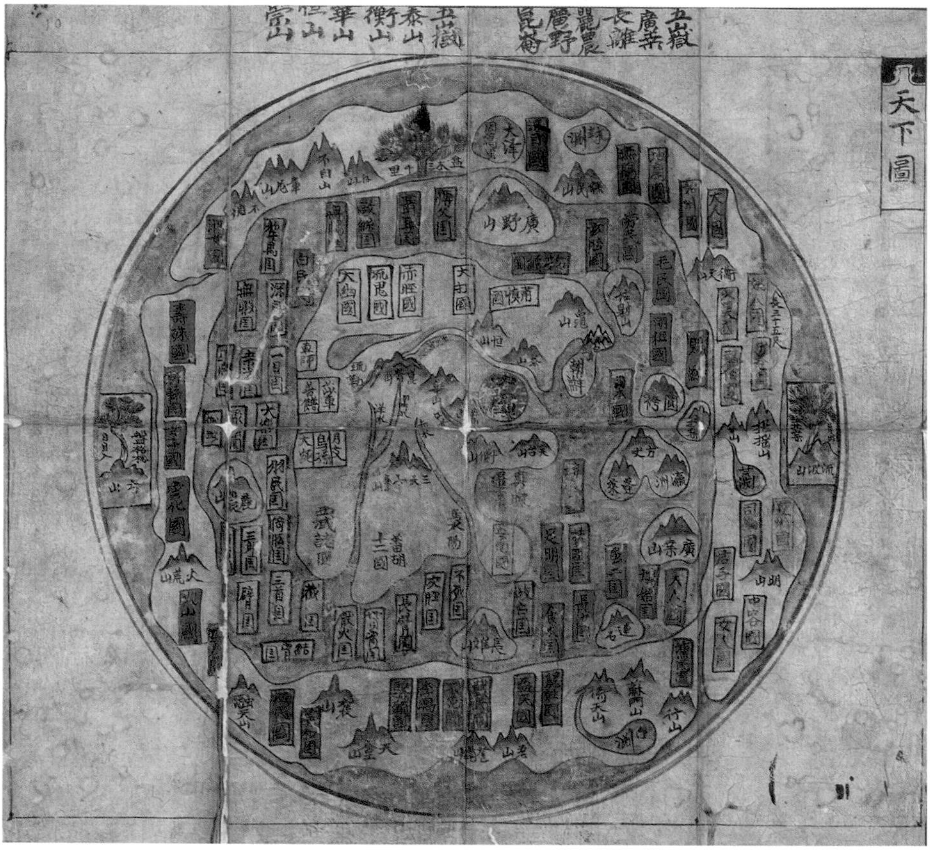

Fig. 5.28. Ch'onhado (Korean map of all under the heavens). Presents Korea, China, and their East Asian neighbors as surrounded by rings of exotic and mythical lands and people. The entire image is encased by an encircling ocean. Mid-18th century. Courtesy: Library of Congress, Washington, DC. Geography and Map Division, 77.2a.

(beyond them) is the boundless ocean, where the sea and sky blend their colors, and the passing ships sail only by means of the south-pointing needle. This has to be watched closely by day and night, for life or death depend on the slightest fraction of error."[96]

The famed Mingtang "luminous hall"—a cosmological temple of sorts designed for the performance of imperial rites, as well as an architectural symbol of virtual government—is based on the nine-palace or *jing* (i.e., three-by-three-grid) form. Later Han texts describe the temple: "Its base was square, the shape of the earth, its round roof resembled the heavens, and the whole structure was surrounded by water that represented the four seas."[97]

Although the *gaitian* concept and its more than one millennium of influence on Chinese cosmographic thought is well documented, there are no "world map" examples of it. Even though Chinese scholars began producing detailed district and provincial maps from the late Han period onward, it wasn't until the proselytizing Christian missionaries of the early Renaissance arrived on their shores that the Chinese began to produce maps of the "whole world" as opposed to just China and its provinces.

From the Korean world, we get the popular "wheel-maps" of the Ch'onhado (map of all under the heavens) tradition (fig. 5.28).[98] These maps depict the known world—with a distinct emphasis on China and Korea—surrounded by not one but two encircling oceans and are now the subject of considerable dispute. Scholars are split between those who see the tradition as an authentic recension of an ancient Korean representation of the world and those who see it as a modern (i.e., eighteenth /nineteenth century) hoax.[99] Ch'onhado supporters claim that the map is based on the work of the third century BCE naturalist philosopher Zhou Yan and specifically on the following passage: "As for the sea around China in the four directions, I call it 'the tiny sea' [bihai]. Beyond that sea there is a great continent that rings it, and beyond the great continent there is a vast ocean sea circling around. It is only here [that one comes to the edge] of the earth."[100] There is more work to be done to complete the picture on Sino-Islamic carto-geographic connections.[101]

Though the preceding survey is far from comprehensive, it is clear even from this limited series of examples that the Encircling Ocean is a metaform in the most fundamental sense. It is an integral part of water-based epiphanies and cosmogonies that establish water as the source of the world.[102] Eventually, it comes to exist independently as the basic outline of all premodern imagines mundi and is used as a useful signifier of power and dominion.

In the next chapter, I will expand the iconographic analysis with a look at what was once the standard answer to the question of how the notion of an Encircling Ocean came to hold sway over medieval Islamic mapping, namely, the Greek, Roman, and medieval European mappamundi traditions.

6

Classical and Medieval Encircling Oceans

Ancient and Classical Greek Cartography

The fundamental issue with all ancient Greek cartography is that we have no extant examples of any maps.[1] As a result, the exact form that these maps took is based on conjecture and reconstruction from long lost sources.[2] As Germain Aujac explains in the *History of Cartography* series, "We have no original texts of Anaximander, Pythagoras, or Eratosthenes—all pillars of the development of Greek cartographic thought. In particular, there are relatively few surviving artifacts in the form of graphic representations that may be considered maps. Our cartographic knowledge must therefore be gleaned largely from literary descriptions, often couched in poetic language and difficult to interpret. In addition, many other ancient texts alluding to maps are further distorted by being written centuries after the period they record; they too must be viewed with caution because they are similarly interpretative as well as descriptive."[3]

Even the earliest manuscript attributed to Claudius Ptolemy, the second-century Hellenistic geographer heralded since the Renaissance as the "father of cartography," is dated to the mid-thirteenth century—more than a millennium after Ptolemy's death.[4] Among scholars of Greek cartography, the question of whether Ptolemy ever drew maps, or whether he just wrote about drawing them, is a matter of considerable debate.[5] From the fourteenth century until the rise of Renaissance mapmakers (such as Mercator, Bleu, etc.), there was a wave of map manuscripts of European provenance all claiming Ptolemy as author. The maps in these so-called Ptolemy manuscripts vary from one to another so widely in form and content that the attribution of these manuscripts to a single author is nothing short of astonishing. Visual evidence to the contrary aside, the Renaissance-spawned myth that Ptolemy is the father of cartography still endures.[6]

In the absence of extant maps, our understanding of Greek cartography is heavily dependent on textual sources. One of our best and earliest textual references comes from the ancient period of Greek history through the Homeric tradition (ca. eighteenth century BCE) and one of our most beloved *Iliad* characters, Achilles—specifically his shield. According to Homer, Hephaestus fashioned a huge shield for Achilles with concentric rings of metal—gold in the center, surrounded by two rings of tin, and two of bronze—with a three-ringed metal rim. Homer tells us that on the front plate Hephaestus created a schematic depiction of the earth and the universe.[7] Arrayed around the edge of the shield was the vast and mighty river of the Encircling Ocean, described thus:

> First of all he forged a shield that was huge and heavy,
> elaborating it about, and threw around it a shining
> triple rim that glittered. . . .
> There were five folds composing the shield itself, and upon it
> he elaborated many things in his skill and craftsmanship.
> He made the earth upon it, and the sky, and the sea's water
> . . . .
> . . . and the strength of Orion
> and the Bear, whom men give also the name of the Wagon,
> . . . .
> and she alone is never plunged in the wash of the Ocean.
> . . . .
> He made on it the great strength of the Ocean River
> which ran around the outermost rim of the shield's strong structure.[8]

So pervasive is the influence of this shield on the history of cartography that Denis Cosgrove suggests, tongue in cheek, that this shield should be considered the cornerstone of the European cartographical imagination![9] And Yāqūt al-Rūmī, the late twelfth-/early thirteenth-century compiler of the opus magnum of geographical dictionaries—*Muʿjam al-buldān* (Dictionary of Places)—informs us that Achilles's shield was discussed by medieval Islamic scholars as well.[10]

The dearth of extant maps means that we have to rely on textual sources for information. For the classical period there are but a few obscure snippets, such as the one from Aristophanes's fifth century BCE comedy *The Clouds*. An anonymous student introduces the farmer Strepiades to the location of Athens and Sparta on a map:

> **Student** (pointing to a map): Now then, over here, we have a map of the entire world. You see there? That's Athens.
>
> **Strepiades:** That, Athens? Don't be ridiculous. Why, I can't see even a single lawcourt in session.
>
> **Student:** Nonetheless, it's quite true. It really is Athens.
>
> **Strepiades:** Then where are my neighbours of Kikynna?
>
> **Student:** Here they are. And you see this island squeezed along the coast? That's Euboia.
>
> **Strepiades:** I know that place well enough. Perikles squeezed it dry. But where's Sparta?
>
> **Student:** Sparta? Right over here.
>
> **Strepiades:** That's much too close! You'd be well advised to move it further away.
>
> **Student:** But that's utterly impossible.
>
> **Strepiades:** You'll be sorry you didn't, by god.[11]

From this reference we get a clear, if comical sense, of the spatial location of places in relation to one another.[12] From the will of Theophrastus (third century BCE), a student of Plato and Aristotle, we learn of the commissioning of a painted world map on wooden panels for the Lyceum that is unfortunately no longer extant.[13]

With his caustic critique about simpleton mapmakers in *Histories*, Herodotus (ca. 484–425 BCE) fills out the picture a little more: "For my part, I cannot but laugh when I see numbers of persons drawing maps of the world without having any reason to guide them; making, as they do, the ocean-stream to run all round the earth, and the earth itself to be an exact circle, as if described by a pair of compasses, with Europe and Asia just of the same size."[14] Herodotus held that Europe was twice the length and breadth of Asia and without a confirmed terminus. Therefore he objected to the notion of an Encircling Ocean since he thought that the landmass of Europe extended where the ocean was drawn. Aristotle bolsters Herodotus's critique by arguing in his *Meteorologica* that those who drew *periodos ges* (circuits of the earth) were illogical.[15] In this thinking, Aristotle went against the teachings of his mentor, Plato, who also speculated on the shape of the world and invented a fanciful creation myth in *Timaeus* using information from Solon. In a dialogue with the usual suspects, Socrates and friends, Timaeus explains that

God placed two other elements of air and water, and arranged them in a continuous proportion—fire:air::air:water, air:water::water:earth, and so put together a visible and palpable heaven, having harmony and friendship in the union of the four elements. . . . And as he was to contain all things, he was made in the all-containing form of a sphere, round as from a lathe and every way equidistant from the center, as was natural and suitable to him. . . . All that he did was done rationally in and by himself, and he moved in a circle turning within himself, which is the most intellectual of motions. . . . And so the thought of God made a God in the image of a perfect body, having intercourse with himself and needing no other, but in every part harmonious and self-contained and truly blessed.[16]

This idea of the sphere as the perfect form was picked up by both Christian and Muslim Neo-Platonists and, as Lilley argues, can be seen as the basis for the importance of the circular form in medieval images, including maps.[17] Plato does not mention anything specific about the Encircling Ocean except to say that Okeaynus (the Encircling Ocean) was part of the elements mixed into the original cosmogony.

In his opus *Geography*, which preserves the works of many ancient and classical Greek authors, Strabo discusses the matter of the ocean in a logical, matter-of-fact manner: "We may learn both from the evidence of our senses and from experience that the inhabited world is an island; for wherever it has been possible for man to reach the limits of the earth, sea is found, and this sea we call 'Oceanus.' And wherever we have not been able to learn by the evidence of our senses, there reason points the way."[18] Without extant images, however, we cannot tell for certain if theories of the shape, form, and geographical structure of the earth, such as those of Strabo, were translated into actual maps or not. The most frequent references to maps in the ancient Greek context are to those known as the *periodos ges* or "Ring-Around-the-World" images possibly made on wooden panels or bronze tablets (*pinax*; pl. *pinaces*). At issue is the meaning of the word *pinaki*, which is the dative singular of *pinax* and depending on the context can mean "on the pinax," "for the pinax," and so on. *Pinax* has a wide range of meanings, including "board," "plank," "writing-tablet," "votive tablet," or "board for painting," and could therefore refer to a picture, plate, or even a public notice board.[19]

Periodos ges (pl. *periata gaies*), as these "Ring-Around-the-World" maps are sometimes called, is not a clear reference to a map either. The same name is used as the title for textual accounts of the world initiated by the sixth century BCE Ionian scholars of Miletus—Anaximander and Hecataeus—and continued by Herodotus's fifth century BCE contemporary Democritus and the fourth century BCE scholars Eudoxus and Dicaearchus of Sicily. As indicated by Dionysius's popular poetic version, the tradition of writing *periodos ges* continues well into the second century CE. Thus, *periodos ges* can just as easily be translated as a "journey around the world"—an ancient travel account akin to the medieval Islamic *riḥla* (travelogue) tradition. Discerning the difference between a textual reference to a journey and a reference to a physical map is yet another example of how challenging it is to study ancient Greek maps.[20]

Given all the references to the shape, description, and size of the earth, *pinaces*, and *periata gaies*, and many others that I have not mentioned in the interest of brevity, it is indeed surprising that no ancient Greek maps have survived. What these textual sources indicate is that from the sixth/fifth century BCE onward ancient Greek society was growing accustomed to quotidian ways of seeing and thinking about the world, if only in theoretical form. Precisely what forms these maps took remain undetermined. How much of the world did they represent, and what were the roots of their inspiration? These questions have formed the basis of some of the most intellectually stimulating work on the history of cartography, but without extant visual examples, no one can know the precise nature of the influence of Greek ideas on Muslim and medieval European cartography.[21]

The matter is complicated not only by the lack of extant ancient Greek maps but also by the low survival rate of other key geographical works. While there is no evidence to suggest that the Arabs knew about the work of Strabo or that they consulted it, there is no question that the Arabs knew and preserved Ptolemy's *Almagest*, hence its Arabized title. But the *Almagest* is primarily a work of astronomy while Ptolemy's *Geography* was one of latitudinal and longitudinal listings of the location of eight-thousand-plus places. Ptolemy's work likely influenced the development of the voluminous Islamic zīj tables tradition,[22] but the question of influence on Islamic mapping is not an open-and-shut case as has been asserted in the past.[23] Although Masʿūdī asserts that the *Ṣūrat al-Ma'mūniyya* was more beautiful than the maps of Ptolemy and Marinus, none of these maps are extant.[24] In the absence of copies of either, we are left in the dark as to the exact nature of this influence. Zuhrī claims that his work is a copy of the *Ṣūrat al-Ma'mūniyya* but his work does not contain maps either, and Zuhrī's usage of the Iranian *kishwar* system (see previous chapter for details) has puzzled scholars who suggest that the *Ṣūrat al-Ma'mūniyya* may instead have been a mixture of the Iranian and Ptolemaic systems.[25] It is just as likely that Masʿūdī and Zuhrī were referring to a KMMS-type world map that may have had its roots in the period of the Abbasid caliph Ma'mūn (r. AH 197–218/813–833 CE). Another possibility is that they were using the word *ṣūrat* in a textually visual sense to refer to images—the way that the ancient Greeks used *periodos ges* and *pinax*. Either way, we cannot look solely to the Greeks for influence on the KMMS world map. Tibbetts goes so far as to suggest that the KMMS maps were "produced independently as a reaction against work dependent on Greek and other foreign agencies."[26]

Roman Cartography

The case of Roman cartography needs to be considered in the context of Greek cartography. Influential Romans, including the emperors, actively patronized Greek

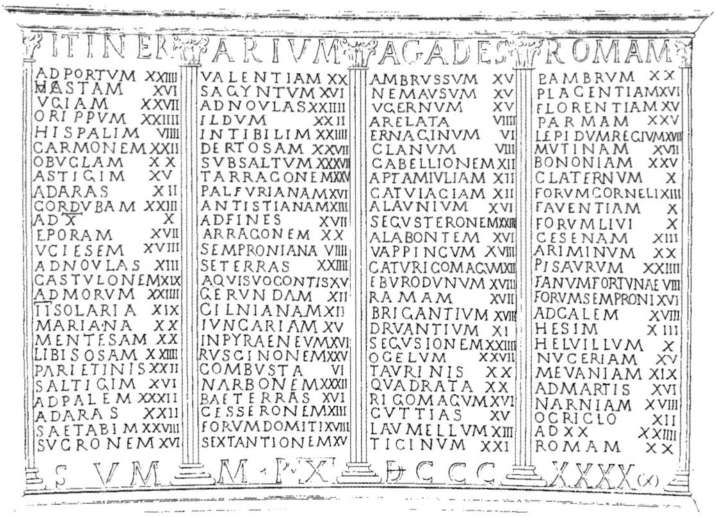

Fig. 6.1. *Itinerarium a Gades Romam.* Tabular itinerary showing a list of stops and distances from Cádiz in southwestern Spain to Rome. Circa 330. Line drawing of text on one of the silver beakers from Bagni di Vicarello: *Corpus Inscriptionum Latinarum* XI 3281, Academia Litterarum Borussica, Leipzig/Berlin 1862–1943. Original beaker is in Rome: Museo Nazionale delle Terme, inv. No. 67497.

scholars. It is generally acknowledged, however, that the Romans approached mapping from a more practical vantage point—that is, for cadastral and architectural surveys.[27] The extant record of possible Roman maps is almost as scant as in the case of the Greeks. What little exists speaks of a limited mapping tradition.[28]

Map-wise what the Romans were most interested in laying out were their roads, and it is therefore not surprising to learn that they produced itineraries in large volume, at least from the fourth century CE onward when pilgrimage to the Levant became popular (fig. 6.1). Usually a bare list of stops and distances indicating *mutatio* (change of mount) and *mansio* (overnight stops), these itineraries took a number of forms. Some were etched on pillars of marble, others jotted down in manuscripts, and still others etched around silver beakers. There are many extant examples of these itineraries, including famed ones such as the octagonal column from Tongeren (Atuatuca Tungrorum) in present-day Belgium, the Antonine Itinerary (Itinerarium Antonini), the Patara monument, and pilgrimage accounts such as that of Theophanes from Hermopolis Magna in Egypt to Antioch in Syria (322–324).[29] What they attest to is that

> It was the mind's eye which ancient chorographers stressed, since written texts—"la carte écrite"—created no less visible, spatial images than maps. . . . Written itineraries could be more easily memorized than maps, because they were sequential. . . . Space itself was defined by itineraries, since it was through itineraries that Romans actually experienced space, that is, by lines and not by shapes. . . . Movement was more important

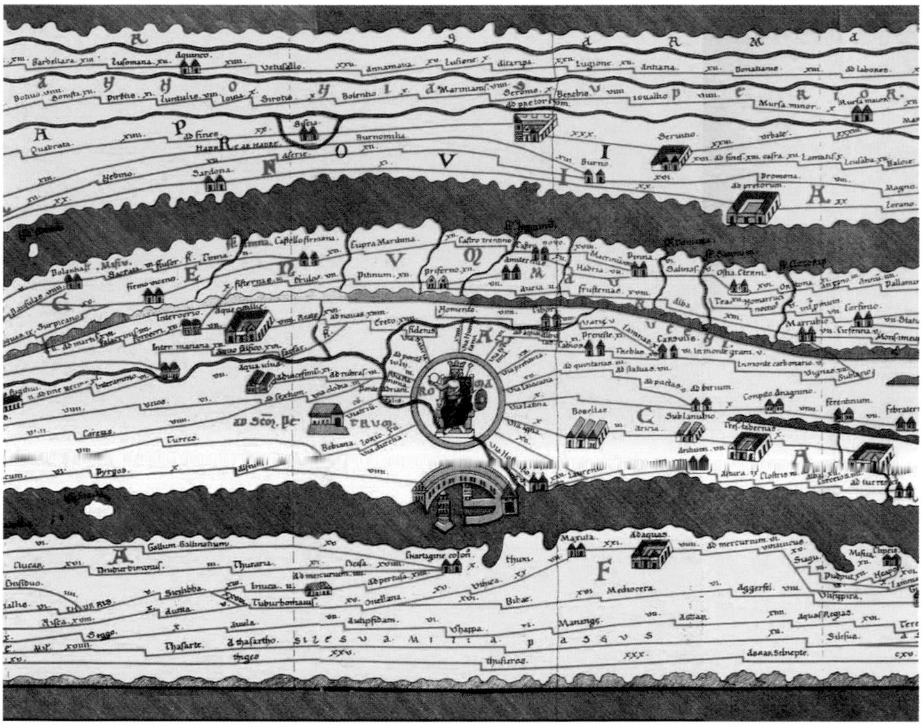

Fig. 6.2. Close-up of Roma from Konrad Miller's 1887 / 1888 line drawing of *Tabula Peutingeriana*. This is an illustrated *itinerarium* showing the road network of the Roman Empire. It covers Europe, North Africa, and parts of Asia. Parchment scroll in eleven sections. Earliest copy dates to the thirteenth century. 0.34 × 6.75 m. Österreichische Nationalbibliothek, Vienna, Codex Vindobonesis 324, segment 9 (of 11).

> than the form. . . . The essence of itineraries was space between topographic points—towns, *vici*, *mansions*, and so on—which followed in linear succession. It reflected the Roman political view of a world made up of a network of *civitates* and urban space, between which lay nothing except curiosities for the traveller.[30]

The only visual example we have of this Roman zest for routes is the famous Peutinger Table (fig. 6.2). It too suffers from the Ptolemy syndrome (discussed earlier in this chapter) in that the earliest extant copy only dates back to the thirteenth century. Richard Talbert, an ardent proponent of the authenticity of the Peutinger map, while acknowledging the legitimate reservations of skeptics, devotes a book to documenting the history of this map. Talbert argues that although the Peutinger map is clearly a medieval production, content analysis suggests that the original copy was done no later than 700 and most likely stems from the period of Diocletian's Tetrarchy (ca. 300)—in tandem with the pilgrimage boom.[31]

The map is composed of multiple sheets and focuses on displaying routes and realms of and for the Roman Empire's network of roads. It does not contain an Encircling Ocean—unless the green background behind the edges of the landmass is intended to be an indication of the ocean encasing the land. It does, however, contain an image of Roma encased in a double-ringed circle that many scholars believe would have been the center of this map if all its pieces were extant. The crowned female figure seated in the center of this double-ringed circle holds a ball in her right hand that could be interpreted as an orb representing Rome's dominion over the world. In true medieval European illustration style, as we shall see in the next section, this figure fits in with the portrayals of God and earthly rulers encased by symbolic double-ringed circles. Could this circular form be a resonance of the Achaemenid double-ring motif (discussed in the previous chapter) bequeathed by the Romans to European illuminators? This is a riddle that is unlikely to be resolved unless maps extant from the Roman period are uncovered.

The same dilemma holds true for the Agrippa world map said to have been commissioned during the reign of the Emperor Augustus. Some suggest that Agrippa never completed this map, while others say it wasn't really a world map but rather a self-aggrandizing view of only the Roman Empire as the world.[32] Agrippa's world map is sometimes linked with the Hereford world map of the late thirteenth century (fig. 6.10) on the basis of a sketch of Augustus Caesar in the lower left-hand corner issuing an edict for the Hereford map. The connection is tenuous at best. If there is any validity in it, it would suggest that the Romans upheld the concept of an Encircling Ocean.

This interpretation of the presence of the Encircling Ocean in Roman world maps is reinforced by Eumenius's panegyric on the rebuilding of a rhetorical school in Gaul in which he mentions a wall map that is on display in the school:

> In [the school's] porticoes let the young men see and examine daily every land and all the seas and whatever cities, peoples, nations, our most invincible rulers either restore by affection or conquer by valor or restrain by fear. Since for the purpose of instructing the youth, to have them learn more clearly with their eyes what they comprehend less readily by their ears, there are pictured in that spot . . . the sites of all locations with their names, their extent, and the distance between them, the sources and mouths of rivers everywhere, likewise the curves of the coastline's indentations, and *the Ocean, both where its circuit girds the earth and where its pressure breaks into*.[33]

There are some extant images such as Roman landscape wall murals that can be interpreted as maplike images with "curves of coastlines" and "mouths of rivers" and mosaics involving routes and places, such as the famous Madaba mosaic map. None

Fig. 6.3. Quartered earth within a circular heaven of Recension A of the *Corpus Agrimensorum Romansorum*. Occurs in a Roman land-surveying manual. Blue gouache and red and black ink on parchment. Courtesy: Herzog August Bibliothek Wolfenbüttel. Cod. Guelf. 36.23 Aug 2°,43b.

of these, however, show the whole world, so I am excluding them from this discussion on the Encircling Ocean, although they are certainly relevant to the heritage of Roman mapping.

The most interesting material from the perspective of the Encircling Ocean form are the images in a set of Roman surveying manuals known as the *Corpus Agrimensorum Romansorum*. In addition to dealing with land surveying, this manual addresses elements of world cosmography. Particularly intriguing is the image of the quartered earth within a circular heaven (fig. 6.3). The *Corpus Agrimensorum Romansorum* is also plagued with dating and provenance issues. The only extant copy has been dated variously to between the fifth and seventh centuries and would thus properly fall within the orbit of medieval not Roman painterly technique.[34]

From the twilight years of Roman dominion during late antiquity and the early medieval period come three geographic traditions destined for popularity in the later medieval era: Macrobius (ca. 240–320), Orosius (ca. 383 to post-417), and Isidore (ca. 560–636). These traditions spawn a plethora of medieval European maps from the ninth to the fifteenth century (as discussed in the next section). Since no manuscripts are extant from the period of the original authors and the earliest extant maps date only from the ninth century onward, these maps are discussed as part of the medieval European canon not Roman.[35] In "Seeing Like a Roman," Whittaker asserts that while "the conceit of looking down on high, like Icarus or Zeus, was common among geographers," it "was of limited use in describing the real world of travellers or administrators."[36]

Here, for lack of any other significant extant maps, ends the discussion of Greek and Roman aspects of Encircling Ocean iconography. Given the impossibility of clear visual connections between Greco-Roman cartography and medieval Islamic maps, it is perplexing that so much significance has been attributed to the influence of the Greco-Roman tradition on Islamic mapping. A shift in this thinking is an important step forward in the study of the history of Islamic cartography. In contrast, the medieval European mapping tradition is rich in extant images, world maps and others, replete with Encircling Ocean motifs. Yet these are rarely referenced in discussions on Islamic cartography.[37] What follows is an attempt to redress the narrative.

Medieval European Mappamundi[38]

Although there are differences between the forms of the Islamic and medieval European mapping traditions, such as the Islamic emphasis on the Indian Ocean, and the sizes and shapes of the continents, the one feature that stands out as a distinct mark of parity between the two mapping traditions is the Encircling Ocean.[39]

The best known of all the models of medieval European world maps is the Isidorian *Orbis Terrarum* tradition that comes from the encyclopedic tradition of *Etymologies* by Bishop Isidore of Seville (ca. 560–636). Better known as a T-O map, it is so-called because of the shape of the Christian cross implied by the "T" encased within an O-shaped world (fig. 6.4). The image derives from the Mediterranean as the leg of the "T" with the Bosphorus/Tanais/Don[40] on the northern end and the Nile on the southern end as the top of the "T." The "T" is surrounded by a circular band that signifies the Encircling Ocean.[41] The circle encases a schematic image of the old world (*oecumene*) depicting the continents of Asia, Africa, and Europe oriented eastward in the direction of Jerusalem—hence the origin of the phrase "to orient a map."[42] There are in excess of seven hundred examples of T-O maps, which by the fifteenth century were appearing in the first printed versions of Isidore's *Etymologies*. The problem is that we have no copies of either *Etymologies* or *De natura rerum* contemporaneous with Isidore. The earliest dated Isidorian manuscripts that include T-O maps date from the eighth/ninth century, but even on this dating, there is no definitive consensus.[43] Isidore does, however, emphasize the importance of the Encircling Ocean in his textual description of the world. In his discussion of "De oceano" in Book 13 of "De mundi et partibus" (The Cosmos and Its Parts), he says that "Greek and Latin speakers so name the 'Ocean' (*oceanus*) because it goes around the globe (*orbis*) in the manner of a circle (*circulus*), [or from its speed, because it runs quickly (*ocius*)]. . . . This is what encircles the edges of the land, advancing and receding with alternate tides, for when the winds blow over the deep, the Ocean either disgorges the seas or swallows them back."[44] In keeping with Isidore's description, most illustrators of Isidorian manuscripts emphasized the Encircling Ocean in their T-O map renditions.[45]

Fig. 6.4. Typical Isidorian T-O map. The "T" stands for the Christian cross and places Jerusalem at the center of the world. The "O" enclosing the entire image represents the Encircling Ocean. Late 11th century. Ink on parchment. 6.3 cm. British Library Board, London. Ms. Royal 6 C I, fol. 108b. © British Library Board / Robana / Art Resource, NY.

These Isidorean T-O maps come in many variations including what has come to be known as the Y-O version (fig. 6.5). In this variation the mouth of the Bosphorus, made up of the Marmara Sea, and the Aegean combine to create a Y shape. Another variation is the so-called list T-O in which a long list of place-names is placed on each of the continents (fig. 6.6).[46] It is reminiscent of Roman itinerary lists (see fig. 6.1).[47] In both of these cases the Encircling Ocean persists as the outlining signifier of the world. With the onset of the Crusades in the late eleventh century, examples of the T-O model multiply and evolve into a visual manifestation of Christian zeal feriozibus terre .

Incipit liber quartusdecim9.

Terra est i media
mūdi regioneʒ po
sita omnibus pti
bus celi i moduʒ
centri eqli inter
uallo consistens
que in singulari
numero totuʒ ortem significat pluā
liʒero singulas partes cuius nomi
na diuersa dat racio. Nam terra di
cta a superiozi parte qua teritur. hu
m9 ab iferioze vel humida terra. vt
ß mari . tellus autem. quia fructus
eius tollimus. hec a ops dicta. eo ꝙ
opem fert frugibus . Eadem a arua
ab arado a colendo vocata. Proprie
autem terra ad distinctonem aque
arida nuncupatur. sicut scriptura

Oriens
Asia
Sept.
Sem
Meri.
tripalus
I tanai fluui9
nibus fluuius I
Europa Jafet
Occi
meoth
mare magnū
Africa Cham dens

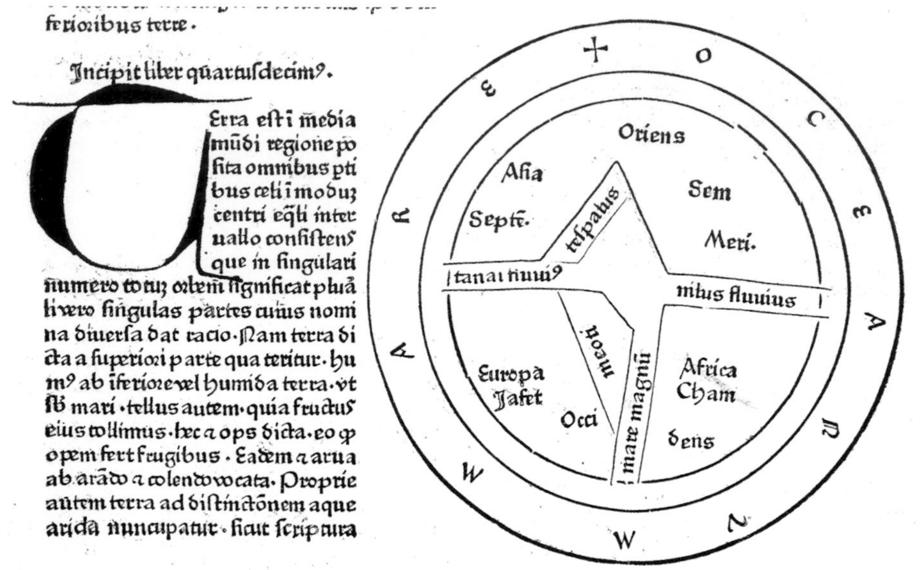

Fig. 6.5. Example of a modified T-O map called a Y-O map. A more elaborate version of Isidore's original schema with the Sea of Azov. 11.1 cm. From Isidore of Seville, *Etymologies* Cologne. 1478. This item is reproduced by permission of The Huntington Library, San Marino, California. HEH 89025.

and a desire for world dominion. See, for example, figure 6.7, which shows the T-O map in a twelfth-century rendition of Isidore's *Etymologies* in which the four directions are indicated by crosses.

The Encircling Ocean in these later medieval European T-O renditions is dotted with islands, resembling late medieval Islamic world maps of the Idrīsī-type (see the next chapter for details). They often include multiple encircling bands that signify both the encircling ocean and the cosmos. The same occurs in eleventh- and twelfth-century examples from Bede's *De natura rerum* and Lambert's *Liber floridus*. Both show multiple encircling bands of the sun, moon, and planets surrounding the basic T-O model of the earth (fig. 6.8).[48] This multiple encircling format resembles the prehistoric, Buddhist, Indic, and Iranian typologies discussed in the previous chapter.

Even with the transition to complex, multilayered mappamundi in the later Middle Ages, the T-O stamp remains the underlying hallmark.[49] Sometimes the T-O stamp is more obvious, at other times less so. The renowned mappamundi the Psalter world map of the mid-thirteenth century (fig. 6.9) and the Hereford mappamundi of the early fourteenth century (fig.

Fig. 6.6. Typical list T-O map from Lambert's *Liber floridus*. Reminiscent of Roman itineraries, this type of T-O map lists names of key places in the space provided for each continent. It also occurs in Isidorean manuscripts. Circa 1121. Gouache and ink on parchment. Courtesy: The University of Ghent. Ms. 92, fol. 19a.

6.10) provide prime examples of this late medieval process of elaboration.[50] Although varying greatly in size—by surface area the Hereford map is some 150 times larger than the Psalter map—they both share the same T-O base overlaid by an intricate wealth of detail. One striking feature of the evolution of the Isidorian T-O model into the full-fledged medieval Western European mappamundi is the increased religious signifiers that dot the surface of the map, which do not occur on the Islamic maps. These include features such as church-like place markers and references to key biblical events in the form of Noah's Ark, the twelve tribes of Israel, biblical monsters, and Earthly Paradise in particular.[51]

Both the Psalter and the Hereford mappamundi use a double-ringed band to

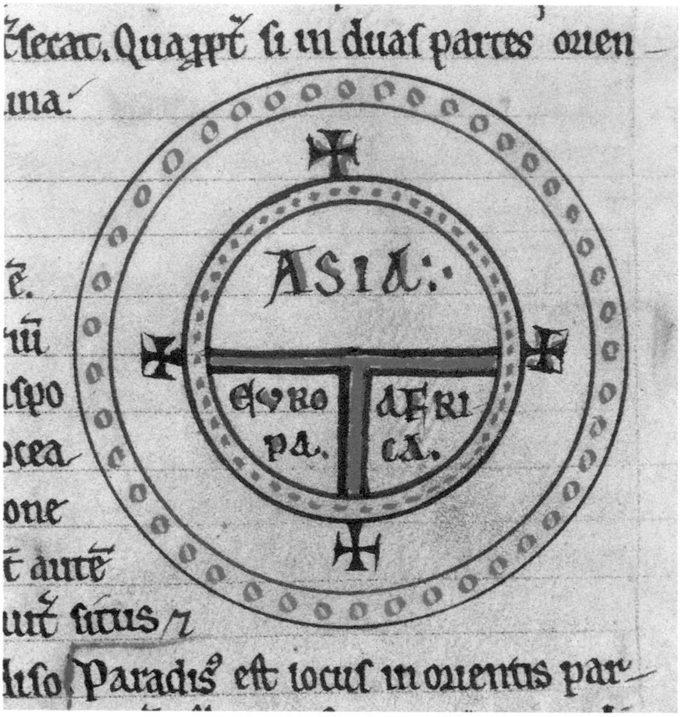

Fig. 6.7. Evolution of the T-O map during the Crusades from a copy of Isidore's *Etymologies*. Crosses indicating the four directions betray signs of crusading influence. This T-O map is unusual in that it indicates two sets of encircling oceans. Circa 12th century. Ink on parchment. Courtesy: British Library Board / Bridgeman. Ms. Royal 12 F. IV, fol. 135v.

separate the profane world from the sacred garden of Paradise. In the Psalter map, Paradise is a circled space on the main landmass at the top of the map. In the Hereford mappamundi, Paradise is no longer part of the main landmass. It is depicted instead as a double-ringed island within the Encircling Ocean (fig. 6.10). Complete with Adam, Eve, and the sacred rivers, its relocation to the Encircling Ocean emphasizes the extreme otherness and inaccessibility of its sacred space.[52] In both cases, God and his angels sit above and beyond the sacred encircling circles. The distance between God and earth is stressed in the Psalter map where we see that in his left hand God is holding an orb with a miniature T-O motif while with his right he is blessing it. A similar double-ringed marker is used in other places on medieval maps to separate sacred spaces from the profane. Most notable of these is the walled city of Jerusalem. In the case of the Hereford depiction, the cross breaks the plane of the earth and rises toward heaven from the center point of the circle (fig. 6.11). Examples abound, especially from the Crusading period, of similar depictions of circular Jerusalem sacrality (fig. 6.12).[53]

It is not only cities that find their way into double-ringed circles; as in the Sassanid examples discussed in the previous chapter, people of import do so too. This fits

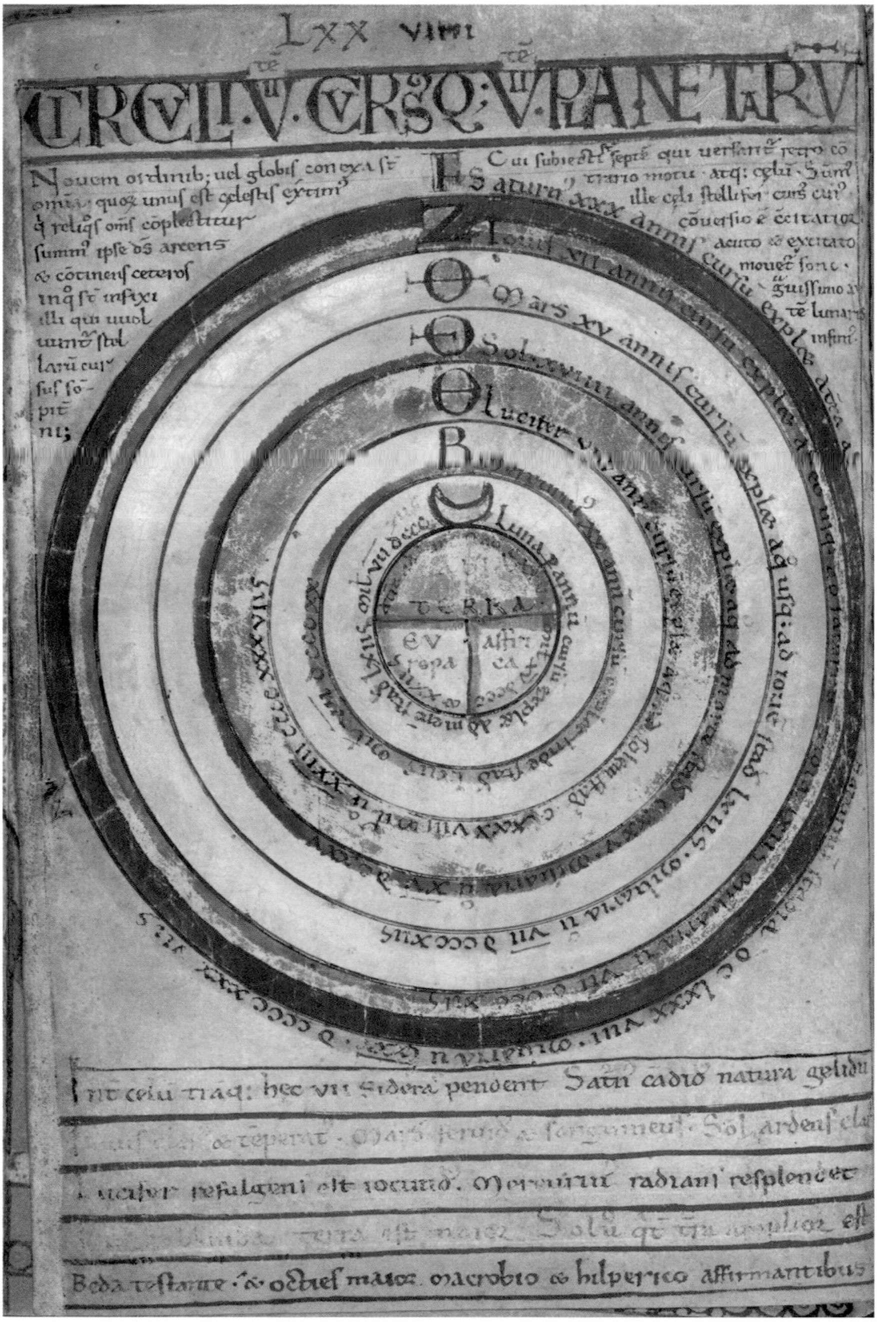

Fig. 6.8. Planetary circles around Terrum Orbis from Lambert's *Liber floridus*. *Liber floridus* is the earliest medieval encyclopedia containing a universal history that eventually supersedes Isidore's work. It is lavishly illustrated and contains a variety of T-O maps. The copy at Ghent is an autograph copy in the hand of the author. Circa 1121. Gouache and ink on parchment. Courtesy: The University of Ghent. Ms. 92, fol. 94v.

Fig. 6.9. Psalter world map. Among the most famous of the medieval European mappamundi, this tiny but extremely detailed world map depicts Christ towering over the world flanked on either side by two angels. In spite of the significant expansion of information from the simple Isidorean T-O map, it is still encircled by an double-ringed Encircling Ocean now dotted with a plethora of islands. Circa 1265. Gouache and ink on parchment. Diameter 9 cm. British Library Board, London. Add. Ms. 28681, fol. 9a. © British Library Board/Robana/Art Resource, NY.

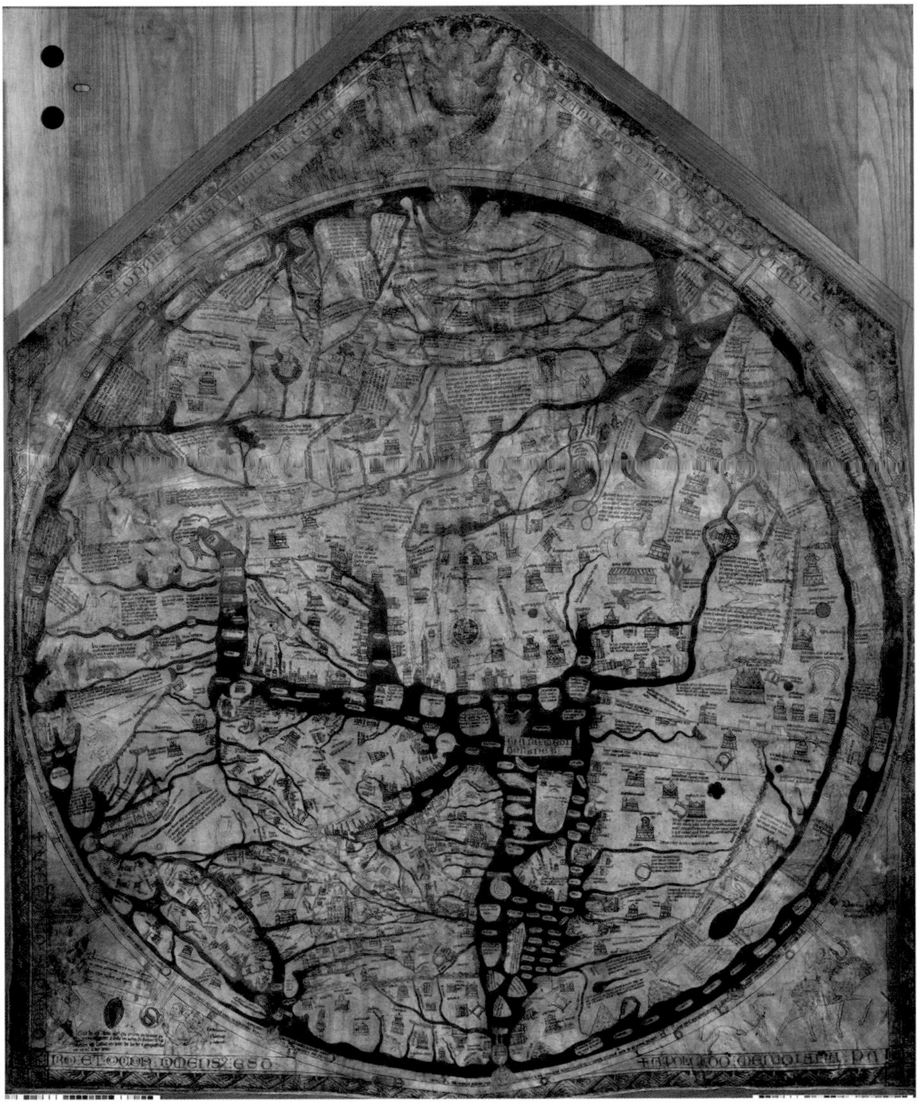

Fig. 6.10. Hereford mappamundi. One crucial difference between this and the Psalter map, which is some 150 times smaller but shares the same T-O base, is that in the Hereford mappamundi, Paradise is no longer shown on the easternmost edge of Asia. Instead it is detached and indicated as one of the many islands in the Encircling Ocean. Circa 1300. Gouache and ink on parchment. 158 × 133 cm. Courtesy: The Mappa Mundi Trust and Dean and Chapter of Hereford Cathedral.

Fig. 6.11. Island of Paradise with encircling ring on the Hereford mappamundi. A double-ringed motif encircles Jerusalem suggesting that it was used to separate sacred spaces from the profane. Circa 1300. Gouache and ink on parchment. Courtesy: The Mappa Mundi Trust and Dean and Chapter of Hereford Cathedral.

with the well-established Christian iconographic practice of surrounding the heads of God, Christ, prophets, and angels with halos. So ubiquitous are examples of these that there is no need to prove this, but it does raise the possibility that the halo—like the Ahura Mazdā symbol—originates in the basic double-ringed Encircling Ocean marker for the world. Depictions of rulers and religious figures within double-ringed circles date back at least to the silver plates of the Sassanid period (see fig. 5.13 of the previous chapter). The images from Lambert's medieval encyclopedia *Liber floridus*—the earliest autograph edition—depicts the ruler (specifically Caesar Octavian Augustus but in fact an allusion to the ruler of the region in which the manuscript was made) with a T-O orb in one hand and a sword symbolizing earthly dominion in the other (fig. 6.13).[54] Note the parallels with God in the Psalter map holding a T-O orb in his left hand (see fig. 6.9).

As Gurevich reminds us, "The highest position is occupied by the monarch appointed by God. Just as the world (macrocosm) is directed by God, and the human body (microcosm) is directed by the soul, so is the body politic directed by the monarch, whose relation with his subjects might be compared to the relation between the head and the limbs. The power of the monarch does not depend on the will of the subjects. The monarch is subject to one God alone, whom alone he serves (*rex-*minister Dei)."[55] It is no exaggeration to say that the *Liber floridus* is replete with images employing Encircling Ocean forms, suggesting that the double-edged circle as a symbol of separation of the sacred from the profane was a well-established device by the twelfth century.

Where medieval European maps craft their own unique dictum is with the "T" in the "O" standing as it does for the connection between a crucified Christ-man and the

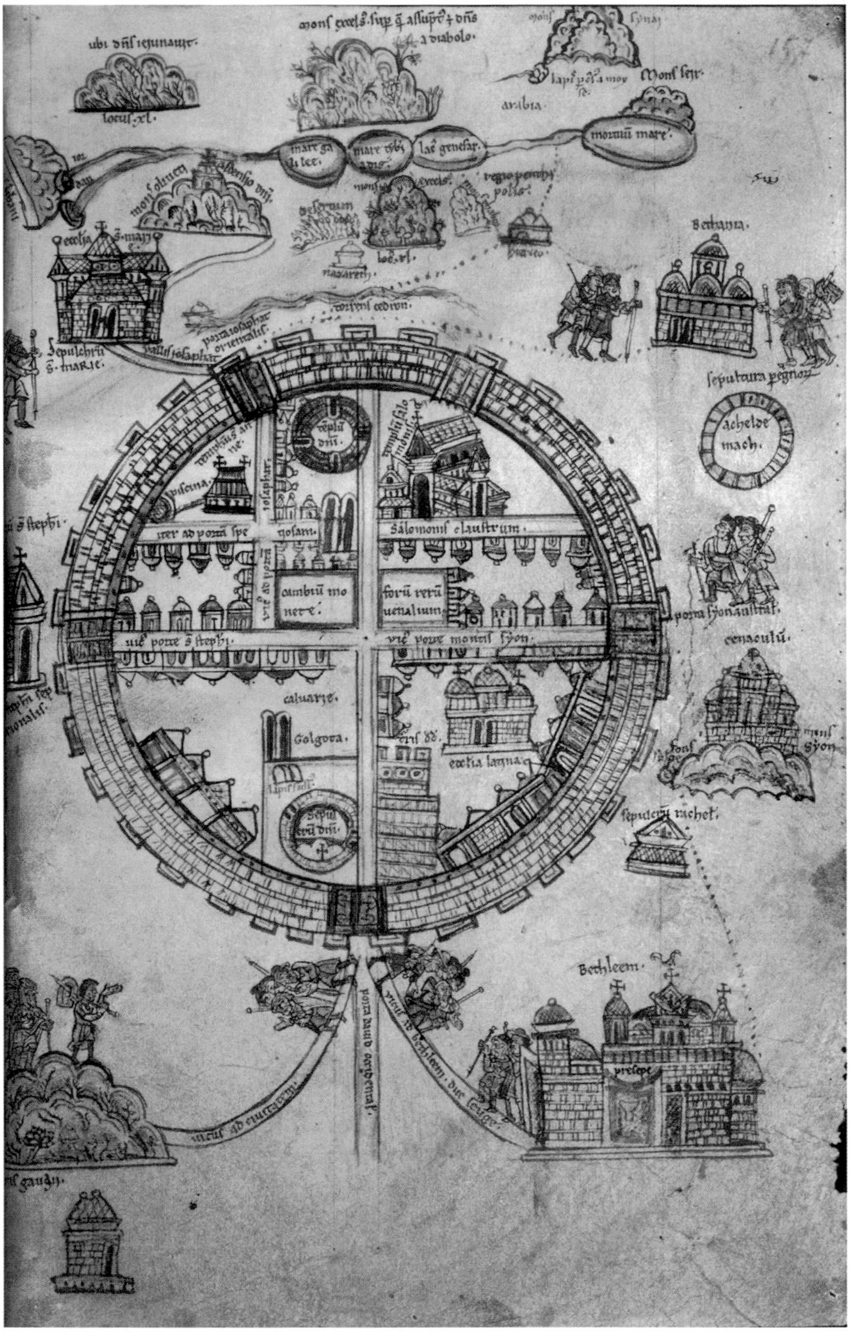

Fig. 6.12. Jerusalem encircled with a double ring. Indicates an imagined symmetry between the shape of the earth and an idealized earthly Jerusalem, a form that gained popularity in the wake of the crusaders' withdrawal. Circa late 12th century. 31.3 × 21.0 cm. Courtesy: Bibliothèque royale de Belgique, Brussels. b. Regia No. 9823-24, fol. 157a.

Fig. 6.13. Ruler encased by circle of power holding T-O map from Lambert's *Liber floridus*. Symbolic use of the double-ringed motif as a metaphor for power reinforced by the encircling ocean encasing the T-O map held by the ruler. Circa 1121. Gouache and ink on parchment. Courtesy: The University of Ghent. Ms. 92, fol. 138b.

cosmos.[56] This connection is articulated in an illustration from a twelfth-century copy of a Hildegard of Bingen manuscript. In this image of the world, encircled by multiple ripples of an encircling ocean, a distinct T form of Christ as man connects with God and Christ the Son of God. In this way the sacred firmament is connected to the earth through the corporeal body of Christ (fig. 6.14). His veins represent the course of the sacred waters that are shown cycling through the body, beginning and ending in the Encircling Ocean.[57] The connection between this rapturous Bingen vision of the world and T-O maps is apparent when we compare it to an image of a T-shaped figure with outstretched arms gripping the "O" of the Encircling Ocean in

Fig. 6.14. Macrocosm, microcosm, and winds from Hildegard of Bingen's *Liber divinorum operum*. World encircled by multiple ripples of the encircling ocean is connected to God through a "T" formed by the body of Christ. 13th century. Courtesy: Biblioteca Statale di Lucca. Cod. lat. 1942, fol. 9r. Photo credit: Scala/Art Resource, NY.

Fig. 6.15. Floriated "T" with human figure gripping the "O" of the Encircling Ocean from Isidore's *Etymologies*. The Encircling Ocean is stressed with six points highlighting its importance for framing the world and humans within it. 13th century. Blue, green, and yellow gouache and red and blank ink on parchment. Courtesy: Biblioteca Medicea Laurenziana, Firenze. Ms. Conv. Soppr. 319, fol. 90b.

a floriated "T" for "Terrat/es" from a thirteenth-century illuminated copy of Isidore's *Etymologies* (fig. 6.15).[58]

Sources of the European Mappamundi Encircling Ocean Motif

What are the sources of the medieval European T-O model with its single and multiple encircling bands? There are two likely candidates, the Macrobian and Beatus traditions, heralding from the late antiquity and the early medieval period respectively. The former is a legacy of the twilight of Roman dominion and the latter has

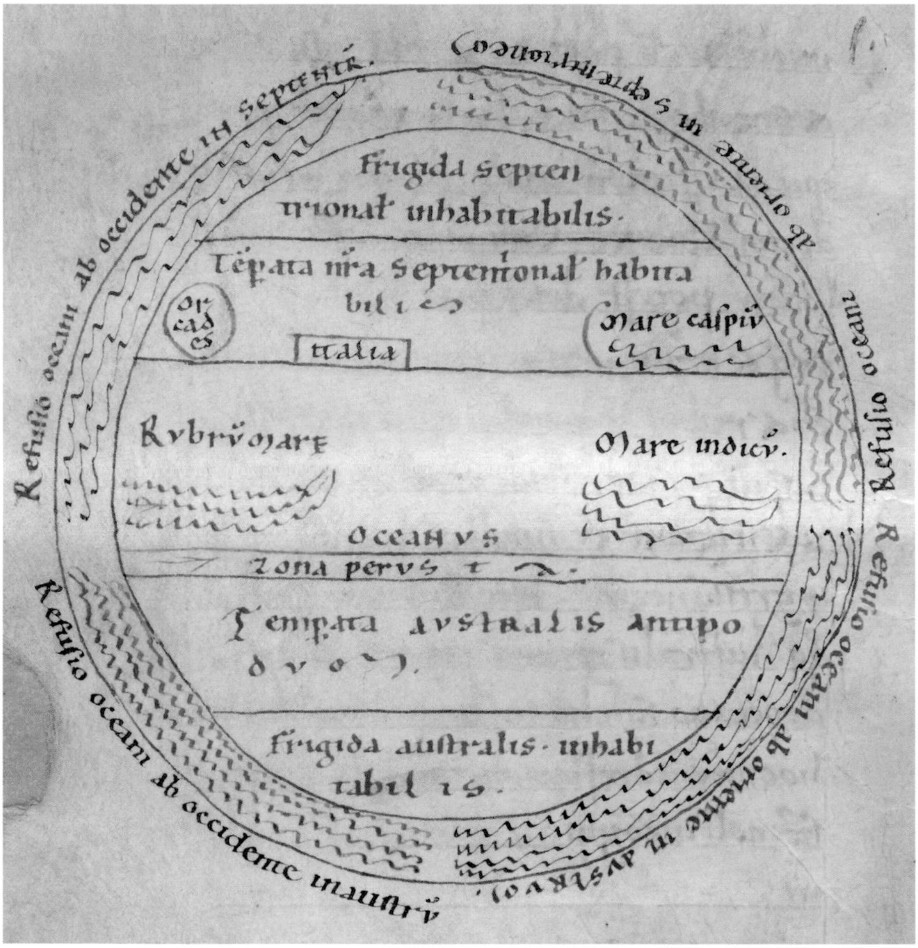

Fig. 6.16. Zonal image of the world from Macrobius's *Commentary on the Dream of Scipio*. Circa 9th or 10th century. France. Red and black ink on parchment. Courtesy: Bibliothèque royale de Belgique, Brussels. Ms. 10146, fol. 109b.

Visigothic roots. One developed during the heyday of the Patristic period (ca. 400–700 CE); the other grew out of adversity and a time of war and conflict between invading Germanic tribes and invading Muslims, giving rise to apocalyptic views of the world. Both models employed an Encircling Ocean motif and highlighted its significance through space and prominent embellishments.

The Macrobian tradition, based on the early fifth-century commentary by Macrobius Ambrosius Theodosius on Cicero's *Dream of Scipio* (51 BCE), takes a visual

Fig. 6.17. Mozarab world map from *Beatus Liebana—Facundus*. Mozarabic map depicts the world in an oblong shape with dramatic markers for Jerusalem and Paradise. It is surrounded by an ultramarine Encircling Ocean filled with large fish outlined in white. 1047. Gouache on parchment. 31.2 × 28 cm. Courtesy: Biblioteca Nacional de España, Madrid. Ms. Vitr. 14-2, fols. 63b–64a.

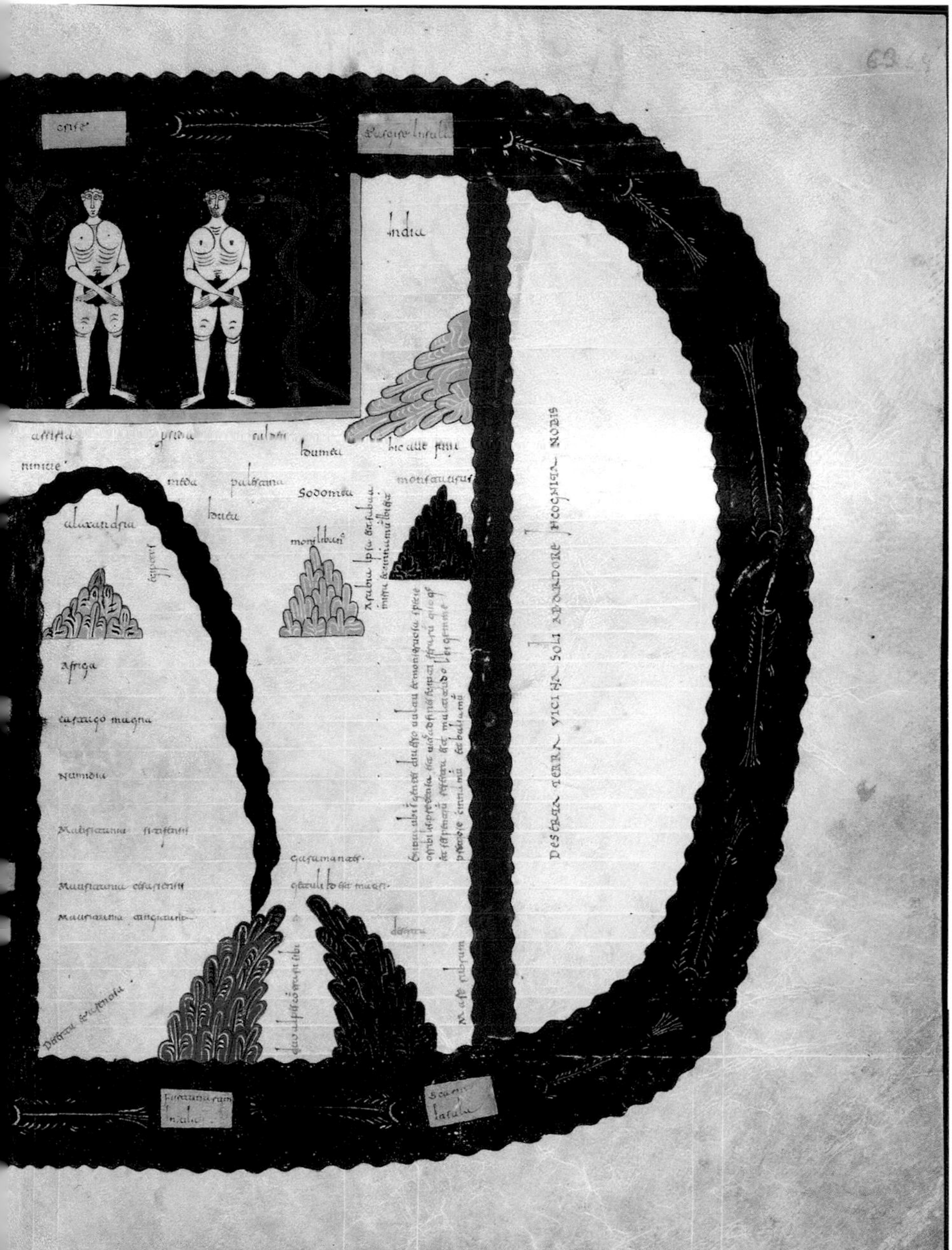
India
Africa
Numidia
DESERTA TERRA VICINA SOLI AB ARDORE INCOGNITA NOBIS

approach to Scipio's "God Trick" of looking back at the earth from outer space while listening to the music of the spheres. Alfred Hiatt emphasizes the importance of the ocean in the maps of the Macrobian tradition. He argues that "the primary purpose of the image was to illustrate the direction of ocean flows, the formation of seas, and the relationship of the known world to unknown but hypothesized regions."[59] This emphasis is clear from the earliest extant Macrobian map dated approximately to the ninth/tenth century (fig. 6.16).[60] The copyists of this French rendition reinforce the importance of the Encircling Ocean in the Macrobian vision by highlighting the ocean with squiggly lines indicating waves and currents.

The Visigothic Beatus tradition begins toward the end of the eighth century after the lightning Muslim invasion of the Iberian peninsula, when it was rumored that the events of the Apocalypse as written in the biblical book of Revelation were about to unfold. To this end, in 776 CE, the monk Beatus prepared his *Commentary on the Apocalypse*. To date we know of twenty-six elaborately illustrated copies of this manuscript. It is important to note that none of the copies date to the time of Beatus.[61] The Beatus maps come in two key models: square and oval. Both models display a world surrounded by a prominent Encircling Ocean.[62] The example shown in figure 6.17 is from the Beatus of Liebanna and is of Mozarab provenance. Among other things, this example of a Beatus world map shows the influence of local Andalusi Mozarab artists,[63] and parallel elements from this rendition can be found in other Islamic maps. (See, for instance, fig. 7.1 for parallels in the concept of gigantic fish that populate the waters of the Encircling Ocean.)

Like the illustrated *Liber floridus* tradition, the Beatus manuscripts too are filled with images incorporating sacred double-ringed circles. In the "Vision of God Enthroned" in the 1220 CE Las Huelgas manuscript God is seated within a circle surrounded by the prophets, apostles, and saints.[64] A similar image is that of the Lamb of God (fig 6.18) symbolizing Christ in a Heavenly version of Jerusalem surrounded by the angels.

In his book *City and Cosmos*, Keith Lilley expounds on the multiplicity of double-ringed circles in everything from T-O maps to images of the cosmos and renditions of Jerusalem on earth and in the heavens. Lilley's answer to this widespread phenomenon is the influence of Neo-Platonic thought—a key influence in medieval Islamic culture as well.[65]

> What we can deduce from comparing images of the Christian world with contemporary depictions and descriptions of earthly and heavenly cities is that they share a sacred geometry. Particular shapes—circles and squares, the cross and cardinal axes—connected city and cosmos, urban form and cosmological form were analagous, and their shared geometrical forms conveyed a common symbolic meaning. Both formed part of a hierarchy of concentrically ordered spaces from center to edge, and both were marked with

Fig. 6.18. Apocalyptic vision of Heavenly Jerusalem with the Lamb of God from Maius's *Morgan Beatus*. Sacred hierophantic circles abound in the medieval European illustrated manuscript tradition. Circa 940–945. Taìbara, Spain. 38.1 × 27.9 cm. Courtesy: Morgan Library and Museum, NY. Ms. M.644, vol. 1, fol. 87a.

> the sign of the cross, symbolizing Christ himself. These Christian urban and cosmological imaginings, with their common structuring "geometric schema," were drawing upon Neoplatonic and Aristotelian sources to elucidate and conceptualize the form and the order of the cosmos. . . . They forged a symbolic link between city and cosmos in the medieval imagination.[66]

Lilley reinforces his argument with images such as the well-known one from the *Bible Moralisée* (fig. 6.19), which depicts God as architect, with compass in hand, measuring the earth with its distinctive blue Encircling Ocean band.[67] The four-petal enclosure encasing God in the likeness of Christ of the Bodleian version resembles Indic cosmographic forms discussed in the previous chapter.[68]

In an attempt to reconcile certain passages in the Bible with the known world, medieval European scholars proposed a variety of innovative interpretations that resemble other premodern religio-cultural traditions. Among the key passages of the Hebrew Bible that refer to the Encircling Ocean are the following:

> **Genesis 1:6–7:** "And God said, 'Let there be a dome in the midst of the waters, and let it separate the waters from the waters.' So God made the dome and separated the waters that were under the dome from the waters that were above the dome. And it was so."

> **Genesis 1:9–10:** "And God said, 'Let the waters under the sky be gathered together into one place, and let the dry land appear.' And it was so. God called the dry land Earth, and the waters that were gathered together he called Seas. And God saw that it was good."

> **Psalm 148:2:** "Praise him, you highest heavens, and you waters above the heavens!"[69]

One gets from the Bible a sense of a primordial ocean from which the earth was created—an idea that was prevalent in the ancient period (see chapter 5 for a detailed discussion of the Judaic vision of the world in the cosmos). These verses mention not one but multiple Encircling Oceans. Evidence that medieval scholars wrestled with this idea is indicated by their rationalization of the phenomenon. Isidore, for instance, says: "Our dwelling-place is divided into zones according to the circles of the sky."[70] Explanations such as these help us link the medieval European picture of the primordial Encircling Ocean with what came before and what was transmitted after (see chapters 5 and 7).

One of the most common presumptions was that all the waters of the earth must be interconnected, resulting in a "congregation of waters" emanating from a single source. This theory is bolstered by Genesis 2:6: "A spring rose out of the earth, watering all the surfaces." On the basis of this, St. Augustine posited that it was via this spring that water cycled back and forth from the rivers to the ocean, and that there

Fig. 6.19. Architect of the world from *Bible Moralisée*. Mid-13th century. Reims, France. Gouache and gold on parchment. 24.4 × 26 cm. Courtesy: Oesterreichische Nationalbibliothek, Vienna, Austria. Cod. 2554, fol. 1b. Photo credit: Erich Lessing/Art Resource, NY.

must have been a subterranean passage through which the water passed.[71] (See reflections of these concepts in Judaic conceptions of the world and the cosmos as discussed in the previous chapter, in particular figure 5.16.) Bernard of Clairvaux portrayed Christ as an unending source of virtue and knowledge among men, in the way that "the sea is the source of fountains and rivers." Likewise, Hildegard of Bingen assumed that the water of wells, springs, and rivers is derived from the ocean, which surrounds the earth.[72] It is an idea that is reminiscent of the Mazdean concept of a single water source recycled between Mt. Hara and the Vurukasha Sea (refer to the discussion in the previous chapter and fig. 5.15).[73]

Similarly, the waters of the universe, within which the earth was said by the Bible to be contained, were a matter of speculation. Ambrose suggested that they were "intended to cool the axis of the universe, overheated by its perpetual rotation." Others argued that it was intended "to screen the earth from the fiery heat generated by the stars and the sun." Ambrose went so far as to suggest by analogy that "if the earth can hang in the center of the universe without support so also can the waters hang unsupported above the firmament."[74] Isidore extolled the virtue of water in a like manner:

> The element of water rules over all the rest, for water tempers the sky, makes the earth fertile, gives body to the air with its exaltation, ascends to the heights, and claims the sky for itself. Indeed, what is more amazing than water standing in the sky? And it is not enough that it reaches such a height, but it snatches a school of fish with it and when poured out becomes the cause of all growing things on earth. It brings forth fruits and trees, produces shrubs and grasses, cleans away filth, washes away sins, and provides drink for all living creatures.[75]

All this water, above, below, and beyond, necessitated a concept of a ball-like earth bobbing on the surface of an all-encompassing body of water like the Encircling Ocean. So ubiquitous, in fact, is the Encircling Ocean in medieval European motifs—mappamundi and otherwise—that in her analysis of the Hereford mappamundi, Naomi Kline devotes an entire chapter in her book *Maps of Medieval Thought* to the subject of "the circle as a conceptual device."[76] She asserts that "that mappae mundi . . . functioned in a way similar to medieval *rotae*, flat, circular wheels that made understandable, in a simplified graphic manner, concepts that explained the way the world worked. . . . In effect medieval mappae mundi share with medieval *rotae* the circular format as containers of diverse information. Our consideration of the medieval mappae mundi is based upon the idea of the circle as a shorthand device for assisting and learning and memory and situates medieval mappae mundi within the medieval realm of wheels of memory."[77]

In an extensive archaeology of Encircling Ocean–like images, Kline shows that the concept of a world within a multiplicity of heavenly spheres was connected to the medieval European imagining of the cosmos. She elucidates her thesis by marshaling the evidence of multiple *rotae*-type cosmographic images that were found in monastic schoolbooks such as William of Conches's *Summa magistri Wilhelmi de Conches* and Gauthier of Metz's *Le Romounce del ymage du monde*.[78] Indeed, Kline's entire *Maps of Medieval Thought* is a contemplation of the primacy of double-ringed circles in medieval Christian illuminated manuscript art.[79] She does not, however, make the connection between the Encircling Ocean and the symbolic stamp of the imago mundi[80] as the origin of this medieval Christian obsession. Nor does she take the route of Lilley and others who see it as a Timean reflection of Neo-Platonic influence.[81]

Fig. 6.20. Macrobian-like world map from al-Sijzī's *Tarkīb al-aflāk*. The Northern Hemisphere has a faint latitudinal grid inked in red. Circa 12th century or earlier. A reader's note on the opening folio indicates a date of 846/1442. Diameter 13.25 cm. Courtesy: Leiden University Libraries. Cod. Or. 2541, fol. 21b.

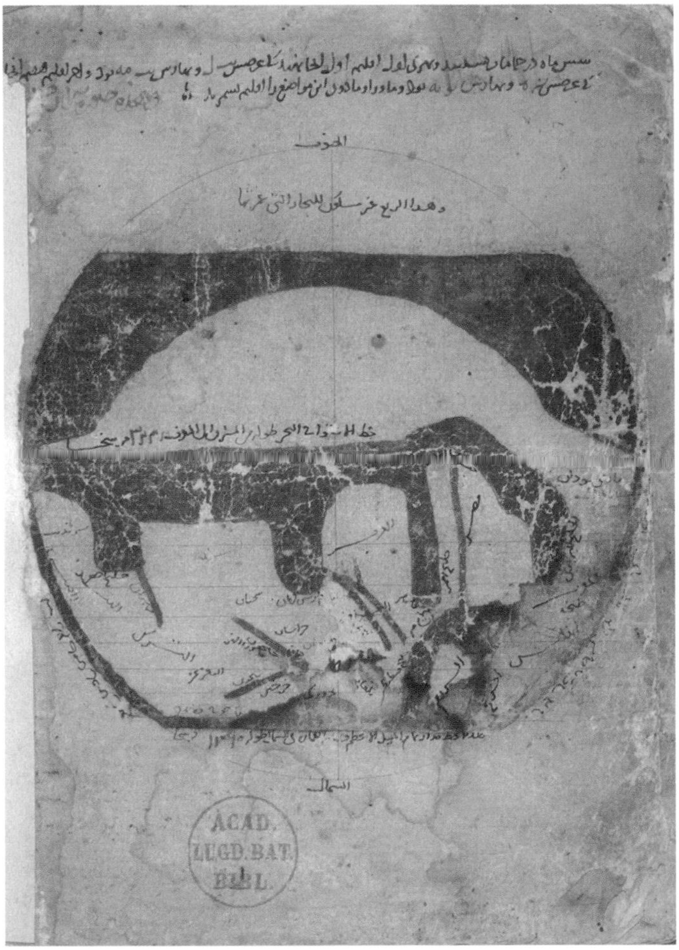

Islamic and Medieval European Mappamundi Connections

Just like the medieval Islamic world maps, all medieval European mappamundi contain the same dominant feature: an Encircling Ocean that frames the image of the world—ergo imago mundi. Did medieval European maps influence the Islamic ones or vice versa? Or were they mutually exclusive?[82] The Mozarabic maps of the Beatus tradition, discussed above, hint at a cross-pollination of decorative motifs at least within the artistic milieus of Islamo-Christian Spain. There are also some Islamic maps reminiscent of medieval European Macrobian models that show a fourth southern continent.[83] One such example is found at the end of an incomplete Sijzī astronomical manuscript, *Tarkīb al-aflāk* (Composition of the Spheres), housed in the Oriental manuscript collection in the library at Leiden (fig. 6.20), recently redated to between 1041 and 1165.[84] The map is labeled in Arabic even though the manuscript is in Persian. It shows a fourth continent separated from an overextended Africa by a large body of water. The main landmass is broken up into distinct segments for North Africa, Egypt, the Arabian Peninsula, Sind, Sarandīb (Sri Lanka), which is unusually shown as attached to the South Asian mainland, and Central Asia with its customary Turkic tribes and requisite rivers (Jayhūn and Sayhūn), ending with the landmass of al-Andalus at the northwestern edge.[85]

In addition to the unusual world map from the Persian astronomical manuscript discussed above there is another surprising T-O map from an eighth- or ninth-century Isidorian manuscript that points to Islamo-Christian transcultural cartographic connections:[86] a basic Isidorian T-O world map labeled in Arabic (fig. 6.21).[87] For the first time we have proof of a definitive, early medieval tie between the two cartographic traditions.[88] The Arabic notations placed on the map in a rough list T-O fashion (compare with fig. 6.6) record the names of key Muslim places on each of the three landmasses, such as Mecca (Makka) and Medina (Yathrib) on the Levantine side, the Berbers and the Ḥabasha (Ethiopians) on the African flank, and Gog and Magog along with al-Rūm (Byzantium) and al-Andalus (Islamic Spain) on the European side. Equally interesting are the notations in the Encircling Ocean, which give the dimensions of this ocean on three flanks. At the top of the circle, we read in Arabic that the earth is collectively 24,000 farsakhs, and in Latin that it is 180 furlongs.[89]

The Arabic commentary ranges from the names of the landmasses to a listing of place-names to ecumenically related symbolic narratives. Aside from the heavily annotated map, there are marginalia comments in Arabic scattered throughout this manuscript beginning early on folio 4v and occurring throughout the manuscript to the final colophon folio of 163v. This map confirms something that historians of medieval Islamic history have long sensed but could not prove: namely, that Arabic speakers in western parts of the Islamic world not only knew about medieval European maps but interacted actively with them. It suggests that the Arab reader and owner of this manuscript took an active interest in the geographical material of this manuscript. He marked folios 70–150 with names of places in Arabic that match the Latin text, and he took enough of an interest in the T-O map to add his own personal notations.

We can further deduce that at some point this manuscript returned to Latin hands. The notation above the top of the "T" in Latin was written *after* the Arabic notations. Furthermore, the Arabic notations themselves are in two different hands: one for the internal regions of the map and the other for the external Encircling Ocean band listing the measurements. We cannot, however, be sure of the date either of this manuscript, its T-O map, or the Arabic notations. Latin experts have pegged this manuscript vaguely as sometime between the eight and the ninth centuries. Oddly,

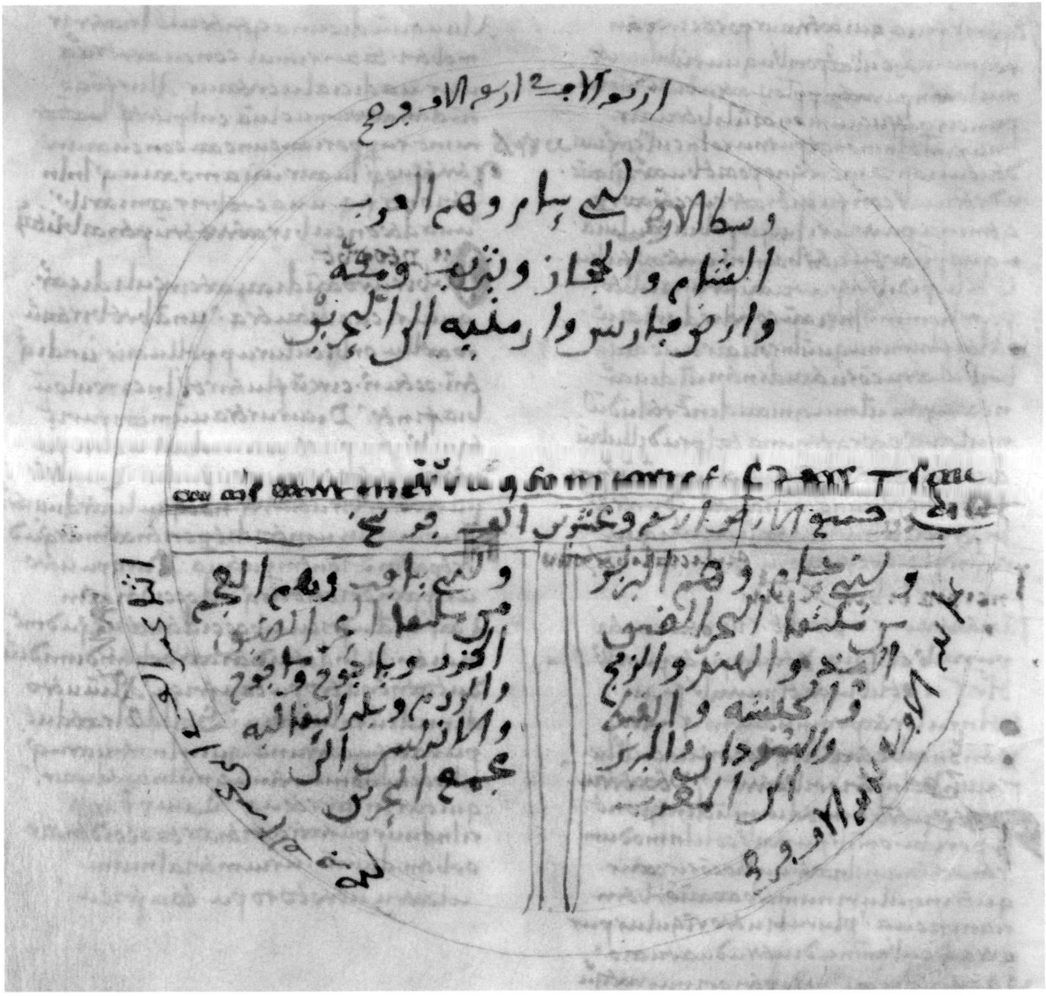

Fig. 6.21. T-O map with Arabic notations from a copy of Isidore's *Etymologies*. Circa 8th–9th century. Ink on parchment. Courtesy: Biblioteca Nacional de España, Madrid. MS. Vitr. 014/003, fol. 116v.

until recently historians of medieval European cartography mentioned neither this map nor this manuscript in their accounts of medieval European mappamundi.[90]

This map is nothing short of stunning both for specialists in Islamic cartography and medieval European cartography and for the field of history of cartography in general. It implies an early intersection between the medieval European and Islamic mapping traditions (at least in the Maghrib).[91]

Transcultural Cartographic Drops

At the XVIIIth International History of Cartography conference in Athens, in July 1999, Christian Jacob gave a visionary paper on the subject of transcultural influences on Hellenistic Greek cartography in which he asked: "If I pour some drops of Chinese cartography on Greek Hellenistic cartography, how does the latter react? What does the Chinese case reveal when applied to the understanding of Hellenistic cartography?"[92] Sitting in the audience, I found myself thinking: "If you add additional drops from the Indian, Iranian, Mesopotamian, Egyptian, and Central Asian cartographic traditions, then you have Islamic maps!"

This broad yet necessarily cursory transcultural survey of the Encircling Ocean in this and the previous chapter reveals that it was a widespread motif best classified as a metaform—nurtured by the observation that no matter what direction one travels in, one eventually reaches the sea. It emerges as an important component of water-based hylogonies—namely, cosmogonies that are based on water as the source of life and death of the world.[93] The Encircling Ocean becomes the basic outline stamp of all premodern *imagines mundi* and turns into a useful signifier of power and domination for ruling groups seeking an easy stamp of world dominion.

What this iconography seeks to prove is that the "Encircling Ocean" feature of Islamic maps cannot be exclusively pegged to the Greeks, Ptolemy, or Neo-Platonists. Rather, it is the most common of transcultural metamotifs of premodern world maps. My purpose is not to deny the connections between Greek and Islamic cartography but only to set them in a metacultural perspective. Greek influence is present, but so are Indian, Iranian, Central Asian, Chinese, Egyptian, Roman, and medieval European influences.

The Muslims retained the Encircling Ocean in their maps as a metacartographic feature, but to understand how they thought about it, we need to examine their textual descriptions of this timeless ocean. Did the Muslims share the ancient Iranian, Indian, and early Semitic view that all the water in the world stemmed from a single source linked to the Encircling Ocean? Did they believe in one or multiple Encircling Oceans? Can the primary texts help us to determine the specific cultural sources for their ideas on the ocean, or are they as metaculturally diverse as the images? Only a close reading of some of the key Arabic and Persian sources can help us fully address the significance of the Encircling Ocean in the Islamic cartographic imagination.

7 The Muslim Bahr al-Muhīt

The Qur'ān is the natural starting point for an analysis of the Encircling Ocean in the Islamic context. But when it comes to cosmographical and geographical details, the Qur'ān is vague. As guest editor Ahmet Karamustafa puts it in the *History of Cartography* volume *Cartography in the Traditional Islamic and South Asian Societies*, "The Qur'ān, the single most important source of Islamic culture, does not contain a systematic cosmology. No single Qur'ānic verse addresses the structure of the universe directly, and materials of cosmological import that appear in the Qur'ān are as a rule devoid of descriptive detail and do not lend themselves to comparative analysis."[1] Similarly, the Qur'ān does not contain a systematic geography. What we get instead are reminders of God's mercy and wrath set against a generic terrestrial background. Take, for instance, the most expansive passage in the Qur'ān involving seas:

> It is your Lord Who drives on ships at sea for your sake, that you may seek His bounty. To you He has ever been Compassionate.

When calamity touches at sea, they are nowhere to be found—those you call upon instead of Him. Man is truly ungrateful.

Are you so confident He will not, once more, send you back to sea, where He will unleash upon you a devastating hurricane, drowning you for your ingratitude, whereupon you will find none to plead your case before Us?

We honoured the progeny of Adam and carried them on land and sea.

We provisioned them with delicacies and preferred them far above many whom We had created. (17:66–70)[2]

In these verses, the seas form no more than a vague background reference. Mention of them occurs only to remind us that we are at the mercy of God. Occasional mention is made of specific telluric forms, such as the *barzakh* (barrier/hurdle). The word occurs three times in the Qur'ān. One of these speaks of the separation between this world and the next and is regarded as the primary meaning of *barzakh* by Muslim theologians and thinkers (23:100).[3]

The remaining two verses (25:53 and 55:19–20) refer to two seas on either side of the *barzakh*:

It is He Who merged the two seas,
This one fresh and sweet water,
That one salty and bitter.
Between them He erected a barrier, an impassable boundary. (25:53)[4]

And

He has brought the two seas together, but as they meet,
Between them is a barrier they do not overrun. (55:19–20)[5]

When approaching these verses from a geographic perspective, this barrier is usually taken to refer to the area of Suez and the separation between the Mediterranean and Red Seas. This interpretation is reinforced by another verse about the two seas (al-baḥrayn) involving Moses and his youthful attendant seeking the place where the two seas meet (majmaʿ al-baḥrayn) (18:60) and is thus taken as a reference to the parting of the Red Sea.[6] Burnham insightfully points out that Andalusi geographers had a different reading of these verses and the *majmaʿ al-baḥrayn*. They saw the Straits of Gibraltar as the *barzakh* and the "confluence of the two seas" as a reference to the joining of the waters of the Mediterranean and the Encircling Ocean.[7]

There are two additional Qur'ānic verses that refer to the two seas: 35:12 stresses the differences and similarities between the two seas: "The two seas are not alike: this one is fresh and sweet water, tasty to drink, that one salty and bitter. From both you eat flesh that is soft, and you extract jewelry to wear. Therein you can see ships ploughing through the waves, that you may seek of His bounty—perchance you will give thanks." While 27:61 casts the two seas as part of the full creation of the earth: "Is it not He Who made the earth be at rest, and made rivers run through it, and erected mountains therein, and built a barrier between the two seas? Is there to be another god with God? Truly, most of them have no understanding."[8]

In contrast to geographers, religious scholars interpret these verses about the two seas and the barriers that separate them metaphorically. Based on the verse that follows 25:53, which speaks of kinship, blood, and marriage, Ahmed Ali, for instance, adheres to the view that the two seas are representative of the differences between men and women and the inevitable barrier between them.[9] Sufis, in particular Ibn 'Arabī (d. AH 638/1240 CE), hold that these passages about barriers and the two seas reinforce the paradoxical nature of the *barzakh*, that it is, on the one hand, an insurmountable barrier and, on the other hand, a means for bringing two sides of the coin together—such as, the here and the hereafter, being and nothingness, and, pertinent to our interest, earth and the cosmos.[10]

As is the case with all verses in the Qur'ān, meaning resides in interpretation and the perspectives, aims, and backgrounds of the interpreters. Different theological schools follow their own interpretations, which inform the Muslim geographers who then apply these interpretations to their geographical discourses. The Qur'ān's veiled geographical content acted as a stimulus that encouraged geographers to learn more in order to explicate the meaning of obscure Qur'ānic verses, and in doing so they built up a picture of the world around them that necessarily relied on information from other cultures to fill in the gaps.

The Throne verse, 11:7, is an excellent example of this form of geographical interpretation: "He it is Who created the heavens and the earth in six days, and His Throne was upon the waters, so as to test you: who among you is the best in works." Much ink has been devoted to an understanding of this enigmatic verse and the nature of God's throne (kursī). Its aquatic location imparts to all seas the special attribute of creation, a reading that is reinforced by other verses, such as 21:30, which discusses the creation of "every living thing from water," and 25:54, the verse following a *barzakh* reference that says, "It is He Who, from water, created man." These verses hark back to ancient Egyptian, Mesopotamian, Indian, Greek, and Hebrew ideas of water as the *fons et origo* of the world, as discussed in the previous chapters, and become an important component of the representational fabric of the Muslim Encircling Ocean in both textual and visual contexts. It is not surprising to find that many a medieval Islamic geographer identified the water that held God's throne as the Baḥr al-Muḥīṭ.[11]

"It is God who created seven heavens, and of earth their like, between them the Command descending, that you may know that God is powerful over everything and that God encompasses everything in knowledge" (65:12).[12] This Qur'ānic mention of the seven earths and seven heavens is another transcultural ancient resonance. It is discussed at length later in this chapter.

Ḥadīth traditions of stories involving the Prophet are another major source of information for Muslims. Like the Qur'ān, ḥadīth do not dwell on geographical details. Instead they focus on morals relating to daily life and piety as emulated by the Prophet Muḥammad. When the sea is invoked, it is usually employed metaphorically to illustrate interminability or ultimate sacrifice—thus the sins of the most grievous sinner are compared to the foam of the sea ("His sins will be forgiven even if these are as abundant as the foam of the sea")[13]—or to signify complete obedience ("By God whose control is my life, if you order us to plunge our horses into the sea, we would do so").[14] One who drowns is considered a martyr,[15] and the Prophet sees Muslim fighters sailing the sea atop thrones like kings in a dream.[16] Muḥammad designates seawater pure enough for ablutions, and fish that die in it are purified by its water and are therefore permitted as food.[17]

But the sea is not all good. It also has malevolent overtones in its association with Dajjāl (the Muslim Antichrist) who resembles the beast in Revelation 13:1. Dajjāl is the apocalyptic opponent of Jesus who is destined to appear at the end of the world. When Dajjāl is on the verge of defeating his Muslim opponents, Jesus will descend and fight him off. In defeat Dajjāl will seek out the sea and be killed near it. Ḥadīth tell of how Tamīm al-Dārī, a companion of the Prophet, was shipwrecked near an island in the Mediterranean where a strange creature called al-Jassāsa was held in chains. In the question-and-answer session that follows with the Muslims, the creature identifies itself as none other than the dreaded Dajjāl and promises to come one day and destroy Muslim towns.[18] This ḥadīth briefly mentions the Baḥr al-Shām (Sea of Syria) and the Baḥr al-Yaman (Red Sea) and Constantinople in the context of the appearance of Dajjāl who will be defeated upon the descent of Jesus. We are told that Dajjāl will disappear "as salt dissolves itself in water"—the only other reference to water in this ḥadīth.[19] Mecca and Medina are mentioned regularly as are places in Egypt, Syria, and Iran, while oases, lakes, and islands are also mentioned by name. But details on these places are sparse and vague and serve only as a distant background to the morality of the tale.

The impression that one gets from the Qur'ān on geographical matters is that God knows everything (wa allāhu a'lam) but that he keeps this knowledge to himself:

> He has the keys of the Unknown.
> No one but He has knowledge;
> He knows what is on the land and in the sea.

Not a leaf falls without His knowledge,
Nor a grain in the darkest (recess) of the earth,
Nor any thing green or seared
that is not recorded in the open book (of nature). (6:59)[20]

This does not mean that humans are not free to explore. On the contrary, the Qur'ān leaves open for investigation the "book of nature" and repeatedly encourages Muslims to "travel and see" (3:137, 6:11, 12:109, 16:36, 22:46, 27:69, 29:20, 30:9, 45).[21] This Qur'ānic injunction to "travel and see" was picked up and stressed by ḥadīth. There are a large number of ḥadīth extolling the virtues of traveling for the purpose of learning. Among the most famous is one reported by Tirmidhī who makes it clear that "those who go out in search of knowledge will be in the path of God until they return."[22] Indeed, the canonic body of ḥadīth owes its very existence to the indefatigable *ṭalab al-'ilm* (students of knowledge) who traveled for the sake of collecting ḥadīth and acquiring religious knowledge.[23] As Houari Touati puts it in the opening line of his book *Islam and Travel in the Middle Ages*, "Muslim men of letters of the Middle Ages were mad for travel."[24] And Sam Gellens insightfully remarks,

> This civilization—really a network of variegated societies united by their commitment to the *shari'a*—was one which in the fullest sense owed its vibrancy to constant movement. Travel in all its myriad forms—pilgrimage, trade, scholarship, adventure—expanded the mental and physical limits of the Muslim world, and preserved and nourished the various contacts that Muslims perennially maintained with one another. . . . Travel in its broadest definition ensured the unity of the Muslim community, but likewise encouraged appreciation of one's home. Travel bound them together and simultaneously stimulated the appearance of their local and regional identities.[25]

It is for this reason that examination of all aspects of the world and its peoples along with the quest for religious knowledge was the impetus of early Muslim scientific investigations. Touati distinguishes between aural and visual phases of this travel for the purpose of collecting knowledge. In the earliest centuries *samā'* (hearing) holds sway over *'iyān* (seeing), and it is for this reason that eighth-century Arabic philologists (in the manner of anthropologists) went to the desert to hear the Bedouin for themselves, and why the *muḥaddithūn* (ḥadīth scholars) traveled great distances in search of ḥadīth with high *isnād* (chain of transmission of a ḥadīth) and the *ṭalab al-'ilm* suffered the privations of travel, lack of food and sleep, and even blindness to gain their *ijāza* (certificates) from ḥadīth authorities of repute. By the ninth century, *ijāza* morph into *wijāda* (certifications of expertise without a master). Similarly, travel writers had two routes: "I come, I listen, and I write" or "I go, I see, and I write," with the latter gaining ground in the later Middle Ages and culminating in the unique *riḥla* (travelogue) genre pioneered by Abū Bakr ibn al-ʿArabī (d. 543/1148) and perfected by Ibn Jubayr (d. 614/1217). This affected geographers too who went from being real seekers of geography as active travelers to armchair geographers and compilers who relied on copying information from previous geographers. Travel was the gel of medieval Islamic society not because it sought out "otherness" as the Western European tradition did but "sameness"—for the purpose of creating a single *mamlaka* (realm) in which Muslims could feel at home anywhere.[26] Ḥadīth, such as the one cited below, reinforced this by extolling the virtues of travel for knowledge as an easier path to heaven.

> I heard the Messenger of Allah say: "Whoever follows a path in the pursuit of knowledge, Allah will make easy for him a path to Paradise. The angels lower their wings in approval of the seeker of knowledge, and everyone in the heavens and on earth prays for forgiveness for the seeker of knowledge, even the fish in the sea. The superiority of the scholar over the worshipper is like the superiority of the moon over all other heavenly bodies. The scholars are the heirs of the Prophets, for the Prophets did not leave behind dinar or dirham rather they left behind knowledge, so whoever takes it has taken a great share."[27]

The adventures in travel that resulted from this perceived religious mandate to search for knowledge, and from it the forays into human and physical geography, along with visual representation, are among the guiding beacons that informed the authors of Islamic geographical manuscripts and cartographic images. The start of the KMMS series sometime in the tenth/eleventh century fits with Touati's observation of the switch in emphasis from audio to visual, that is, from *samāʿ* to *ʿiyān*.

It is from this perspective that we must view the work of the geographers, the historians, the scientists, and others in their commentaries on the composition of the world. Muslim societies did not emerge in a vacuum: therefore their output must be seen in light of the works of antiquity that crossed the paths of the traveling scholars. Some of these influences are better documented than others. This holds especially true for an enigmatic and impossible to fully survey body of water such as the Encircling Ocean.[28] The importance of Greek influence in the realm of science and geography is repeatedly emphasized over Chinese, Indian, Persian, Mesopotamian, Egyptian, and Hebraic contributions.[29] One has only to take a look at Ṣāʿid al-Andalusī's *Ṭabaqāt al-umam* (Classifications of Nations), for example, to gain an understanding of the kind of information that was circulating in the medieval Middle East in the eleventh century pertaining to the cultural dispositions and scientific explorations of a wide variety of people. Ṣāʿid al-Andalusī divides the world into those nations that did not take an interest in science and those that did. Among the former—oddly—he lists the Chinese and the Turks, while among the latter he lists the achievements of the Indians, the Persians, the Chaldeans, the Greeks, and even the ancient Egyptians, along with the achievements of the Muslims, as well as the contributions of the Banū Isrā'īl (i.e., the Jews).[30] In the preface to his translation of Masʿūdī's *Kitāb al-tanbīh wa-al-ishrāf* (Book of Instruction and Supervision), Carra de Vaux informs us that this tenth-century historian-geographer, who had journeyed as far as India to the east and Andalus to the west and Zanzibar in the south and the Caspian in the north, was well acquainted with Buddhism, Mazdeanism, Christianity, and Judaism:

> He is a very intelligent and well-informed philosopher. His mind is open to all the systems, from the philosophies of legendary sages to the many doctrines of the sects of his own time. An historian of religions, he greatly furthered his own research, he knows Mazdaism, Sabeism, Buddhism, and has extensive information on Christians and Jews. In the course of his travels, he himself questioned doctors and scientists of different nations, Jews, Persians, Christians, Kurds and Karamantes; he debated or dialogued with them, bringing to the interviews as much his own kindness as of his curiosity, as much intelligence as an absence of fanaticism.[31]

"For," as Masʿūdī puts it and Tarif Khalidi transcribes, "there can be no comparison between one who lingers among his kinsmen and is satisfied with whatever information reaches him about his part of the world, and another who spends a lifetime in travelling the world, carried to and fro by his journeys, extracting every fine nugget from its mine and every valuable object from its place of seclusion."[32]

Other geographers were equally well versed in transcultural traditions. The polymath Bīrūnī (362–c. 440/973–c. 1048), for instance, visited India and spent much of his life fascinated with India and Indian traditions and wrote one of the earliest and most comprehensive books on medieval India—*Kitāb fī taḥqīq mā lil-hind min maqūla* (Book of the Verification of What Is Said about India)—which is discussed in chapter 5. A century earlier, the Zoroastrian convert Ibn Khurdādhbih (ninth century), one of the earliest geographers of the routes and realms tradition, who spent time in both Ṭabaristān and Jibāl, must have had access to a broad range of transcultural information as director of the postal and intelligence services (ṣāḥib al-barīd wa 'l-khabar) for the Abbasids. Touati credits Ibn Khurdādhbih's contemporary Yaʿqūbī (ca. late third century/ninth century) with initiating the "geography through travel" revolution. Whereas Ibn Khurdādhbih relied on the travel reports of voyagers and merchants, Yaʿqūbī wrote based on his own travel experiences, which were far-flung and lengthy of duration. He went to Armenia at a young age; from there he made his way to the court of the Tahirids in Khurasan, and after that spent time in India. Later, Yaʿqūbī was named a functionary in Egypt and frequented the court of the Rustamids in Tāhart in the central Maghrib (present-day Algeria).[33] As he puts it himself, "In the flower of my young age, at a time when I possessed a full acuity of mind and a great vitality and intelligence, I did my utmost to know the history of the world, as well as the distances that separate one state from another. This penchant came to me as a consequence of uninterrupted voyages that I made, beginning in my childhood, and that long kept me far away from my native soil."[34]

The fathers of the KMMS tradition tell us in their works that they traveled extensively to collect and verify data. Iṣṭakhrī (fourth / tenth century) claims to have visited Arabia, Iraq, Khūzistān, Daylam, and Transoxiana, as well as Sind, if we are to believe Ibn Ḥawqal's account of the meeting that took place there between him and his mentor at which Iṣṭakhrī handed over the KMMS baton.[35] Along with the Iraqi and Iranian provinces of al-Jazīra, Iraq, Khūzistān, and Fārs, Ibn Ḥawqal (mid to late fourth century / tenth century) traveled extensively in Central Asia (Khwārazm and Transoxiana), the Mediterranean (including Spain and North Africa, up to the southern edge of the Sahara, and Egypt), and Armenia and Ādharbayjān.[36] Of all the KMMS carto-geographers, Muqaddasī (late fourth century / tenth century) is considered the most widely traveled and the best writer due to his organization, rhymed prose, and sophisticated arguments. He claims to have visited every region in the Islamic *mamlaka* and left no princely library unexamined. "His entire work bears witness to a vast travel experience, which gives his work its picturesque aspect and a scintillating brilliance. . . . Muqaddasī accumulates his adventures like fantastic mini-novels."[37] Based as he was in a strict definition of the *mamlakat al-Islām* (realm / world of Islam), Muqaddasī is also the only KMMS carto-geographer who did not include a map of the whole world and his maps are the most schematic of all.[38]

The entire edifice of eye-witnessing geography stems, according to Touati, from the inspiration of one the greatest belletrists of all times, Jāḥiẓ (d. 255 / 868 or 869) the master of *adab* (etiquette, urbanity, culture, and, specifically, literary style) who argued that "the human mind is perfected only by voyages, the knowledge of lands, and the frequentation of people of all social conditions." For geographers who took up the Jāḥiẓian gauntlet,

> the voyage as experience was the organizing principle of Muslim geography of the tenth century. . . . Because the voyage becomes an empirical mode of investigation, it guides the collection of information—a store of information that is necessarily considerable, given the project that it is destined to serve—but also its verification and rectification. Once that information has been collected and confirmed, it is organized, thus gaining in coherence and rationalization. Cartography, in return, draws profit from this collection of information by having available more ample and richer material. Because the geographer is by definition a voyager, and even a globe-trotter, geography does not hesitate to borrow its writing models from the travel narrative.[39]

Descriptions of the Encircling Ocean from Geographers and Historians

Nowhere is the voyaging geographer more challenged than by the enigmatic Baḥr al-Muḥīṭ that, by its world-encircling and cosmos-connecting nature, defies total exploration. The geographers present a conflicted Manichean view of this world-encircling sea. Mirroring the Qur'ānic mention of *al-baḥrayn*, we are told that it is the only sea that contains both sweet and brine waters—merged or flowing side by side depending on the interpretation of *maraja al-baḥrayn* in verses 25:53 and 55:20[40]—and that it simultaneously houses the throne of God and the islands of the devil, Iblīs. The crossing of this multivalent encircling sea is dangerous and forbidden to ordinary people because it separates the here of the mundane earth from the heavenly cosmos of the hereafter. Only exceptional humans like Dhū al-Qarnayn (Alexander the Great), Khiḍr (the mythical green man), King Soloman, and the perfect Sufi who has attained *fanā'* (nirvana of a kind) can attempt such a crossing.

It is composed of a series of radical opposites that Burnham refers to as a "conceptual malleability." It is, on the one hand, the finite end of the world and on the other infinite because no one can determine if or where it ends. The sense conveyed in geographical texts is either that it is infinite and connects with the cosmos as part of the seven encircling seas or that it skirts the mountains of Jabal Qāf that encircle and stabilize the earth. It is the transitional body between the mundane world of humans and the cosmos of the divine. It is water, fire, and fog; dark, forbidding, and light; primordial and life-giving while simultaneously encapsulating death and the most terrible forms of evil.

> The encircling sea is a destabilizing concept because, in the way it embodies origin, transition, and connection, it undermines the integrity and stability of the other meta-geographical concepts that constitute the earth . . . belying the separateness of climes, sea, and sphere, and threatening to collapse the edifice of ideas that have been erected on them. . . . The encircling sea embodies the creative tension between distinction and obliteration. . . . Even as it undermines the distinctions upon which the climes and the sphere rest, it remains the substance from which the earth and the sky were created, and the site upon which God drew the first distinctions [and] upon which all of creation is based. The encircling sea as a generative concept and a site of origins brings the tension inherent in its destabilizing nature full circle, back to its own waters.[41]

Because of its capricious nature, references to the Encircling Ocean in the geographical texts range from sparse to immensely detailed; from banal empirical statements to fabulous accounts. Some geographers barely mention the Baḥr al-Muḥīṭ, and if they do, dispense with it quickly and vaguely, as if they do not wish to destabilize their geographic accounts with the unfathomable nature of this sea. Others embrace the enigma of this all-encompassing sea and incorporate fantastic tales of the dread that lies within—complete with mythical signs warning the incautious traveler of the dangers that lie ahead. Stretching unlimited across all horizons, the Encircling Ocean is thus the paradoxical source of life and death. In the Islamic picture of the world, it reigns supreme as the ultimate, boundless, uncharted, unknowable, and, hence, the most intimidating and enigmatic sea on earth.

Although al-Baḥr al-Muḥīṭ is the most common name for the Encircling Ocean in Arabic texts, it is also known by a variety of other names. Among the most common alternatives are al-Baḥr al-Ẓulūmāt (Sea of Darkness) and al-Baḥr al-Akhḍar (Sea of Green). The name al-Baḥr al-Ẓulūmāt is used to designate the terrible, out-of-control part of the ocean where no sailor can venture because of its impenetrable darkness, further emphasized by some geographers with the name of al-Baḥr al-Muẓlim (Sea of Darkness), al-Baḥr al-Aswad (Sea of Black), or even al-Baḥr al-Zift (Sea of Pitch).[42] Al-Baḥr al-Akhḍar is a harder name to define. On the one hand, it permits the splitting of the Encircling Ocean into two halves: the dangerous half signified by black, and the good, safe section close to the shore of known landmasses intimated by green.[43] On the other hand, some scholars suggest that the name al-Baḥr al-Akhḍar is related to God's throne situated in the Encircling Ocean (see discussion above) and the emerald mountains of Jabal Qāf (discussed later in this chapter) that are said to surround the Encircling Ocean and separate it from the cosmos. There is no consistency to the manner in which geographers and other medieval Islamic scholars use these different names. For Bakrī, al-Baḥr al-Aswad is a reference to the eastern part of the ocean near China (also referred to as al-Baḥr al-Ṣīn), whereas for Masʿūdī, the same eastern portion of the Encircling Ocean is labeled as Aswad's opposite: al-Baḥr al-Akhḍar.[44] Sections of the Encircling Ocean are also referred to in the geographical texts by cardinal directions and the landmass along that section of the ocean. The western part of the Encircling Ocean that washes along the shores of the Maghrib is sometimes referred to as al-Baḥr al-Maghrib (Sea of the West), while the eastern section of the ocean that borders the lands of China is referred to as al-Baḥr al-Ṣīn (Sea of China). Zuhrī often uses the term al-Baḥr al-ʿAẓam (Great Sea) but his naming convention is not followed by other geographers.[45] The term al-Uqiyānūs (Ocean) is only used by the late medieval geographer Bakrī. Ṣīnce al-Baḥr al-Muḥīṭ is the one name that is used universally by all medieval Islamic authors to refer to the Encircling Ocean and since it is the name used to indicate the Encircling Ocean on maps, it is the name that I will use in this chapter.[46]

Authors with more of an administrative bent provide a factual view of this sea. Such is the approach of the quintessential Abbasid bureaucrat-scholar Qudāma ibn Jaʿfar (d. 337/948), who, in his tenth century encyclopedic *Kitāb al-kharāj wa-ṣināʿat al-kitāba* (Book of the Land-Tax and the Craft of Writing), takes a perfunctory and dismissive approach to the Baḥr al-Muḥīṭ: "No one knows what goes on in it except close to the shores of the Maghrib and Ḥabasha and al-Andalus. . . . It is called 'Muḥīṭ' because ships do not circulate in it and I do not know of one good report of its condition."[47] Four centuries later, the famous fourteenth-century Maghribi historian Ibn Khaldūn (733–808/1332–1406) says something similar: "Nothing of the sort exists [of charts] for the Surrounding Sea. Therefore, ships do not enter it, because, were they to lose sight of shore, they would hardly be able to find their way back to it."[48]

On its unnavigability, there is near unanimity. The anonymous tenth-century Persian geography *Ḥudūd al-ʿālam* (Regions of the World), compiled 372/982–983, reiterates this view, with one crucial difference: the islands are highlighted as stepping stones across short distances of this sea: "This sea surrounds the Earth like the Horizon and ships cannot work in this sea and nobody has crossed it and it is unknown where it ends. And on the whole stretch of the inhabited zone the people see that sea but cannot cross it in a ship, except for a very short distance from the inhabited places."[49] This is a view that even the presumed initiator of the most mimetic maps (see chapter 3), Idrīsī (d. 560/1166), reinforces: "As to the dark ocean, nobody knows what is behind it and no human being possesses any trustworthy report of it, because of the difficulty of sailing over it, the darkness of its light, the towering of its waves, . . . the tyranny of its animals, and the vehemence of its winds."[50]

Overarching this immense body of water with no known limits is a brooding sense of negativity, of darkness, gloom, danger, and death. For Masʿūdī (d. 345/956), the Herodotus of the medieval Islamic world,[51] there is no respite in this sea of gloom: "In fact, no one travels it; one does not find cultivated land, nor does one find human beings; one does not know the spread or the end, one is ignorant of its extremity; and it is called the sea of darkness/gloom, the Green Sea, or the Surrounding sea."[52] A putrid green color and stinking smell is attributed to it as a sign of the pall of death that lingers over it. "It is the uttermost sea, dark, relentlessly green, and lifeless."[53] None is worse than the northwestern end of this ocean, where Masʿūdī compares the sea to a giant sea-lung.[54] True to form, the historian Ibn Khaldūn attempts to provide a rational explanation for this phenomenon: "Moreover, the air of the Surrounding Sea and its surface harbors vapors that hamper ships on their courses. Because of the remoteness of these (vapors), the rays of the sun which the surface of the earth deflects, cannot reach and dissolve them."[55] A later supplier of cosmographic cum geographic tales wrapped in a generous dose of fantasy, Ibn al-Wardī (d. 861/1457), concurs, suggesting further that "its darkness is caused by the great distance from the place of sunrise and sunset." So distant and enigmatic is the Encircling Ocean for Ibn al-Wardī that he presents it as "a sea to which no coast is known. Its depths are not known to anyone but God."[56]

No animals, at least not normal ones, could survive in such a hostile, death-ridden environment. Even birds did not dare to cross it. Satan makes his home here on the Isle of Sah, waiting to ensnare the unsuspecting traveler. Worse still, the fish in this

Fig. 7.1. Picture of the world from Ibn Zunbul's *Qānūn al-dunyā* (Law of the World). Encyclopedia on wonders of the world including geography, astronomy, and dream interpretation. 970/1563. Ottoman. Gouache on paper. Diameter 20 cm. Courtesy: Topkapı Saray Museum, Istanbul. Revan 1638, fol. 4a. Photo: Karen Pinto.

sea are said to be many days long in size, and they are considered to be creatures of the devil because of the way in which they eat their own and devour men in an instant.[57] "The only life the Ocean knows, perhaps, is of horror: fish here are 'several days' long, Satan languishes in his prison on the island of Sah, and the tomb of Solomon, lord of all sages, evokes their presence in this landscape where man does not venture forth."[58]

Images of Satan and the horrific fish of the Encircling Ocean are not represented on the KMMS maps. But an unusual sixteenth-century manuscript called *Qānūn al-dunyā wa ʿajāʾibuhā* (The Order of the World and Its Wonders), by Ibn Zunbul, dated to 1563 and copied by Sheikh Aḥmad Miṣrī, does contain an image of the world in which all else has been blanked out save for the mighty Encircling Ocean and its enormous, terrifying fish (fig 7.1).[59]

Born also in this sea is the dragon-like monster Tinnīn, creator of terrifying storms, whose tail stretches all the way to the Caspian, and who terrorizes Muslim sailors especially in the eastern Mediterranean.[60]

> What is most frightening, however, is the monster, or hydra or dragon, the Tinnīn. Found only in deep water, and only in the Ocean, in the Mediterranean, particularly in the eastern Mediterranean, where the sea widens next to the mountains, and finally in the Caspian Sea. . . . The Tinnīn is born in the depths of the Ocean; on God's command angels draw him from the abyss like a cloud that pulls on him like a magnet on iron. Reluctant at first, the monster eventually comes out, raises his head to the sky, and the devastating hurricane stops in this nightmare scene struck by the tail of the irritated animal.[61]

Ibn al-Wardī elaborates the picture further by incorporating fabulous, jinn-inhabited floating cities, containing fortresses and palaces that disappear and reappear: "In this sea [i.e., the Encircling Ocean] there is the throne of satan (God's curse be upon him). And in it there are cities that float on the face of the water, in which there are the people of the jinn. In it are fortresses and palaces which float on the face of the water, disappearing and reappearing and other wondrous images and strange forms that disappear inside the water."[62]

Mas'ūdī stresses, at length, in both his *Murūj al-dhahab* (hereafter *Murūj*) and *Kitāb al-tanbīh wa-al-ishraf* (hereafter *Tanbīh*), that sailors have been forewarned since antiquity from carelessly passing through the mouth of the Mediterranean by the forbidding statues of Hercules that command the juncture: "At this point of the junction of these two seas, the Mediterranean and the Ocean, is a lighthouse of copper and stone built by the great king Hercules [Hirqil]; it is covered with inscriptions and [surmounted] by statues that [appear] to say by gesture: 'There is neither road nor route behind me' for those who would want to enter from the Mediterranean into the Ocean. . . . It is said that this lighthouse had not originated at this spot, but on an island of the Encircling Ocean, situated near the coast."[63] The location of these mythical pillars of Hercules varies among authors. In his *Tanbīh*, Mas'ūdī expands on his description and specifies that they are situated on the island of Cadiz.

> In the Exterior Sea, at the side of Spain, there is another island called Qādis [Cadiz], situated facing Shadhūna [Medina Sidonia]. . . . On this island is a big lighthouse, which is an edifice, one of the marvels of the world and on top of it is a column surmounted by a bronze statue. The statue is quite big and quite elevated and can be seen in Shadhūna and even farther. . . . They call them the Aṣnām Hirqil [statues of Hercules], [because] they were erected during ancient times by Hirqil [Hercules], the mighty king, in order to indicate to whoever saw them that they could not advance any farther. This is what was

> said by the inscriptions very clearly traced on their breasts in a certain type of ancient characters and the gestures of the hands of these statues pointed to the inscriptions for those who may not have read them. This was done to salute the voyager and to prevent him from going farther and getting lost in this sea. These statues are famous from the days of antiquity. It is even so in our time, that is to say, in the year 345. The ancient philosophers have spoken of them, as well as the scholars that occupied themselves with the shape of the world and the description of the world.[64]

Some five centuries later, Ibn al-Wardī picks up on and expands on these reports suggesting that there was a color code involved in the message of these forbidding statues. In a curious twist to the tale, Ibn al-Wardī asserts that these pillars were not built by Hercules but rather by the Yemeni king Abraha al-Ḥumayrā', who is associated in early Muslim lore with the birth of the Prophet Muḥammad in the Year of the Elephant:[65] "In it are idols which are erected by Abraha al-Ḥumayrā' standing on the face of the sea. These consist of three idols: one of them is green and it signals with its hands as if it is addressing the riders of the sea, commanding them to return; the second is red, signaling to itself and addressing the riders of this sea that it is dead and not passable; and the third, white idol points its finger to the sea as if referring to all those who perished there."[66]

Although these statues are not represented on the world maps, the Mediterranean maps always contain a triangular mountain-island called Jabal al-Qilāl at the mouth of the Mediterranean, where it meets the Encircling Ocean (i.e., the Atlantic). This island has never been definitively identified.[67] I believe it should be read as a symbol for the pillars of Hercules on the maps. The island is often adorned with intimidating symbols resembling the geographic descriptions. Figure 7.2 is an example of a KMMS Mediterranean map with the feature of Jabal al-Qilāl highlighted.

That a massive, seemingly unlimited ocean skirting along the edges of the earth, where man loses acquaintance with the world, should appear mysterious, dangerous, and forbidding comes as no surprise. What is surprising is that woven into this portrayal of bleak negativity is a positive perspective. The source of this conflicted portrayal is the ancient transcultural idea that the ocean is the primeval source of all life.[68] This concept is reinforced by the Qur'ān:[69]

> Do not these unbelievers see
> that the heavens and the earth were an integrated mass,
> then We split them and made every living thing from water?
> Will they not believe even then? (21:30)[70]

If all life is created from water, then it cannot be only bad. This led some geographers to reconcile the conflict. Before his discussion of the harmful vapors that inhabit this sea (as quoted above), Ibn Khaldūn first notes: "The water withdrew from certain

Fig. 7.2. Map of the Mediterranean with Jabal al-Qilāl (red square) from an abbreviated copy of al-Iṣṭakhrī's *Kitāb al-masālik wa-al-mamālik* (Book of Routes and Realms) with a symbol for the mythical pillars of Hercules that guard the mouth of the Mediterranean in all KMMS maps. This version is decorated with dark-red inverted crescents. 589/1193. Mediterranean. Gouache and ink on paper. 34 × 26 cm. Courtesy: Leiden University Libraries. Cod. Or. 3101, fol. 33.

parts (of the earth), because God wanted to create living beings upon it and settle it with the human species that rule as (God's) representative over all other beings."[71]

In this alternate view, "mountains are the coagulated billows of the ocean"[72] and the primeval water supported the throne of God.[73] Ṭabarī, for instance, relates that with the first breezes, red and white foam formed on the surface of the Ocean exactly on the spot of the future sanctuary of God and this was the point of origin of the earth.[74] In overtones that remind us of the Indic notion of the world as a lotus flower (see chapter 5), the thirteenth-century encyclopedist and historian Nuwayrī tells us about the creation of the Ocean: "When God intended to create the water, he created a green hyacinth, fixed its length, breadth, and height. Then he regarded it with a majestic look; then it became a small quantity of water that was in constant motion. This apparent undulation and motion was only a trembling caused by the fear of God."[75]

Thus water, specifically the Ocean, was not the enemy at the time of creation; rather it supported the preexistent throne and pen,[76] and it was obedient to God's commands to the point of trembling under his gaze.[77] When and how did it turn into a menace and the enclave of Satan and his mischief-making genies? This is not specified in the sources. All that we can discern are the different layers of meaning. Even in the days of the terrifying Encircling Ocean of the medieval Muslim geographers, it is possible to see traces of the earlier obedient primeval Ocean lurking along the edges of Paradise, ever promised but unattainable by mortal man. Muqaddasī, Masʿūdī, and Waṣīfī (an early author of the *ʿajāʾib* tradition often quoted by tenth-century Arab authors) all locate the cupola of Paradise (sometimes emerald, sometimes gold) on the other side of the menacing dark-green Encircling Ocean.[78] It is from this cupola that the great rivers of Paradise are said to issue, and it is by following these rivers, specifically the most glorious of them all, the Nile, that one may find Paradise.[79] Resonating with overtones of Egyptian cosmography is the unusual ḥadīth cited by Muqaddasī in which the source of the Nile can only be found by "riding the [terrible] animal that is confronting the sun . . . [attempting] to devour it." This is how Paradise, situated on the other side of the Ocean, containing the head of the Nile, can be reached.[80]

The angels, meanwhile, abide not only in Paradise but on the mighty Ocean itself, for as Masʿūdī tells us, it is an angel who stamps on the sea and thereby causes the ebb and flow of the tides. "According to some interpreters of the law, the ebb and flow were caused by an angel that God had destined for this purpose and set at the ends of the seas; this angel would dip his foot or toe into the sea, causing the sea to rise, and when he would pull it out, the water would return to its place, and this would be the ebb."[81] In his *Aḥsan al-taqāsīm fī maʿrifat al-aqālīm* (Best Divisions for Knowledge of the Earth), Muqaddasī cites a similar story about the mythical green man, Khiḍr, meeting an angel who informed him that the breathing of gigantic whales cause the ebb and flow of the tides in the seas.[82] Illustrated versions of Qazwīnī's *ʿAjāʾib al-makhlūqāt* depict this tide-controlling angel and also show a world perched atop a bull and a fish that is held up by an angel (fig. 7.3).[83] This symbolic pictograph of

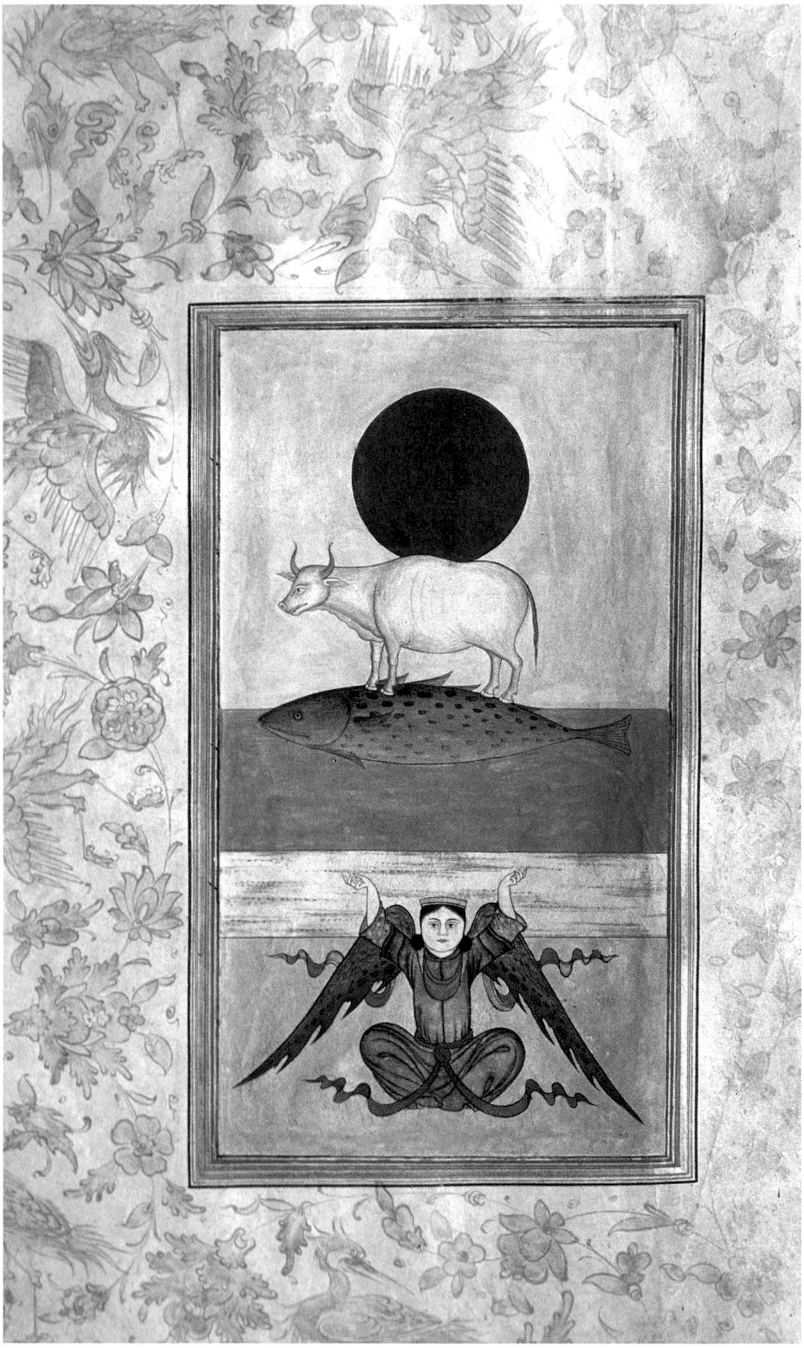

Fig. 7.3. Image showing the world held up by a bull standing on a fish swimming in the Encircling Ocean with an angel holding up the Ocean from a Turkish copy of al-Qazwīnī's *ʿAjāʾib al-makhlūqāt*. 1198/1784. Gouache on a background of delicate gold foliation on highly polished, thin paper. Dimensions 23 × 19 cm. Courtesy: Topkapı Saray Museum, Istanbul. Hazine 409, fol. 105b. Photo: Karen Pinto.

the world is a recurrent image in late medieval illustrated *ʿajāʾib* manuscripts. It is a motif that appears to be linked to Mesopotamian, Iranian, Indian, and Hebrew cosmographic myths (see chapter 5).[84]

In addition, there are the islands that dot the Encircling Ocean, such as the Jazāʾir al-Khālidā—"Eternal Islands" (a reference to the present-day Canary Islands, which the ancient Greeks set as 0° for longitude calculations). Almost every geographer mentions these islands in the context of the Encircling Ocean. Masʿūdī notes that "when the sun sets at the edge of China, it rises upon these Jazāʾir al-Khālidā in the Atlantic Ocean and that when it sets here it rises in the confines of China."[85] They are located in the westernmost part of the Encircling Ocean and are blessed because they produce corn and fruit from uncultivated soil. One author even suggests that they have an abundant source of gold.[86] They are sometimes conflated with the Jazāʾir al-Saʿāda, the "Islands of Bliss." From Bakrī (d. 487/1094), we get one of the best descriptions of the blessings and happiness of these islands that dot the surface of the perilous encircling ocean:

> Over against Tanja are situated the islands called by the Greeks Furṭunātash [Fortunatas, i.e., the happy ones]. They are called thus because their trees and shrubs produce all sorts of delicious fruits without having been planted or cultivated, and their ground bears corn instead of grass and different sorts of aromatic plants instead of thorns. They are separated one from the other, though at short distances. It is said that once it happened that the wind drove a ship ashore on one of these islands. When the sailors went ashore they found different sorts of fruit-trees and spice-trees and various precious stones. They took of them what they could and returned to Spain. When the king asked them where they had got this, they told their story. Then he provided ships and let them sail, but they did not reach an island for they perished because of the high billows and the vehement wind so that none of them returned.[87]

Not that these islands are without their own internal dangers. Idrīsī and Qazwīnī on the authority of Abū Ḥāmid al-Andalusī refer to terrible wild beasts that roam the island of Laghūs. We are also told that some of the Jazāʾir al-Saʿāda house Christian magicians.[88]

Like the island of Hercules, some of these preparadisical islands, in particular Masfahān and Laghūs, contain statues warning travelers that there is no safe passage beyond. Not that these statues deterred the fearless. Qazwīnī mentions Dhū al-Qarnayn's (Alexander the Great's) mythical journey across the Encircling Ocean to reach the Islands of the Blessed and Paradise beyond. We are told that Alexander did not succeed and many of his sailors drowned in the stinking sea, but it is said that he did get a glimpse of Paradise through the clouds and mist.[89]

These and other inaccessible islands dot the surface of the Encircling Ocean like tempting stepping-stones to the inaccessible Paradise beyond. The anonymous Iranian author of the *Ḥudūd al-ʿālam* takes the matter further by including north of the Islands of the Blessed the mysterious Island of Men and its counterpart the Island of Women.[90] "On the former the inhabitants are all men, and on the latter women. Each year for four nights they come together for the purpose of procreation and when the boys reach the age of three years they send them to the Island of Men. On the Island of Men there are thirty-six big rivers that fall into the sea, while on the Island of Women there are three such rivers."[91]

The anonymous author of the *Ḥudūd al-ʿālam* balances the picture by including in the eastern end of the Encircling Ocean Jazīrat al-fiḍḍa (Silver Island) on which there are many *sāj* (teak) and *ābanūs* (ebony) trees, as well as silver mines. We are told that this island is heavily populated, prosperous, and that it belongs to China. It is not classified as either good or evil.[92]

These islands appear for the first time on the maps of Idrīsī, the cartographer best known for some of the earliest extant mimetic maps (see figs. 3.1 and 9.9 as well as discussion of Idrīsī's work in chapter 3).[93] Although the islands aren't named on the Idrīsī world maps—a sign perhaps of their inaccessible, unconquerable status[94]—there is no other explanation for the nine islands in the Atlantic end of the Encircling Ocean and the three on the Pacific end. (Note that the version of the Idrīsī world map in the *Book of Curiosities*—see fig. 9.9—shows more islands in the Atlantic section of the Encircling Ocean, including the island of England [Inqilṭirra]).[95]

Idrīsī also begins the tradition of accentuating the Encircling Ocean by subtly ringing it with a delicate series of mountains, which are none other than a representation of the mythical mountains of Jabal Qāf. Linked by some to the Qurʾānic idea that God placed mountains on the earth to stabilize it and prevent it from moving (13:3, "It is He who stretched the earth and placed upon it stabilizers and rivers," as well as 16:15 and 78:6–7), the Jabal Qāf do not make an appearance on Islamic world maps until the late twelfth century. This late emergence suggests that they cannot be explained as a direct consequence of the Qurʾānic verses but should be seen as an indication of transcultural metacartographic influences and, in particular, the rise in importance of Sufi (mystical) concepts of the other world in which the Qāf mountains played a central role. André Miquel argues that the Qāf mountains are a direct borrowing from ancient Iranian traditions and the Hara-berezayti mountains that were said to ring the Vurukasha Sea—the source of all water in pre-Islamic Iranian cosmologies.[96]

The Qāf mountains can also be linked to the Hindu Purāṇic concept of the Lokaloka mountains that encircle the world and separate the visible from the invisible, forming the frontier of humanity. (For examples of this visual link, compare figs. 3.1 and 7.4 of world maps encircled by Jabal Qāf mountains with the earlier Indic examples in chapter 5 along with Mandala images.) These mountains become prominent in the medieval Islamic imagination through Sufi work, exemplified by the mid-twelfth-century allegorical tale *Manṭiq al-ṭayr* (Conference of the Birds) of the Persian poet Farīd al-Dīn ʿAṭṭār (d. ca. 1193). *Manṭiq al-ṭayr* tells the story of birds

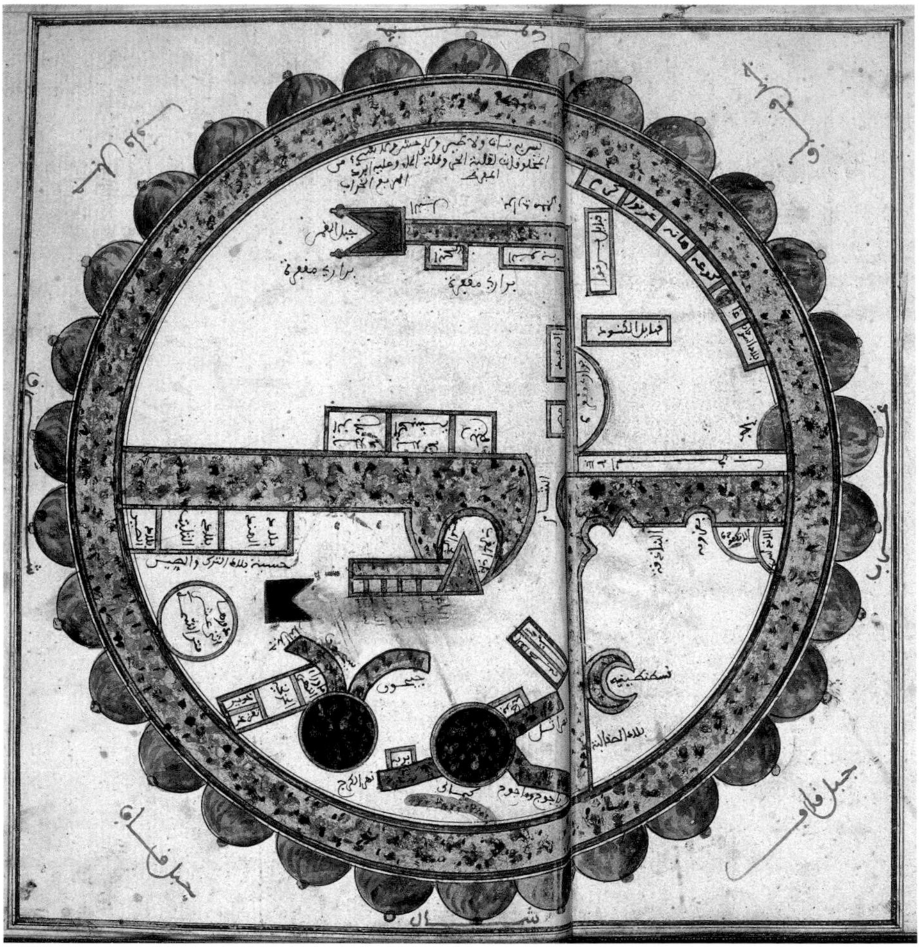

Fig. 7.4. Dramatic example of Jabal Qāf from an Ibn al-Wardī manuscript. This is one of the most striking examples of the emerald green mountains of Jabal Qāf that are shown to ring the world. Circa 16th century. Ottoman. Gouache on paper. Courtesy: The British Library Board, London. Ms. Or. ADD. 9590, fols. 3b–4a. © The British Library Board.

in search of their mythical leader, the Sīmurgh, who existed since the beginning of the world and withdrew to the mountains of Qāf to contemplate and seek wisdom. Thus these mythical mountains that reinforce the separation between the mundane world and the cosmos take on a sense of mountains of wisdom and contentment. The implication is that those who have the spiritual stamina and patience to make it across the Encircling Ocean that separates us from the heavenly cosmos will attain wisdom and peace.

Escape your self-hood's vicious tyranny—
Whoever can evade the Self transcends
This world and as a lover he ascends.
Set free your soul; impatient of delay,
Step out along our sovereign's royal Way:
We have a king; beyond Kaf's mountain peak
The Simorgh lives
. . . .
Do not imagine the Way is short;
Vast seas and deserts lie before His court.
Consider carefully before you start;
The journey asks of you a lion's heart.
The road is long, the sea is deep—one flies
First buffeted by joy and then by sighs;
If you desire this quest, give up your soul
And make our sovereign's court your only goal.[97]

Thought to have been composed of brilliant green emeralds, the Qāf mountains become such a celebrated feature of some late medieval Islamic mappamundi that they overshadow the depiction of the world. Yāqūt's mid-thirteenth-century reference to the Baḥr al-Muḥīṭ as "halo of the moon" can be read as confirmation that by the twelfth century the Encircling Ocean had grown in the Islamic geo-cartographic imagination into more than just a terrifying ocean ringing the world. Reinforced by the wisdom and contentment of the Jabal Qāf, the Encircling Ocean turns under Sufi influence into a symbolic, sacred hierophany. ʿAṭṭār reconciles the Manichean dichotomy of the Encircling Ocean by arguing that its turbulence lies in its feeling of disgrace and deep longing to be one with God.

A hermit asked the ocean: "Why are you
Clothed in these mourning robes of darkest blue?
You seem to boil, and yet I see no fire!"
The ocean said: "My feverish desire
Is for the absent Friend. I am too base
For Him; my dark robes indicate disgrace
And lonely pain. Love makes my billows rage;
Love is the fire which nothing can assuage.
My salt lips thirst for Kausar's cleansing stream.

For those pure waters tens of thousands dream
And are prepared to perish; night and day
They search and fall exhausted by the Way."[98]

By the late seventeenth century, the idea of the ultimate quest for solitude, wisdom, and the ecstasy of unity with God and the cosmos is presented in an impressive Escheresque Safavid miniature that depicts a dervish (mystic) in an imaginary journey flying across the turbulent Ocean among fish, birds, and other animals, toward the central icon of the mythical bird Sīmurgh residing in Jabal Qāf (fig. 7.5).[99]

At the empirical level, the dilemma for the geographers is how the sweet waters of the rivers that feed into the seas turn into the immense saltwater storehouse of the Encircling Ocean. Where does sweet water come from and how does it turn into the salty brine of the seas and ocean? How is it that in spite of all the sweet riverine waters that pour into the seas, the ocean continues to remain salty and brackish? The answer is none other than the age-old "fountain of life" cosmographic theory that, as I have shown in chapter 5, parallels the Iranian tradition of the Vurukasha Sea and Hebraic idea of Tehom.[100] Mas'ūdī and Hamdhānī suggest that the source is a giant underground reservoir—an abyss in the bowels of the earth—which contains as much sweet water as the Encircling Ocean does saline. This reservoir feeds the rivers and ensures the equilibrium of all water on earth.[101]

Actively debated among the geographical scholars is the number of seas and whether the Encircling Ocean should be included in this number. The opinions range from two to seven, with the inclusion of the Encircling Ocean contested. The KMMS authors, Iṣṭakhrī, Ibn Ḥawqal, and Muqaddasī, side with the Qur'ānic references (discussed at the outset of this chapter). According to the KMMS authors, there are but two seas: Baḥr al-Rūm (the Mediterranean) and the much larger Baḥr al-Fāris (i.e., the Persian Gulf/Indian Ocean). Iṣṭakhrī and Ibn Ḥawqal list both of these seas as gulfs of the Encircling Ocean,[102] which implies that there is really only one sea and it is the Encircling Ocean. Neither author explains this discrepancy. The other KMMS author, Muqaddasī, devotes a long passage to a discussion with an imaginary interlocutor on the matter of how many seas there are. Employing Qur'ānic arguments, Muqaddasī confirms the accuracy of the Qur'ān on this matter. The limits of the Encircling Ocean are not known, and therefore, according to Muqaddasī, it cannot be counted; while the other seas (e.g., the Black Sea, the Caspian, the Aral, etc.) are internal seas and therefore not seas at all. Since Muqaddasī is the only KMMS geo-cartographer to not include a world map, he does not have to resolve this matter visually.[103] Ibn Rusta argues for five seas by adding the Encircling Ocean, the Black Sea, and the Caspian to the Mediterranean and Persian Gulf/Indian Ocean. Hamdhānī settles on four: the Indian Ocean, the Mediterranean, the Caspian, and the one that stretches from Khwārazm to Rūm (namely, the Baltic Sea cum Black Sea, which was

Fig. 7.5. "Mystical Journey." Leaf from a Safavid album. Circa 1650. Isfahan. Gouache and black bordered with gold on off-white paper. 12.5 × 25.2 cm. Courtesy: Harvard Art Museums/Arthur M. Sackler Museum, Grace Nichols Strong, Francis H. Burr and Friends of the Fogg Art Museum Funds, 1950.135. Photo: Imaging Department © President and Fellows of Harvard College.

thought at the time to be connected). The anonymous Persian author of the *Ḥudūd al-ʿālam* privileges the Qurʾānic verse of seven seas by adding to the total of four seas adumbrated by Hamdhānī the Aral Sea and the Encircling Ocean divided into two halves of eastern and western seas. The Ikhwān al-Ṣafāʾ (Brethren of Purity) vacillate between four and seven.[104] Masʿūdī wavers between five, six, and seven seas. He includes the Sea of Azov to make six and concludes by emphasizing the unity of all seas that combine to form the Encircling Ocean as the seventh.[105]

Finally, there is the question of the influence of the transhistorical concept of multiple Encircling Oceans (discussed at length in chapters 5 and 6) and the extent to which it influences Islamic carto-geographical thought. It is obvious neither in the maps that I have dealt with thus far nor in the texts that accompany them. Yet, below the surface, in a wide range of sources, the concept lurks. Sura 65, verse 12, contains the only mention of seven earths and seven heavens in the Qurʾān. Even though it is a single reference, it is picked up in some ḥadīth, and these seven seas are individually named in the Qiṣaṣ al-anbiyāʾ (Tales of the Prophets), which were especially popular in the later Middle Ages.[106]

According to Kisāʾī, author of one of the most popular versions of the Qiṣaṣ al-anbiyāʾ, all the seas of the earth derive from the seven seas that encircle the seven earths. He names each of these seas (Nīṭash, al-Aṣamm, Mihrās, al-Sākin, al-Mughallib, al-Mu'annis, al-Bāqī) and specifically notes that "each of them surrounds the sea before it."[107] Even the eminent ninth- to tenth-century historian Ṭabarī discusses the seven encircling seas that separate the seven earths in his opening volume on history "from Creation until the Flood," in the context of the *haykal* within which the physical world as a whole including the planets is contained. This is in turn encased within God's footstool (kursī). Distinctly reminiscent of the Indic vision of the world with multiple encircling oceans discussed in chapter 5, Ṭabarī cites a tradition from Wahb as follows:

> According to Muḥammad b. Sahl b. 'Askar—Ismā'īl b. 'Abd al-Karīm—'Abd al-Ṣamad—Wahb, mentioning some of His majesty (as being described as follows): The heavens and the earth and the oceans are in the *haykal*, and the *haykal* is in the Footstool. God's feet are upon the Footstool. He carries the Footstool. It became like a sandal on His feet. When Wahb was asked: What is the *haykal*? He replied: Something on the heavens' extremities that surrounds the earth and the oceans like ropes that are used to fasten a tent. And when Wahb was asked how earths are (constituted), he replied: They are seven earths that are flat and islands. Between each two earths, there is an ocean. All that is surrounded by the (surrounding) ocean, and the *haykal* is behind the ocean.[108]

Two ḥadīth references in Abū Dāwūd's ninth-century *Sunan* suggest the existence of multiple Encircling Oceans. From Abū Dāwūd's *Kitāb al-jihād* comes the following ḥadīth linked to 'Amr ibn al-'Āṣ, architect of the Muslim conquest of Egypt in the early 640s: "The Prophet (peace be upon him) said: No one should sail on the sea except the one who is going to perform hajj or umrah, or the one who is fighting in Allah's path for under the sea there is a fire, and under the fire there is a sea."[109] Also from Abū Dāwūd's ḥadīth collection comes another verse about encircling oceans that traces back to the Prophet's uncle al-'Abbās ibn 'Abd al-Muṭṭalib:

> I was sitting in al-Baṭḥā' with a company among whom the Prophet of Allah was sitting, when a cloud passed above them. . . . Abū Dāwūd asked: "Do you know the distance between Heaven and Earth?" They replied: "We do not know." He then said: "The distance between them is seventy-one, seventy-two, or seventy-three years. The heaven, which is above it is at a similar distance (going on till he counted seven heavens). *Above the seventh heaven there is a sea*, the distance between whose surface and bottom is like that between one heaven and the next. Above that there are eight mountain goats the distance between whose hoofs and haunches is like the distance between one heaven and the next. Then Allah, the Blessed and the Exalted, is above that."[110]

The KMMS geographers do not address the matter of the heavenly seas. But Ibn al-Wardī makes an opaque statement about the Encircling Ocean, which could be

Fig. 7.6. Angel Rūḥ holding up the cosmos from a Safavid rendition of al-Qazwīnī's *'Ajā'ib al-makhlūqāt*. Late 16th century. Qazvin. Gouache, gold, and ink on paper. 22.5 × 12.5 cm. Courtesy: Ashmolean Museum, Picture Library, Oxford. Reitlinger Gift, EA 1978.2573.

read as a reference to a second encircling ocean: "Its shore on the open side is the dark ocean which encircles the surrounding one as the latter surrounds the earth."[111] Hamdhānī includes an extensive passage on the seven worlds that encircle the earth, of which the fifth is composed of shallow, shimmering water. Hamdhānī specifies that each of the worlds are surrounded by water, which are in turn surrounded by air.[112] Resembling the version in Kisā'ī's *Qiṣaṣ al-anbiyā'*, Qazwīnī's thirteenth-century cosmography highlights the multiple oceans encircling the world:

> And Kaʿb al-Aḥbār said God created seven seas and the first of them is it [which] encircles the earth called "Bunṭus" [Pontus? Name of the Black Sea?] and behind it a sea called "Qabīs" [Cadiz?] and behind it a sea called "al-Aṣamm" [the Deaf One?] and behind it a sea called "al-Muẓlim" [the Dark One] and behind it a sea called "Marmās" [Marmaras?] and behind it a sea called "al-Sākin" [the Calm One?] and behind it a sea called "al-Bāqī" [the Remaining One] and it is the last of these seven seas which encircles them all. Each of these seas encircles that which is in front of it.[113]

Illustrated copies of Qazwīnī's *ʿAjā'ib al-makhlūqāt* manuscript series often contain a depiction of the angel Rūḥ holding up an image of the spheres (aflāk) surrounded by multiple bands that have been interpreted as spheres encircling the earth,[114] but these encircling spheres can also be read as a visual reference to the seas encircling the earth as one of the bands painted blue suggests (fig. 7.6).[115] Copies of Yāqūt's thirteenth-century geographical encyclopedia, *Muʿjam al-buldān*, present a popular medieval view of the spheres with the earth surrounded first by a circle of water (al-mā'), then a circle of air (al-hawā'), then a circle of fire (al-nār), ending with the sphere of the moon (Falak al-Qamar) (fig. 7.7). (For a discussion of this manuscript along with other examples, see chapter 3.)

What these references suggest is that concepts of multiple encircling oceans akin to those found in Indic traditions (see chapter 5) were widespread in the Islamic visual vocabulary from the thirteenth century onward. While the concept finds its way into Qazwīnī's work, it has a belated impact on the KMMS tradition. It surfaces for the first time in late copies of the KMMS world maps from the mid-nineteenth century (see fig. 4.8).

> We know that creation largely depends on the balance intended by God between earth and water. But we also know that creation is the transition between the original chaos and final disruption of the Last Judgment, where everything is destroyed to make way for a new and eternal order. The balance that allows the human race to live during the time allotted to it has not resolved or forever ended the ancient conflict between earth and water. Creation simply calms and always restrains, within the limits set by God, each of both elements which continue to manifest their power, one through eruptions and earthquakes, another through storms and tides. For now, on each side of these shores,

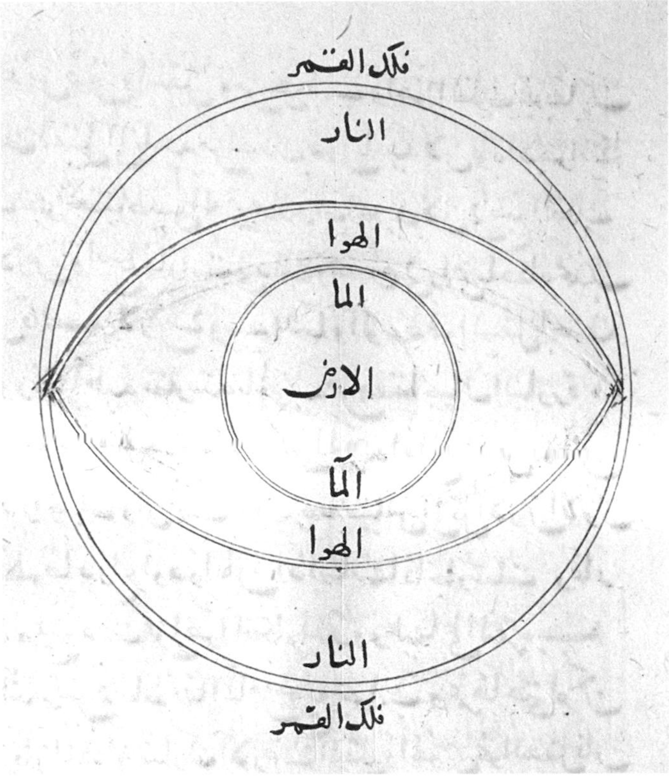

Fig. 7.7. *Aflāk* diagram from Yāqūt's *Muʿjam al-buldān*. 827 / 1424. Red and black ink on paper. Diameter 6.5 cm. Courtesy: Topkapı Saray Museum, Istanbul. Ahmet 2700, fol. 9b. Photo: Karen Pinto.

> victory after defeat, and calm after storm, earth and water in the end concede nothing, each asserting itself. A day will come however when, thrown against each other, land and water will stir in their struggle the foundations of the world. In their static symbolism and their spatial limits, the island and lake epitomize the equilibrium existing today. Yet in their very ambiguity, the land that defies water, sometimes out in the open sea, this water that breaks the continuity of the earth's cover also intimates what formidable confusion will bring our world and our age to an end.[116]

Thus the seemingly benign Baḥr al-Muḥīṭ hovering on the periphery of world maps opens out textually into a complex feature that hovers between two extremes: a hierophany and an antihierophany. It is the realm where reality melds with myth and fantasy, beyond which the cosmic mountains of Qāf beckon souls seeking peace and unity with the maker, and paradise waits. In its many manifestations it presents linkages to a multitude of cultural traditions: The Iranian notion of the Vurukasha Sea, the Hebraic notion of Tehom, the Babylonian notion of primeval chaos, the Indian notion of the multiple Encircling Oceans and worlds, as well as the empirical Greek

Fig. 7.8. Dish with Kufic writing from Samarqand. Bowls such as these ringed with Kufic inscriptions were popular in the ninth and tenth century. Circa 9th–10th century. Earthenware bowl covered with a white slip and painted in a red-and-brown slip under a transparent glaze. Diameter 26 cm. Courtesy: The David Collection, Copenhagen. Inv. no. 55/1974. Photo: Walter Denny.

notions of Oceanos. It hovers on the periphery as an unmistakable mark of the difference between earth and water, between life and death, between the known and the unknown. Its familiar double-edged-ring form turns it into a metamotif that is used to signal the imago mundi. It reinforces the idea that our world is an island marooned in a vast sea struggling to keep the waters at bay.[117]

Limitless Meaning

Metapictures are all like pipes: they are instruments of reverie, provocations to idle conversation, pipe-dreams, and abstruse speculations. Like pipes, metapictures are "smoked" or "smoked out" and then put back on the rack. They encourage introspection, reflection, meditations on visual experience. Their connection to history, politics, contemporaneity is equivocal, for they clearly serve (like puzzles, anagrams, conundrums, paradoxes) the purposes of escapist leisure, consumptive and sumptuary pleasure, a kind of visual orality in which the eyes "drink" in and savor the scopic field.[118]

W. J. T. MITCHELL, *Picture Theory* (p. 72)

The question of visual orality brings me to the next level of iconographic inquiry: the symbolic. Texts provide one dimension of the picture; symbols provide another. Texts undergird a form whereas symbols layer it from above. Textual meaning is gen-

Fig. 7.9. Soaring double-ringed interior of Sultan Ahmet Mosque, Istanbul. Central dome is painted with a blue band encircling a sixteen-arch and petal form that connects to a central blue medallion containing a Qur'ānic inscription. Early 17th century. Courtesy: Mikhail Markovskiy / Shutterstock. Image ID 75321991.

erally explicit and self-contained, whereas the messages of symbols are expansive and reach toward infinite horizons. Humans have envisioned and depicted themselves as encircled since time immemorial. We are subject to the law of the center versus the periphery, the navel vis à vis the rest of the body, the circumference and its radii. We encircle and are encircled ad infinitum. It is this metameaning that the Muslim Baḥr al-Muḥīṭ encapsulates on a symbolic plane.

If we let go of our rigid definition of what constitutes a map, we can see manifestations of the encircling ocean motifs in other visual forms. Those familiar with Islamic visual vocabulary know that the double-ringed circle dominates the art and architecture of the Middle East. Standing as it does for Islam's prime principle of *tawḥīd* (unity) or the Sufi equivalent, *waḥdat al-wujūd* (oneness of being, unity of existence), manifestations of it as a hierophany abound.[119]

Ranging in place, time, and space from a ninth-/tenth-century Central Asian white slip plate with a distinct yin-yang motif at the center surrounded by a double-ringed ochre band and a Kufic inscription[120] wishing the user knowledge and good

Fig. 7.10. Mamluk brass bowl with encircling fish motif. Base of bowl is decorated with a fish encircling a twelve-petal whorl. The outside of the bowl is decorated with double-ringed medallions. First half of the 14th century. Engraved brass originally inlaid with silver. Diameter 18.4 cm. Source: The Metropolitan Museum of Art, New York. Edward C. Moore Collection, Bequest of Edward C. Moore, 1891, Inv. No. 91.1.534.

health (fig. 7.8), to an example of a soaring double-ringed interior in the early seventeenth-century Sultan Ahmed Mosque in Istanbul, featuring a blue band in the central dome encasing a sixteen-petaled flower form in (fig. 7.9), this double circle can be read as an encircling ocean motif used in tandem with the written word to stamp a universal symbol of the imago mundi within the realms of Islam. A survey of Islamic art and architecture shows that this encircling ocean/imago mundi motif dominates Islamic material culture, from mosques and tiles to brass bowls and banners (figs. 7.10 and 7.11).[121] There is no better signifier for the importance of the hierophany of encirclement for Islam than the phenomenon of ritual encirclement of the holy site of the Kaʿba in Mecca during the annual hajj, when countless circumambulating pilgrims create a vibrant encircling ocean of believers around the navel of Islam (fig. 7.12).

Double-ringed bands resembling the encircling ocean/imago mundi motif are often used to encase Qurʾānic inscriptions and the names of Muḥammad and the first four Rashidun caliphs. Attached is an example from a mid-fifteenth-century manuscript depicting the Prophet's *miʿrāj* (ascent to heaven) during his *isrāʾ* (night journey to Jerusalem). To bypass the problem of depicting the face and figure of the Prophet, the miniature artist depicts instead his name, Muḥammad, encased within a gold double-ringed circle with wavelike designs (fig. 7.13).[122]

The issue of depicting the person of the Prophet is a thorny one. In some of the earliest extant miniatures from the twelfth century the Prophet is depicted with his

Fig. 7.11. Las Navas de Tolosa banner. Early thirteenth-century Almohad banner with a central medallion composed of an eight-pointed star surrounded by a studded encircling ring. 3.3 × 2.2 m. Courtesy: Monasterio de las Huelgas-Museuo de las Ricas Telas, Burgos, Spain. Photo credit: Album/Art Resource, NY.

Fig. 7.12. Pilgrims encircling the Kaʿba during Hajj in Mecca. Courtesy: Zurijeta / Shutterstock. Image ID 173485667.

face uncovered. Over time the approach to illustrating the Prophet changes to a veiling of his face and eventually a total elimination of his body replaced by his name and a description of his figure.[123] Thus a unique form for depicting the Prophet, his closest companions, and previous prophets developed called *ḥilya*. *Ḥilya* literally means "physiognomy, natural disposition, likeness, depiction, or description." It started as a textual form in the early eleventh century in the Arab world that was developed six centuries later by the Ottoman Turkish miniature artists into a stand-alone visual motif. Thereafter the *hilye* (as it is referred to in Turkish) becomes an elaborate art form that is highly sought after and is continued today by top Arabic calligraphers around the world.[124] The earliest version of these *hilye* contain a double-ringed band encasing a textual description of the Prophet's looks and persona. Over time this band morphs into the shape of a crescent—the modern symbol of the Islamic world because of its reliance on the lunar calendar. (Refer to examples of a seventeenth-century *hilye* and a contemporary one in figures 7.14 and 7.15.)

Scholars have suggested that the design of the round city of Baghdad founded

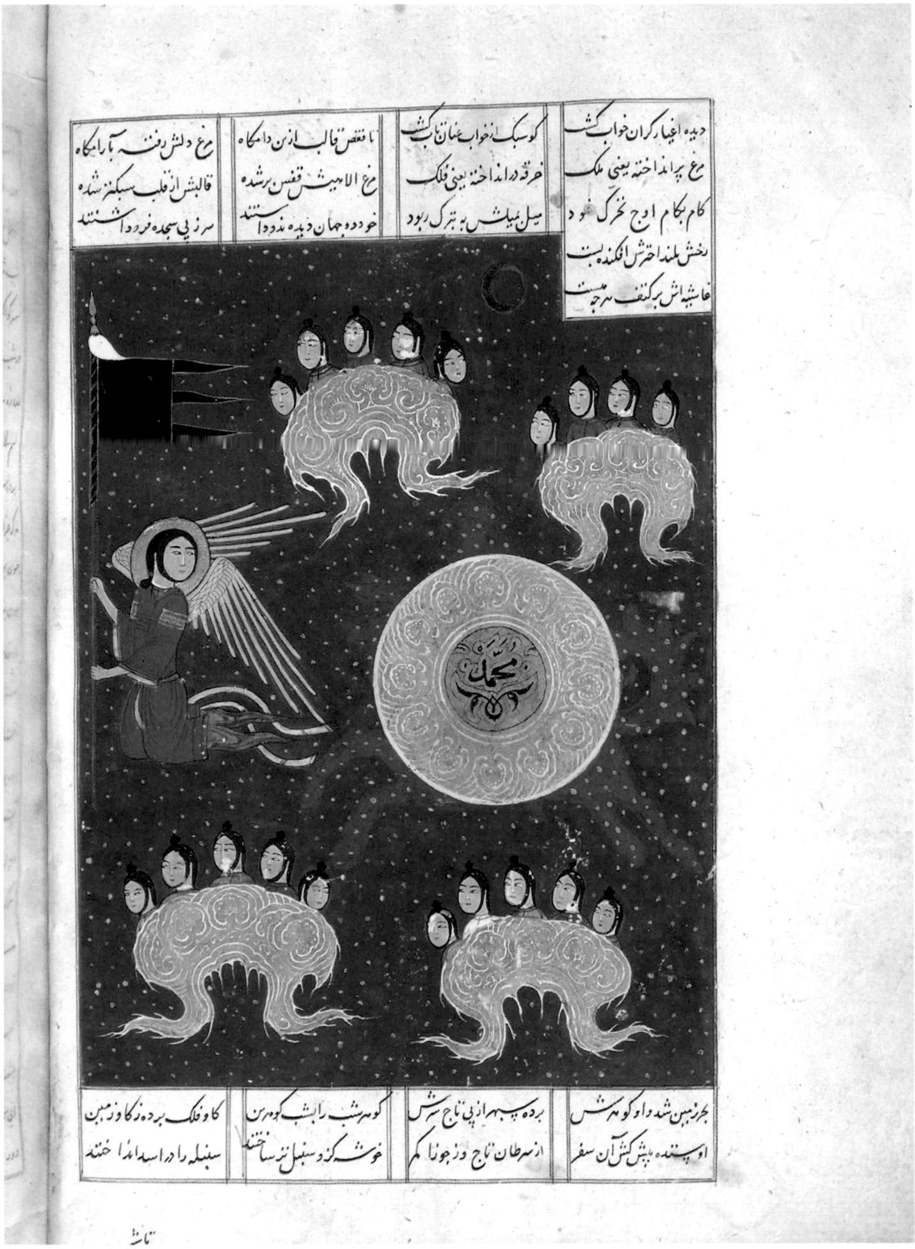

Fig. 7.13. Heavenly ascent (miʿrāj) of the Prophet from Niẓāmī's *Khamsa* (Quintet). 1441. Western India. Courtesy: Topkapı Saray Museum, Istanbul. Hazine 774, fol. 4b.

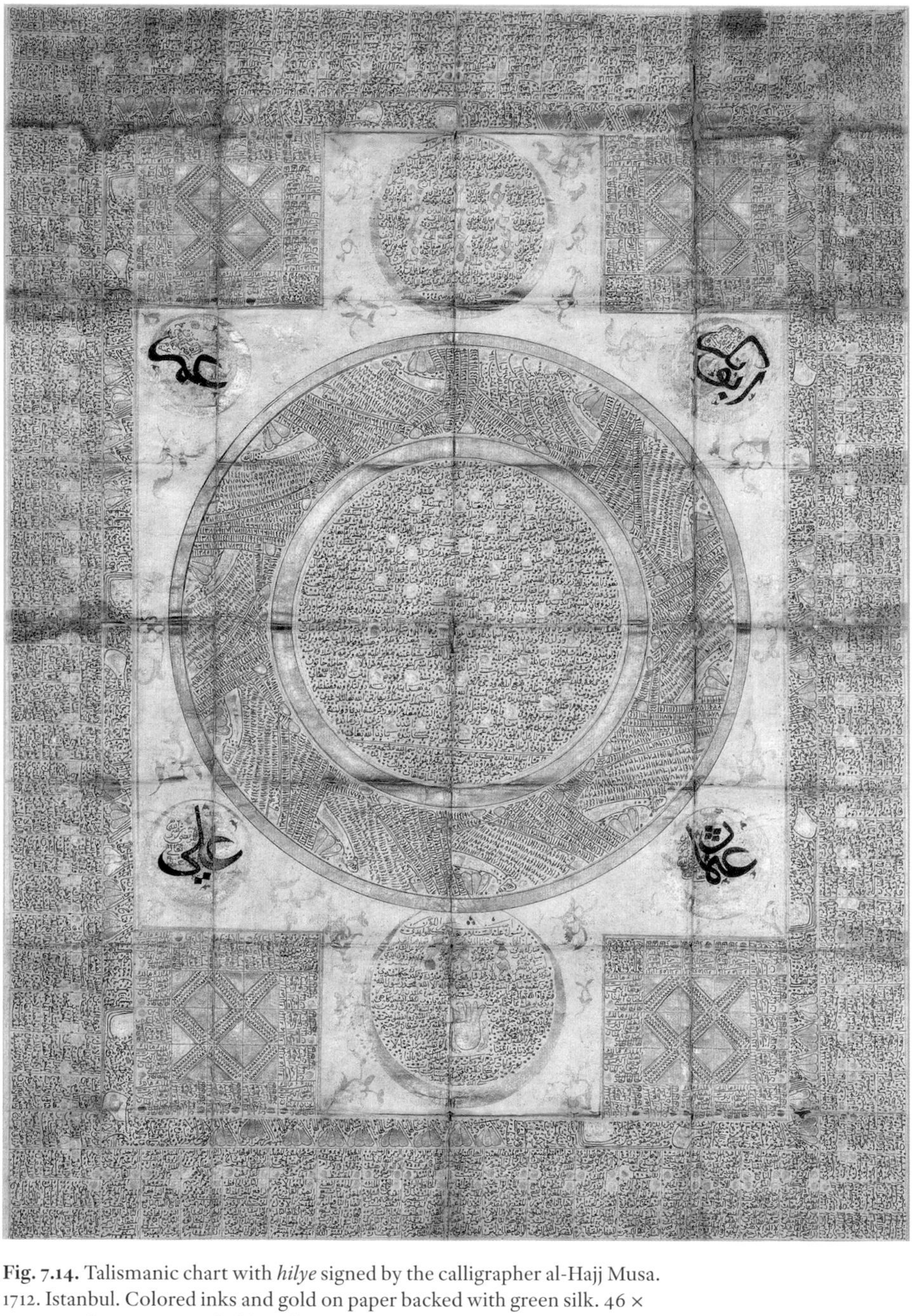

Fig. 7.14. Talismanic chart with *hilye* signed by the calligrapher al-Hajj Musa. 1712. Istanbul. Colored inks and gold on paper backed with green silk. 46 × 34.1 cm. The Nasser D. Khalili Collection of Islamic Art, Mss. 759. Photo © Nour Foundation. Courtesy of the Khalili Family Trust.

Fig. 7.15. Contemporary red *hilye* of the Prophet Muḥammad by the calligraphic artist Mohamed Zakariya. 93.8 × 64.77 cm. Courtesy: Mohamed Zakariya. Photo by Frank Wing.

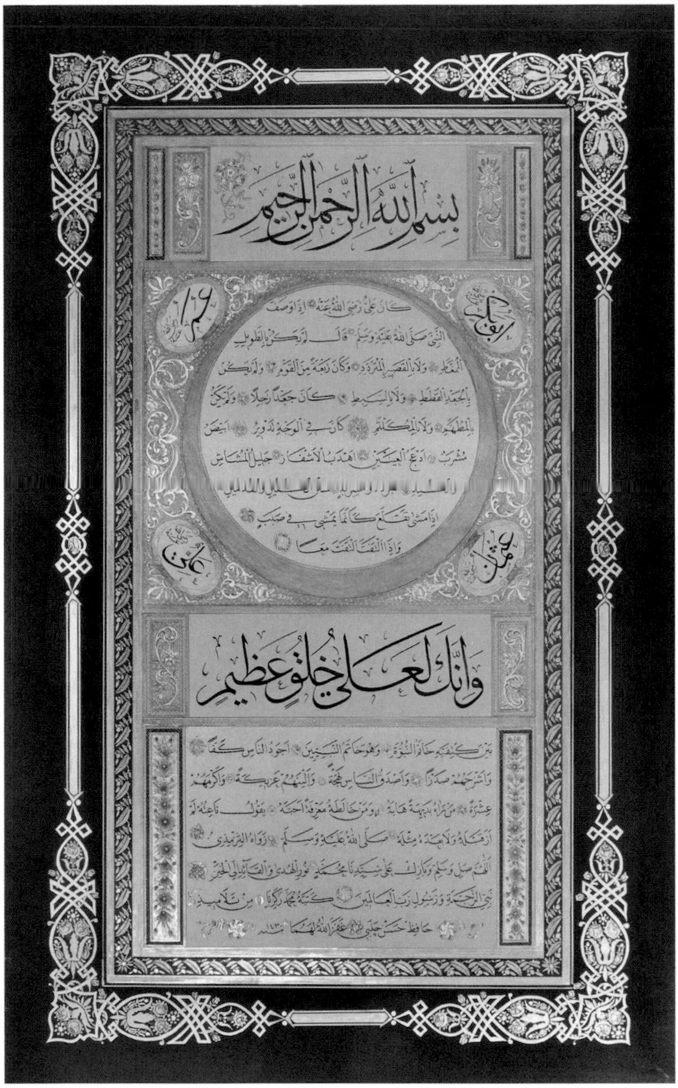

in 762 by the Abbasid caliph Manṣūr was derived from earlier concentric Sassanid city models, such as Hatra, Dārābjird, Isfahan, Takht-i Sulaymān, and Fīrūzābad.[125] Yaʿqūbī's description suggests that Baghdad was built as a microcosmic representation of the world—literally *ṣūrat al-arḍ* (picture of the world), precisely the phrase that is used as a caption for the KMMS world maps.[126] Built along the lines of a min-

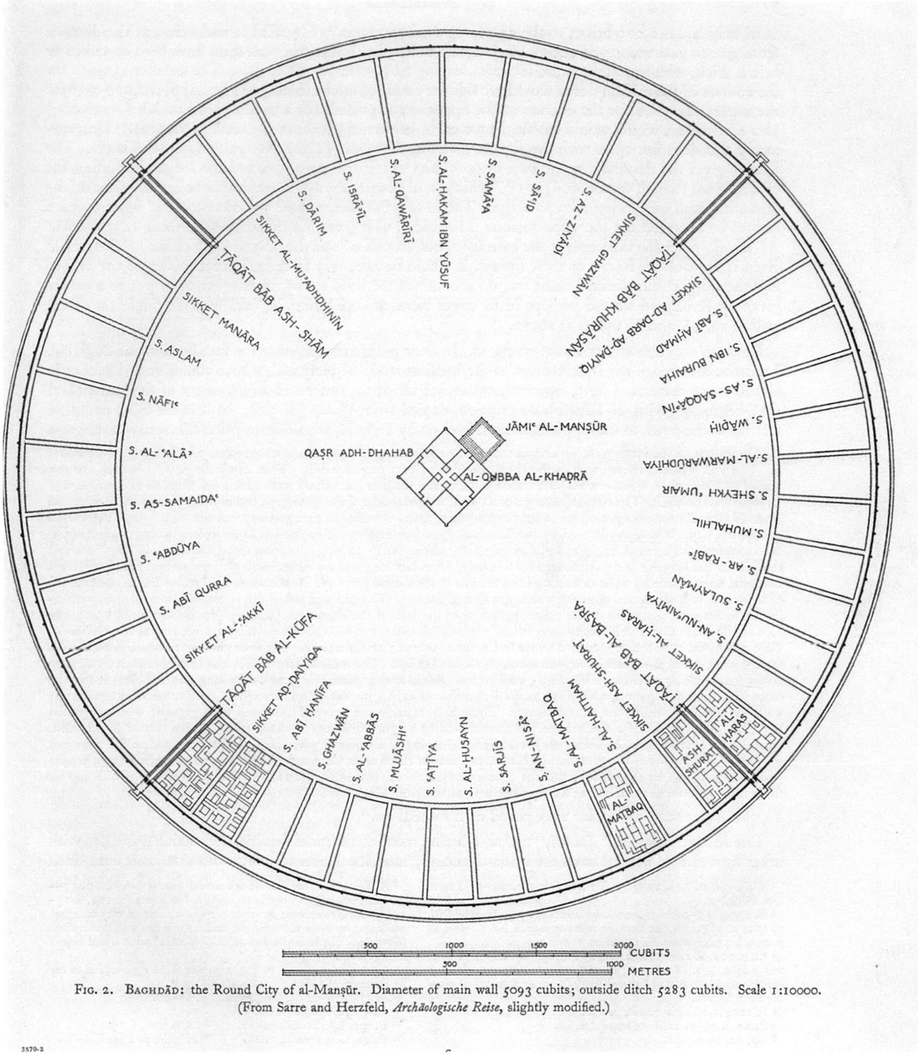

Fig. 7.16. Reconstruction of the round city of Baghdad by K. A. C. Creswell. Courtesy: Ashmolean Creswell Archive and Fine Arts Library, Harvard. K. A. C. Creswell, *Early Muslim Architecture*, fig. 2, negative no. EA.CA.2182.

iature imago mundi with the caliph's palace at the center, the deliberate circular design has been read by some scholars as an Abbasid-cum-Barmakid (the Barmakids were a vizier family from Balkh in present-day Afghanistan that served the early Abbasids from Manṣūr to Hārūn al-Rashīd) attempt to visually stamp Baghdad as the new navel of the world.[127] The architectural historian Creswell first raised this idea

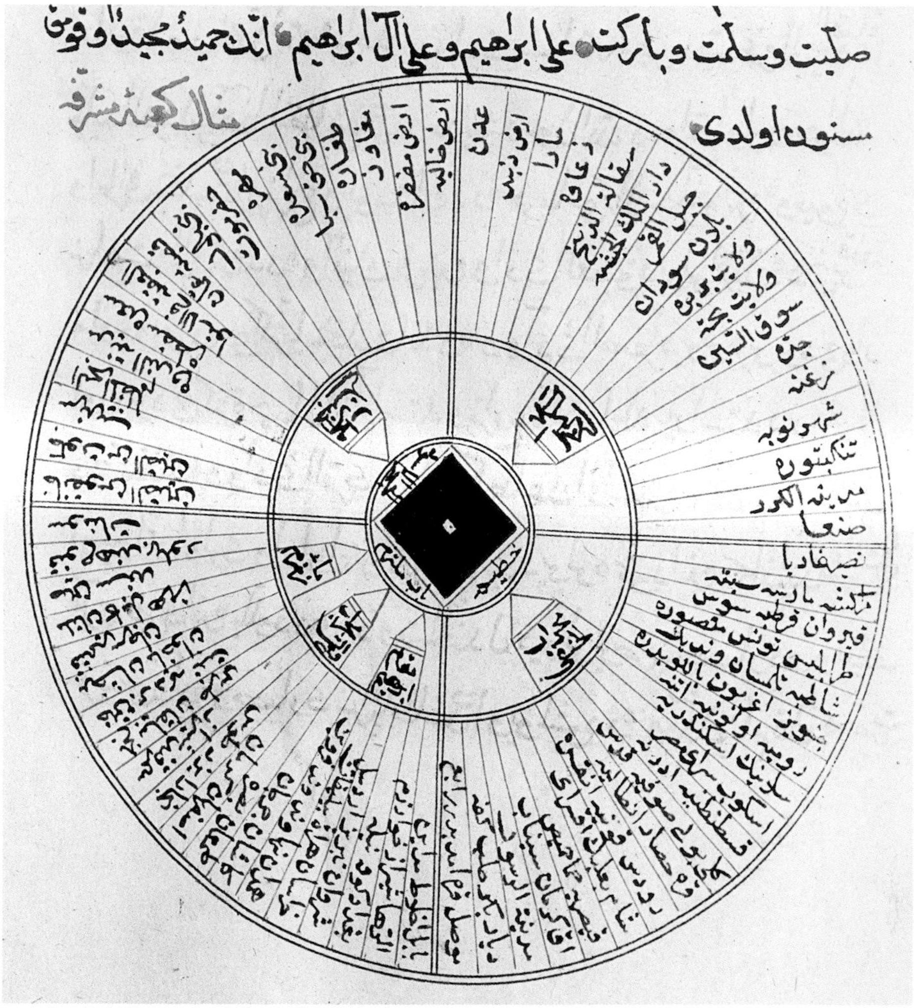

Fig. 7.17. Seventy-two-sector Qibla (direction of prayer) scheme of sacred geography from a Turkish translation of Ibn al-Wardī's *Kharīdat al-ʿajāʾib*. Shows the Kaʿba in Mecca as the navel of the world surrounded by a double-ringed circle indicating the location of places around the Muslim world, such as Bilād al-Sudan, Qusṭanṭīniyya (Istanbul), the Beja, and even a section of the Encircling Ocean (al-Baḥr al-Muẓlim), in relation to the Kaʿba. 1092/1681. Turkey. Red and black ink on paper. 9.7 cm. Courtesy: Topkapı Saray Museum, Istanbul. Bağdat 179, fol. 52b. Photo: Karen Pinto.

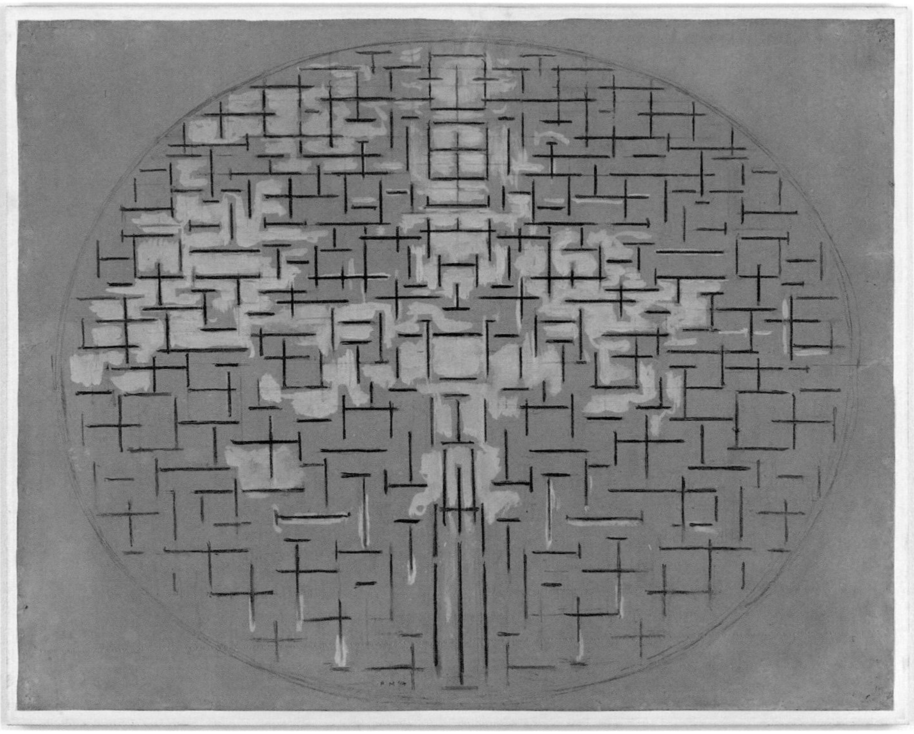

Fig. 7.18. *The Pier and the Ocean* by Piet Mondrian (1872–1944). The metaform of the Encircling Ocean encompasses an entire universe of forms. 1915. Charcoal and gouache on paper. 87.9 × 111.2 cm. Courtesy: The Museum of Modern Art. Mrs. Simon Guggenheim Fund, 34.1942. Digital Image © The Museum of Modern Art / Licensed by SCALA / Art Resource, NY.

in his attempt to re-create the structure of the city.[128] Creswell's architectural plan of Baghdad (shown in fig. 7.16) bears a remarkable resemblance to Islamic Qibla charts (way-finding diagrams and instruments for locating Mecca) (compare fig. 7.16 with fig. 7.17).[129]

Is this all? No, for I am still held by the image. I read, I receive (and probably even first and foremost) a third meaning—evident, erratic, obstinate. . . . As for the other meaning, the third, the one "too many," the supplement that my intellection cannot succeed in absorbing, at once persistent and fleeting, smooth and elusive, I propose to call it the obtuse meaning. . . . *It seems to open the field of meaning totally, that is infinitely. . . . The obtuse meaning appears to extend outside culture, knowledge, information; analytically, there is something derisory about it; opening out into the infinity of language, it can come through as limited in the eyes of analytic reason; it belongs to the family of puns, jokes, useless expenditure. Indifferent to moral or aesthetic categories (the trivial, the futile, the false, the pastiche), it is on the side of the carnival. Obtuse is thus very suitable.*

ROLAND BARTHES, *Image, Music, Text* (pp. 53–55; original emphasis)

The ubiquitous metaform of the Encircling Ocean causes our reading to skid into the realm of limitless meaning. At some point in an iconographic investigation of the Encircling Ocean it becomes clear that the form holds yet another peril for explorers—the metaform encompasses an entire universe of forms from the plate that bears the bodies' nourishment to the domes of every mosque and basilica that offer sustenance to the soul. So as not to end up like Mondrian lost on a pier in the middle of an ocean (fig. 7.18),[130] I leave off the chase for encircling oceans to search instead for the meaning behind a place. The next two chapters seek historical insight through an examination of the contextual fabric surrounding a particular, and fascinating, space on medieval Islamic maps—the land of the Beja.

8

The Beja in Time and Space

The study of history through maps warrants, among other things, a close analysis of place from the point of view of time and space. Graphics of territory—whether real or imagined, macro or micro—imply the collapsing of space from an infinite three-dimensional expanse to a constrained two-dimensional sheet.[1] Like the creation of pictures, this requires a conscious selection of what to include and what to leave out. The production of space on maps revolves around two issues: what places to include and how to create the illusion of an identifiable landscape. It is in the resolution of these issues that the signs of form (mimetic or not), symbol, boundary, label, marker, and place come into active play.

> Not only is it easy to lie with maps, it's essential. To portray meaningful relationships of a complex, three-dimensional world on a flat sheet of paper or a video screen, a map must distort reality. As a scale model, the map must use symbols that almost always are proportionally much bigger or thicker than the features they represent. To avoid hiding critical information

in a fog of detail, the map must offer a selective, incomplete view of reality. There's no escape from the cartographic paradox: to present a useful and truthful picture, an accurate map must tell white lies.[2]

The constraints of space play a vital role in the act of creating any map. The limitations of space require that the map be ordered, and this, in turn, requires that it be distorted. Since the map is a product of a specific time, in a particular milieu, the distortion of space is culturally located. It is in the selections, the distortions, and the omissions that we find telling biases of structure and imagination. In this way the constraints of space turn paradoxically into a window overlooking a set of cultural constructs. Space is informed (if not determined) by time:[3] One without the other would result in a map that is no more than a dot. As Denis Wood puts it, "We cannot have a map without thickness in time unless we can have a map without extension in space."[4]

I seek in this chapter to extract the cultural fabric of time from space by analyzing one site in the ossified matrix of places, which overlays each of the KMMS world maps. Nowhere are the constructs of time more apparent than in the case of world maps, where space is at the highest premium. The ambitious scope of a world map means even less can be included, and the cartographer, whether medieval or modern, situated in a global whirl of events (composed of reproductions and retentions, as Husserl suggests),[5] is faced with the task of choosing that which is most important and its form of representation. It is aspects of this cartographic process of choice, signification, and distortion that I seek to explore through a close analysis of place. The object is to understand the meaning behind the territorial allocation of place and space.

This chapter focuses on two anomalous names in the KMMS world map place matrix: the lands of the Beja and their deserts (Bilād al-Būja and Mafāza al-Būja) in light of the reigning Islamic historiography. While all the places on the medieval KMMS world maps deserve attention, I have chosen these two sites because they deviate from the norm of cartographic selection.[6] Most of the places on the map are familiar sites, but the Beja are virtually unknown. Their prominent presence on the Muslim world maps raises puzzling questions and provides surprising answers for the crucial issues of time, space, and cartographer's choice.

Consistently located on the eastern flank of Africa is a double-territorial ethnonym for an obscure East African tribe known as the Beja (red circle on fig. 8.1).[7] Mention of them in medieval Middle Eastern historiography is rare and, at best, superficial, yet no Islamic world map from the eleventh to the nineteenth century leaves them out. Not only are the Beja privileged with a permanent berth on the Islamic world map; their lands are also the only place on the map signified with a double-territorial marking.

Who were the Beja? Why are they absent in Islamic historiography yet present on the Islamic world maps? The answer emerges from a series of references that hint

Fig. 8.1. KMMS world map marking the lands of the Beja, Nubia, Ethiopia, and the Zanj. The red circle indicates Bilād al-Buja (Lands of the Beja); the yellow circle, Bilād al-Ḥabasha (Abyssinia/present-day Ethiopia); the green circle, the lands of the Zanj (a general reference to eastern Africa); and the purple circle, Bilād al-Nūba (Nubia). 878/1473. Ottoman. Gouache and red and black ink on paper. Diameter 19.6 cm. Courtesy: Sülemaniye Camii Kütüphanesi, Istanbul. Aya Sofya 2613, fol. 3a. Photo: Karen Pinto.

at the oddest reasons for the emphasis—reasons which in turn cause us to question our notions of how and why places make it onto maps. The query prompts surprising answers that can be relegated to the Husserlian domain of retentions, reproductions, and protentions.

> Retentions being the perspective view that we have of past phases of an experience from the vantage of a "now" moment which slips forwards, and in relation to which past phases of our experience of the present are pushed inexorably back. Reproduction of some recollected event, by contrast, involves the temporary abandonment of the current "now" as the focal point around which retentional perspectives cohere, in favour of a fantasied "now" in the past which we take up in order to replay events mentally. Retentions, unlike reproductions, are all part of current consciousness of the present, but they are subject to distortion or diminuation as they are pushed back towards the fringes of our current awareness of our surroundings. . . . Retentions can thus be construed as the background of out-of-date beliefs against which more up-to-date beliefs are projected, and significant trends and changes are calibrated. As beliefs become more seriously out of date, they diminish in salience and are lost to view. We thus perceive the present not as a knife-edge "now" but as a temporally extended field within which trends emerge out of the patterns we discern in the successive updatings of perceptual beliefs relating to the proximate past, the most proximate past, and the next, and so on. This trend is projected into the future in the form of protentions, i.e. anticipations of the pattern of updating of current perceptual beliefs which will be necessitated in the proximate future, the next most proximate future, and the next, in a manner symmetric with the past, but in inverse temporal order.[8]

Out of this inquiry into the ever-present marking of the Beja, the temporal imagination emerges as the dominant architect of space on KMMS world maps. It presents itself as an imagining that is triggered as much by extreme *otherness* as it is by the subtle reflection of *self*.

Why the Beja?

If you pick up any book on medieval Islamic history and scan the index for "Beja" or other variant spellings of the word ("Būja," "Bedja," "Būjah," sometimes even "Baja,"),[9] in at least nine out of ten secondary sources the name will not occur. If the work moves out of the orbit of the central Umayyad and Abbasid lands to a discussion of the Muslim presence in Africa, it will likely include an extensive exposé on North Africa: the Maghrib, Ifrīqiya (Roman Africa),[10] the Berbers, al-Andalus (Muslim Spain), and the early conquests. The main ruling dynasties will be mentioned: the Tulunids (AH 254–292/868–905 CE), the Ikhshidids (323–358/935–969), the

Fatimids (297–567/909–1171), the Ayyubids (564–650/1169–1252), the Mamluks (648–922/1250–1517, of Egypt), the Rustamids (161–296/778–909, of western Algeria), the Aghlabids (184–296/800–909, of Tunisia and eastern Algeria), the Idrisids (172–375/789–985, of Morocco), the Almoravids (454–541/1062–1147, of northwest Africa and Spain), the Almohads (524–668/1130–1269, of North Africa and Spain), the Marinids (614–869/1217–1465, of North Africa), and the Hafsids (627–982/1229–1574). Some texts will also mention the Hilālī incursions,[11] the slave trade, and the gold trade.[12] If the discussion moves beyond North Africa to West Africa, one may also find mention of the gold-producing centers of Ghana, Kanem, Takrūr,[13] Mali, and Timbuktu.[14] Although Ira Lapidus acknowledges that "in the eastern Sudan a mass Islamic society was established at a relatively early date," there has been little in-depth study of the area.[15] When East Africa is discussed, the focus tends to be on the two most important Christian kingdoms in the region: Bilād al-Ḥabasha (i.e., Abyssinia)[16] (highlighted on fig. 8.1 with a yellow circle) and Nubia (highlighted on fig. 8.1 with a purple circle). The lands of the Zanj and its people the Zanjī (highlighted on fig. 8.1 with a green circle) are frequently mentioned in the secondary sources. Both are discussed in the context of East African trading relations with the Muslims, particularly in the context of the slave trade, as well as in the mainstream of Abbasid history where the Zanj gained an early berth for themselves in the medieval Middle Eastern imagination as a consequence of their famous revolt in southern Iraq.[17]

But the Beja are rarely, if ever, mentioned.[18] Lapidus's all-encompassing book *A History of Islamic Societies* affords an excellent example.[19] He casts a wide net in this voluminous sourcebook for Islamic history, ranging from medieval to modern, from the central lands to the periphery. He discusses Africa at length, including an extensive discussion of West Africa, and the Muslim connections. Yet, when it comes to East Africa, although he acknowledges the region's importance for the early Muslims, details are sparse. The details do not increase until we get to the early modern period and the Funj Empire (located south of Ethiopia, around the area of the Blue Nile) of the sixteenth to eighteenth century. For the period prior, Lapidus adopts the standard pattern that dwells on Abyssinia and the Zanj. All that he has to say about the lands in between is that "the Arabs occupied Egypt as far as Aswān in 641, but it was many centuries before Arab and Muslim peoples penetrated Nubia and the Sudanic belt to the south. The first Arab influx occurred in the ninth century, when Egyptians swarmed south to the newly discovered al-ʿAllāqī gold fields between the Nile and the Red Sea. In the twelfth and thirteenth centuries Arab bedouin migrations increased the Muslim presence."[20]

M. A. Shaban, another important historian of the period, known for his *Islamic History: A New Interpretation*, fares a little bit better.[21] He includes two brief references. The first is after a lengthy discussion of the Zanj, when he summarily locates the Beja north of Abyssinia and the lands of the Zanj. The second reference is when he mentions them briefly among the earliest examples of the *iqṭāʿ* (fief/feudal estate system, but different from the European system).[22]

In works devoted to Islam in Africa the situation improves a little. The Beja are at least mentioned, but again the details remain sparse and, more often than not, tangential to the main subject at hand. For instance, in *Race and Slavery in the Middle East* and *Race and Color in Islam*, Bernard Lewis's reference to the Beja is meager and incidental to the Zanj.[23]

Exceptionally, Jay Spaulding, in his article "Precolonial Islam in the Eastern Sudan," devotes almost a page to the subject of the Beja and their Nubian neighbors, noting their sporadic contact with the early Muslim colonizers in Egypt.[24] Spaulding's generous treatment is a mere blip on the historical radar. In a recent book, *The Course of Islam in Africa*, Mervyn Hiskett, in spite of extensive chapters devoted to "the Nilotic Sudan," "Ethiopia and the Horn," and "East and South Africa," mentions the Beja in only one brief phrase: "South of ʿAlwa lay vast expanses of savannah and riverain country roamed over by stateless, Hamitic-speaking, nomadic cattle grazers known as the Beja."[25] Andrew Paul's 1954 *A History of the Beja Tribes of the Sudan* and Yusuf Fadl Hasan's 1966 article "The Penetration of Islam in the Eastern Sudan" are still the only extensive accounts of the Beja and early Muslim colonization in East Africa.

In spite of their stark underrepresentation in the mainstream of modern Islamic and African historiography, there is not a single example of an Islamic world map to be found without the words "al-Būja" (the Beja) marked prominently astride the upper reaches of the Nile between al-Ḥabasha (Abyssinia) and the "Bilād al-Nūba" (Land/Territory of Nubia). It is a striking fact that the Beja are even accorded an additional strip of land between the Red Sea and the Nile, which is referred to on many of the maps as the "Mafāza al-Būja" (i.e., the deserts of the Būja).[26] In this double marking alone the Beja are conspicuously distinguished from every other territorial group on the Islamic world map.[27] See, for instance, the close-up of the region from a fifteenth-century Aqqoyunlu world map (fig. 8.2). This holds true even on a highly stylized nineteenth-century Mughal rendition where the illustrator allocated space to the Beja (fig. 8.3). While the specific location of their name on the map may fluctuate within the prescribed area, one thing is certain—over an eight-century corpus of world maps their presence is rarely overlooked. This suggests that the Beja were an important component of the medieval Islamic conception of the world and that their place on the world map was never questioned, not even in the seventeenth-century Iranian copies, or the nineteenth-century Indian ones.[28]

Why would a place so infrequently mentioned in the historiography, and therefore (one would presume) historically insignificant, receive such a prominent, permanent place on the Muslim world maps? Why would the Beja's territory receive deliberate attention in preference over other, better-known locales in Islamic history

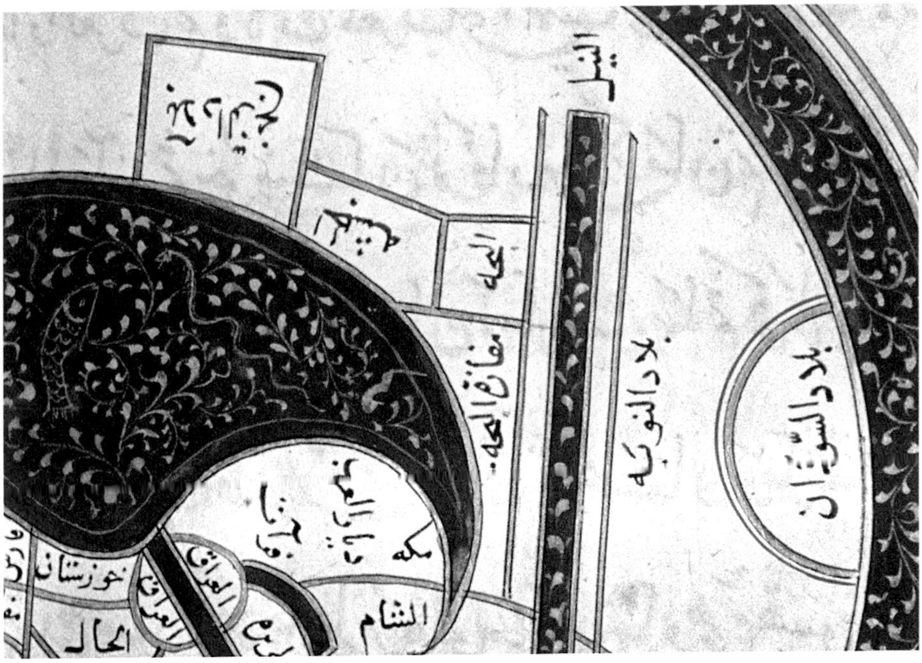

Fig. 8.2. Close-up of Aqqoyunlu KMMS world map showing Bilād al-Buja (Lands of the Beja) from al-Iṣṭakhrī's *Kitāb al-masālik wa-al-mamālik* (Book of Routes and Realms). Shows that the *mafāza* (deserts) are assigned to the Beja. Circa mid to late 15th century. Lapis lazuli (ultramarine) gouache with gold and black ink on paper. Diameter 26.4 cm. Courtesy: Topkapı Saray Museum, Istanbul. Ahmet 2830, fol. 4a. Photo: Karen Pinto.

such as Ifrīqiya, which is not marked on any medieval Islamic map? Why are other African tribes (such as the Zaghāwa, the Kawkaw, the Qumati, the Tuareg, and the well-known Berbers) mentioned in the primary literature but not marked on the maps? The quest for an answer raises an important red flag: that of the possible time period in which the KMMS world maps were first conceived, as well as the time period in which they were frozen.

Who Were the Beja?[29]

Sources tell us that the Beja are a seminomadic people of Hamitic origin. Scholars trace them back to the ancient kingdom of Cush (Kush);[30] others suggest that they are the ancestors of the modern Eritreans, and possibly even the Ethiopians. Until recently "Beja" was presumed to be an Arabic way of naming the troublesome Blem-

Fig. 8.3. Late nineteenth-century KMMS with lands of Beja circled. Mid-19th century. Mughal. Dark-blue, forest green, orange, and yellow gouache and red and black ink on paper. Diameter 19 cm. Courtesy: The British Library Board, London. Or. Add. 23542, fol. 60a. © The British Library Board. (For color version, see fig. 4.8.)

mye of Upper Egypt who harassed the Romans, resulting in the Roman general Maximinus's punitive raid of 453. Lately, the theory linking the Beja to the Blemmye (the headless monsters of medieval European mappamundi) has come under attack.[31] A. J. Arkell, on the other hand, argues that the Beja can be traced back to the "Medju" and inscriptions dating to the Sixth Dynasty of Egypt (2423–2242 BCE), found near the

First Nile Cataract, describing the submission of various tribes in the region to the Egyptian pharaoh.[32] In other words, there is no definitive account of the origin of the Beja.

Muslim geographers consistently locate the Beja textually, as well as graphically, somewhere between the Red Sea and the Nile.[33] Mas'ūdī and Ibn al-Wardī tell us that the Beja occupied the triangle between Abyssinia, Nubia, and Egypt.[34] Some writers present itinerant descriptions. Ibn Ḥawqal, for instance, instructs readers to find the Beja thus: "Starting from Qulzum, in the occidental part of the [Red] sea, we touch an arid desert where nothing grows and one meets nothing but the islands mentioned previously. In the middle of the spread of this desert live the Būja, who have tents made of hair. Their skin is much darker than the Abyssinians, who resemble the Arabs. They have neither villages, nor cities, nor cultivated fields, and subsist from what is brought from the cities of Abyssinia, Egypt, and Nubia."[35]

André Miquel identifies two regions of Beja tribes. One is in a west-to-east direction from the coast, along which lay the caravan route from Quṣ to Qifṭ, which led eventually to the mines of Wādī al-'Allāqī and from there to the port of 'Aydhāb. In this location he also includes the traffic from Yemen and the Hijaz. Miquel identifies the other region as going in a south-to-north direction along the hollow of Baraka, beginning in the plateau of Eritrea and moving toward Aswān and Egypt.[36] According to the "Beja Project," they are still to be found approximately where the Muslim maps mark them: off the Red Sea, south of Egypt, east of Nubia, north of Ethiopia (fig. 8.4).[37]

One thing that the modern map confirms is that, at least in terms of location, the Muslim cartographers placed the Beja accurately on their world maps—demonstrating a degree of accuracy in their knowledge of the Beja tribes and related territories.[38] The mere fact of their existence or their accurate location does not, however, explain the consistent presence of the Beja on the Muslim world maps. To understand the reason for the Muslim cartographers' selection, we have to turn to their texts. These medieval Arabic geographic accounts are among the earliest extant descriptions of the Beja.[39]

The earliest mention of Muslim-Beja contact occurs in the context of the initial Arab campaigns in southern Egypt, following the disastrous encounter with the Nubians (31/651–652), in which Arab forces were decisively stalled.[40] After this resounding defeat at the hands of the Maqurra on the Nile, of which it has been said that "the Muslims never suffered a loss like the one they had in Nubia,"[41] the Arabs turned their attention eastward. On the rebound from Nubian humiliation, 'Abd Allāh ibn Sa'd encountered some Beja tribes, but at that point, in comparison to the mighty Nubians and their famous gold mines, Sa'd regarded the Beja as politically insignificant.[42] It was not long before the Muslims discovered that they had a more easily exploitable treasure trove sitting above the First Cataract of the Nile. At first Beja lands were

Fig. 8.4. Modern map showing the location of the Beja stretching across the borders of Egypt, Ethiopia, and Sudan all the way to the Red Sea. Prepared by Julian Weiss.

invaded to bring the Upper Egyptian border under control and to exact tribute from the chiefs.[43] During the reign of the Umayyad caliph Hishām (r. 105–125/724–743), ʿUbayd Allāh ibn al-Ḥabḥāb negotiated one of the earliest treaties with the Beja.[44] Forays into Beja territory enabled the Muslims to rediscover the ancient mines of gold at Wādī al-ʿAllāqī and emeralds at Qifṭ.[45] Precisely when this occurred is not clear as it is not given by the sources. What is clear is that by the ninth century a major rush for gold and precious stones was under way in Beja territory.[46]

> In their territory, [which lies] between Abyssinia, Egypt, and Nubia, there are emerald and gold mines. These mines extend from around Aswān—at a distance of approximately ten stages in the territory of Egypt—up to the sea, near a fortress named ʿAydhāb. It is here that the tribe of Rābiʿa are grouped—at a point named al-ʿAllāqī, in the middle of the sands and flat ground, with a few hills scattered between this area and Aswān. The products of the mines are directed toward Egypt. They are mines of pure gold without any silver, which are under the control of the Rābiʿa, who are the sole proprietors.[47]

The Beja tribes came to be viewed as a ripe source of slaves, especially for the work of mining.[48] By the early ninth century the Beja were getting restive, probably due to the impositions on their land and people.[49] Thereafter, at regular intervals, the Beja tribes began to break treaties, revolt at the mines, and raid southern Egypt, going as far as Fusṭāṭ. It seems that the Arab tribes of southern Arabia (primarily the Rābiʿa tribe), who had come over in droves to take over the mines, were the primary catalyst for Beja irritation.[50] One sympathetic Muslim observer, for instance, notes: "The Beja who live in this desert are not bad people, nor are they robbers. It is the Muslims and others who kidnap their children and take them to the towns of Islam, where they sell them."[51]

Could the discovery of gold and emerald mines in Beja territory, along with the fact that they were viewed as a good source of slaves, and the frequent occurrence of their troublesome raids, have secured for them a permanent place on the Muslim world maps? Perhaps so, but there were other places, such as Ghana and Kanem in West Africa, that were renowned for their gold mines and slaves and that did not make it onto the Muslim world maps.[52] Certainly, troublesome tribes were abundant in the Muslim lands and peripheries and would not be a sufficient reason to make the cartographic news.

Why mark the Beja on the world maps when the Muslim scholars writing about them held them in such low regard? Ṣāʿid al-Andalusī, for instance, writing in the thirteenth century, does not mince his words: "The only peoples that reject these humane institutions and live outside these rational laws [i.e., royal decrees and divine laws] are a few of the inhabitants of the deserts and the wilderness such as the beggars of Bajah [*sic* Būja], the savages of Ghana, the misers of the Zinj [*sic.* Zanj], and those resembling them."[53]

Four centuries earlier, Iṣṭakhrī said something similar when he opened his treatise on routes and realms by asserting that he will not discuss the Beja and other such uncouth, irreligious people:

> We have not mentioned the land of the Sudan in the west, or the Būja, or the Zanj, or other peoples with the same characteristics, because the orderly government of kingdoms is based on religious beliefs, good manners, law and order, and the organization of settled life directed by sound policy [intiẓām al-mamālik bi'l-diyānāt wa-al-ādāb wa-al-hukum wa taqwīm al-ʿimārat bi'l-siyāsa al-mustaqīma]. These people lack all these qualities and have no share in them. Their kingdoms, therefore, do not deserve to be dealt with separately as we have dealt with other kingdoms. Some of the Sudan, who live nearer to these well-known kingdoms, do resort to religious beliefs and practices and law, approaching in this respect the people of these kingdoms. Such is the case with the Nūba and the Ḥabasha, because they are Christians, following the religious tenets of the Rum. Prior to the rise of Islam they were in neighborly contact with the Byzantine Empire, because the land of the Nūba borders on Egypt, and the Ḥabasha live on the Sea of al-Qulzum [Red Sea].[54]

Iṣṭakhrī goes on to mention the mines of Beja territory, yet he neglects to mention the Beja: "Between them [the land of the Nūba and the Ḥabasha] and the land of Egypt is a desert in which there are gold mines, and they are linked to Egypt and Syria by way of the Red Sea."[55]

An odd set of statements, given that barely fifteen pages later in his text, in the section on the Baḥr Fāris (Persian Gulf/Indian Ocean), Iṣṭakhrī includes an extensive passage on the Beja, their mines, their characteristics, and way of life:

> If you go from Qulzum, the western end of this sea [the Red Sea], extending from it to the empty wasteland, there is nothing until you arrive at the desert of the Būja. The Būja are a people of curly hair, much darker than the Ḥabasha [Abyssinians] who are similar to the Arabs. They do not have villages or cities or grain except what comes to them from the cities of Abyssinia, Yemen, Egypt, and Nubia. Their borders stretch from between Abyssinia to Nubia and Egypt, reaching as far as the mines of gold. Going from these mines in the vicinity of Aswān, Egypt, until a fortress on the sea called ʿAydhāb, it is approximately twenty *marāḥil*.[56] And the place where the people from these mines gather is called al-ʿAllāqī, and it is sandy land, flat without mountains, and the money of these mines is sent to the land of Egypt. And they are mines of gold not silver. And the Būja are composed of a nation of idol worshippers. Then one arrives at the land of the Abyssinians and they are Christian and closer in skin color to the Arabs, [that is,] between black and white.[57]

Ten pages later, in what can be read as an indication of Iṣṭakhrī's fascination with the Beja, he mentions them again. This time Iṣṭakhrī's comments are in the section on Egypt. He describes the approach to Beja lands from Aswān instead of the Red Sea.

> And as for the mines of gold, they are fifteen days away from Aswān. The mines are not in the land of Egypt, rather in the land of the Būja, and they extend until ʿAydhāb. And it is said that ʿAydhāb is not of the land of the Būja, but that it is one of the cities of Abyssinia. The land of the mines is flat without mountains, rather sandy and dusty. And the place in which the people collect is called al-ʿAllāqī. And the Būja have no villages or any prosperity or any richness, rather they are nomads, and they have nobles [among them]. It is said that among the nobles there are none more lowly than theirs. Both their slaves and their nobles, and everything else in their land, reach as far as Egypt.[58]

This textual pattern recurs in Ibn Ḥawqal's *Kitāb ṣūrat al-arḍ*. He begins his discussion of the Beja employing exactly the same words as Iṣṭakhrī (quoted at the outset of this discussion).[59] Then twenty-five pages later, in the section on the Persian Gulf/Indian Ocean, Ibn Ḥawqal inserts one of the most extensive descriptions of the Beja to be found anywhere, comprising seven long pages of a detailed discussion of their lands, the tribal divisions, and the uprisings. In a significant departure from the norm, Ibn Ḥawqal devotes more detail to the Beja than he does to Abyssinia or the lands of the Zanj. Albeit a more balanced appraisal, his discussion is still peppered with disdainful statements: "The Būja are the people of this land. They worship idols and other objects that they consider venerable. . . . They are nomads who breed sheep . . . and their numbers escape all estimates. . . . In the incline of Baraka, live the [Būja] tribes known as Bāzīn and Bārīya—people who fight with bows, poisoned arrows, and lances, but without shields. The Bārīya are known for extracting their incisors and for clipping their ears. They live in the mountains and the valleys where they tend to livestock and cultivate the soil."[60]

The tenth-century Palestinian scholar Muṭahhar ibn Ṭāhir al-Maqdīsī in his *Kitāb al-bad' wa-al-ta'rīkh* (Book of the Creation and the History) echoes a similar sentiment when he deprecatingly notes that the Beja are so uncouth that "there is no marriage among them; the child does not know his father, and they eat people—but God knows best."[61]

Is it possible then that it was not the mines nor the potential for slaves that put the Beja on the Muslim world maps, rather that they were perceived by the Arabs as a reflection of an extreme *other*?[62] The Arabic sources bolster this interpretation and confirm that the Beja were perceived of as a group of people with strange mannerisms and a way of life contrary to that of the Muslims. This is what the curious passage in Ya'qūbī's *Kitāb al-buldān* (Book of Countries) suggests: "The Būja live in tents of hides, pluck their beards, and remove the nipples from the breasts of the boys so that their breasts do not resemble the breasts of women [yanza 'ūna falaka thadai al-ghilmān li-'allā yushbihu thadaihim thadai al-nisa']. They eat grubs and similar things. They ride camels and fight in combat on them just like they fight on horseback, and throw javelins without ever missing."[63]

These sources suggest that it was not the Beja's gold and emeralds and rebellious ways that gave them a berth on the medieval KMMS world maps but rather that it was their extreme alterity that captured the imagination of the Muslim cartographers and propelled them into a permanent berth on the world maps. They are a signifier on the maps of a shocking "uncivilized" people who don't have any laws or religion—"pays sans foi ni loi."[64] André Miquel concurs with this interpretation but adds a spin of his own: "On the one hand, men who mutilate their children and follow a resolutely matriarchal system, on the other, a people that clings, as much as it can, to nomadism and to the desert; a people that follows grass and moves to summer pastures from the sea to the internal valleys; finally, a people that, just as Arabia could not live without the camel, carries with it war and race, an ever-ready sacrifice for the guest that a mad extravagance is forced to honor."[65]

The Beja are, in other words, not just an extreme manifestation of *other*; they are also a paradoxical manifestation of *self*. Or, rather, one of the many selves that make up the Arab Muslim psyche: the primordial, nomadic, desert-based one whence Islam sprang.[66] As Miquel intuitively points out, between the two coasts of the Red Sea, one can perceive a kind of latent fraternity, in which Arabia finds reflections of itself. "The description of the Bedja land puts side by side features that would seem absurd [or abnormal] to an Arab-Muslim mentality, with others that resemble Arabia."[67] Or as Hamdhānī put it, "A piece of Yemen [on African soil]."[68]

But the maps were not produced in southern Arabia by members of the Rābiʿa or Mudar tribes. Rather they are products of the political and intellectual centers of the Muslim world, where the perceived affinities or extreme alterities between the Rābiʿa and the Beja would not have had a significant effect on the prevailing caliphal conception of the world. We must look further to find an answer for the Beja presence on the world maps. In the next chapter, I will look more closely at Arab interactions with the Beja and the resulting variance in representations of the Beja on world maps.

9

How the Beja Capture Imagination

In all the Arab accounts of the Beja there is a curious phenomenon. Extensive references to them and their strange ways, their mores, their mines, their enslavement, and their raids occur in treatises from the late ninth century onward. Iṣṭakhrī wrote in the early tenth century, Ibn Ḥawqal and Maqdīsī in the late tenth century, and Ṣāʿid al-Andalusī in the mid-thirteenth century. The earliest extensive description of the Beja comes from Yaʿqūbī's *Kitāb al-buldān*, composed in AH 276/889–890 CE. The only other major account comes from the late ninth-/early tenth-century historian Ṭabarī who reports events involving the Beja in the year 236/850.[1]

Prior to the mid-ninth century, references to the Beja are extremely rare. One of these occasional mentions comes from Ibn Khurradādhbih's *Kitāb al-masālik wa-al-mamālik*, composed in 232/846. It is a brief, formulaic phrase referring to the bilateral trade agreement (*baqṭ*) made with the Nubians following the stalemate of 31/665–662. It only mentions the Beja tangentially:[2] "In uppermost Egypt[3] are the Nubians, the Beja, and the Abyssinians and ʿUthmān ibn ʿAffān had reached a settlement with the Nubians for four hundred heads per year."[4]

The fact that there is no detail on the Beja in Muslim sources prior to the mid-ninth century is surprising since (as I demonstrated in the previous chapter) active exploitation of the gold mines in Beja territory had begun as early as the mid-eighth century. Why would it take so long for the Beja to make it into the mainstream of Muslim thought?

This suggests that the Beja were not famed for their mines in the Muslim centers of power and learning prior to the mid-ninth century. The exploitation of the Beja mines was, at least until the ninth century, a localized phenomenon not well known beyond the immediate orbit of contact between the Arab tribes of southern Arabia and Upper Egypt. If initial contact with and knowledge of the Beja was only a localized phenomenon prior to the mid-ninth century, then there is no reason why the Beja would make news in the center—where most of the armchair scholars and geographers were based—unless something dramatic happened.[5]

In the first quarter of the ninth century there were a spate of Beja uprisings and raids into Egypt and Muslim towns, reaching all the way to Fusṭāṭ. These raids became so serious and frequent that they eventually came to the attention of the center, specifically the ʿAbbasid caliph Mutawakkil (r. 232–247/847–861), who was forced to step in and take action to redress the situation. Ṭabarī reports that:

> He [i.e. al-Mutawakkil] sought advice concerning the circumstances of the Būja. He was informed: They were a nomadic people, tenders of camels and livestock. Getting to their territory was difficult, and it was inaccessible to troops, for it consisted of desert and steppe. It was a month's journey from the land of Islam to the territory of the Būja, through wasteland, mountains, and barren country, lacking water, vegetation, refuge, or a fortified position. Any government representative who entered Būja territory would have to be supplied with provisions for his entire stay until he returned to the land of Islam. If the extent of his stay was greater than estimated, he and all his comrades would perish. And the Būja would simply overcome them without hostilities. Their land did not remit to the central government land tax or any other tax.[6]

A series of events related to official caliphal intervention in East Africa along with a dramatic visit by the Beja chief to Samarra catapulted the obscure Beja into the mainstream Muslim imagination. The events triggered a flash of acknowledgment that left an indelible stamp in the form of dramatic narratives, fantastic descriptions, and a nomination on the world maps.

Both the geographer Ibn Ḥawqal and the historian Ṭabarī describe, with mostly minor but some major variations, the same series of dramatic events. They tell of the start of the Beja uprisings and raids, the Muslim response, and the journey of the Beja chief, ʿAlī Baba, to the Abbasid capital, Samarra, to meet with the caliph.[7] The major difference between the two accounts is that Ṭabarī presents a terse view from the center of caliphal power, whereas Ibn Ḥawqal provides a detailed, localized account of the Beja uprisings, specific reasons for each one of them, and the various Muslim attempts to suppress them. Ṭabarī, for instance, only says that "from the time of al-Mutawakkil's reign the Būja refrained from delivering [their] tax for several consecutive years," and that the head of the Postal and Intelligence Service, Yaʿqūb ibn Ibrāhīm al-Bādhghīsī, wrote to the caliph to tell him that the Beja had broken the treaty between them and the Muslims and that "they [i.e. the Būja] advanced from their territory to the mines of gold and precious stones that were on the border between Egypt and the territory of the Būja. The Būja killed a number of Muslims employed in the mines for mining the gold and precious stones, and took captive a number of the Muslim children and women. The Būja claimed that the mines belonged to them and were in their territory, and that they would not permit the Muslims to enter them."[8]

Ibn Ḥawqal, on the other hand, provides an elaborate account of the stories behind the early uprisings. It seems the trouble began when in 204/819, a group of Beja abandoned the Muslim chief of Qifṭ, Ibrāhīm Qifṭī, and some of his companions on their way to pilgrimage, leaving them to die of thirst in the desert. Ibn Ḥawqal suggests that the rationale for this was that some of the Beja feared that Qifṭī knew their lands too well: "We definitely must put to death this Muslim who knows our country so well, our camping spots, our watering holes. We cannot trust him."[9]

Word of the deliberate abandonment of Qifṭī and his followers got back to the other Arabs in Qifṭ, and when the Beja chief, Mūhā, made a trip to town with thirty notables to buy supplies, the locals enticed them down to one of the churches and massacred the party. According to Ibn Ḥawqal, in 204/819, the Beja tribes retaliated by marching on the city of Qifṭ, taking seven hundred prisoners, and killing many, which in turn forced the remaining inhabitants to flee to Fusṭāṭ. There the fugitives petitioned for restitution and had to wait seven years before their pleas were heard because the local officials were too busy with other matters. Finally, their cause was taken up by a wealthy private citizen, Ḥakam Nābīghī of the Qays tribe. In 212/827 he executed raids on Beja territory and succeeded in freeing the city of Qifṭ from the Beja stranglehold.

None of this quieted the Beja for long. Twenty years later, in 232/847, the Beja attacked and occupied Onbu, a city in Upper Egypt.[10] This would have coincided with the Abbasid caliph Mutawakkil's ascension to power, and it is possible that this is when he was first informed about the troublesome Beja. Ṭabarī does not specify. All he notes is that Mutawakkil, who was involved in a power struggle with local Turkish leaders at home,[11] did not respond immediately, hoping that the troubles with the Beja would die down. At this point, Ibn Ḥawqal presents a variation. He says that the influx of the Rabīʿa and Muḍar tribes to the area had increased dramatically following the invasion of Muḥammad ibn Yūsuf Ukhaiḍir Ḥasanī into Yemen, resulting in the mass emigration of local inhabitants to Egypt and Beja territory. This brought about more clashes between the southern Arabian tribes and the Beja. Ibn Ḥawqal notes that one flare-up resulted in some Beja insulting the Prophet, and when this reached the ears of Mutawakkil, he reacted: "A brawl flared up between one of their men and one of the Būja, during which the Būja insulted the Prophet. Al-Mutawakkil was informed and he sent to the site a descendent of Abū Mūsā Ashʿarī, called Muḥammad al-Qummī, who was then jailed for homicide. The caliph provided him [al-Qummī] with the men and arms that he asked for and promised him his freedom."[12]

Ṭabarī also wrote that Mutawakkil refrained from responding at first, but as the situation worsened and Beja attacks against the Muslims intensified "to the point that the inhabitants of Upper Egypt feared for their lives and for their children," he appointed Muḥammad ibn ʿAbd Allāh—known as al-Qummī (whom Ibn Ḥawqal claims was in prison for a homicide)—to attack the troublesome Beja.[13]

From this point onward, both texts begin to dovetail. Both concur on Qummī's successful rout of the Beja and provide parallel accounts of the confrontation. Apparently Qummī was greatly outnumbered by the Beja and running short of supplies by the time he reached the heart of Beja territory. He had collected about three thousand men, while the Beja were two hundred thousand men strong with eighty thousand dromedary. When the two armies first met, the Muslims panicked. Qummī urged them to stay and fight for their honor: "There is no escape. Fight for your life and honor and you will be victorious."[14] Undaunted by being outnumbered, Qummī devised a plan to thwart the Beja.[15] According to Ibn Ḥawqal, he surrounded his camp with a network of horses and drums and prepared a series of cloth banners with gilded writing, which he fixed to the top of some spears. At daybreak he proclaimed, "Behold! Soldiers of the Būja, here are missives for you from the Emir of the believers."[16] The banners distracted the Beja who had formed in groups for battle. Curious, some of them broke ranks to take a closer look at the banners. When this happened, Qummī ordered the camels on which his banner men were sitting to rise and indicated that the drums be sounded. Suddenly the Beja found themselves in the middle of a sea of banners with drums pounding. This caused a pandemonium and scattered the Beja forces in disorder. Qummī capitalized on this and routed the Beja army decisively. Many of them were massacred. Others, including the Beja chief, ʿAlī Baba, were taken prisoner.[17]

Ṭabarī provides a similar yet varying account. He indicates that at first the Beja chief was just harassing Qummī's army, exhausting them, waiting for their supplies to run out. ʿAlī Baba only launched an all out attack when he heard that a shipment of fresh supplies had arrived from Baghdad. Like Ibn Ḥawqal, Ṭabarī also notes the use of noise to cause confusion among the Beja ranks during the battle, but he refers to the use of bells, not drums and banners:

> Seeing this [i.e. the arrival of the ships with supplies], ʿAlī Baba, the Būja chief, went on to do battle with the Muslims, rallying troops against them. The two sides clashed and fought violently. The camels upon which the Būja fought were unseasoned and tended to be frightened and alarmed by everything. Noticing this, al-Qummī rounded up all the camel and horse bells in his camp. He then attacked the Būja, stampeded their camels with the clanging of bells. Their alarm was considerable. It drove them over mountains and valleys, totally splintering the Būja forces. Al-Qummī and his men pursued and seized them, dead or alive, until night overtook him. This took place at the beginning of 241 (855–56). Al-Qummī then returned to his camp and could not count the dead they were so many.[18]

Both accounts concur that the Beja were routed and that ʿAlī Baba was taken prisoner, forced to pay the overdue taxes, and dragged off to Iraq for an audience with the caliph Mutawakkil. Ṭabarī provides a dramatic account of this visit. Its long-term effect on the imagination of the people (scholars, chroniclers, and general populace) located close to the center of power can well be imagined:

> ʿAlī Baba appointed his son Laʿis as deputy over his kingdom. Al-Qummī departed with ʿAlī Baba for the gate of al-Mutawakkil, and arrived there at the end of 241 (855–56). He attired this ʿAlī Baba with a silk brocade-lined robe and a black turban covering his camel with a brocade saddle and brocade horse cloths. At the Public Gate, along with a group of the Būja, were stationed about seventy pages, upon saddled camels, carrying their lances, on whose tips were the heads of their warriors who had been killed by al-Qummī. . . . Some(one) of [the respondents] reported seeing ʿAlī Baba with a stone idol in the shape of a young boy to which he prostrated himself.[19]

Stepping back from the texts for a moment, we can imagine the cities of Iraq, especially Basra, Kufa, Baghdad, and Samarra, the new Abbasid capital, of the mid-ninth century, through which ʿAlī Baba and his retinue likely marched. Imagine the throngs of people in the markets and the streets, going about their daily business. Then imagine the sudden appearance of a bizarre procession of the Beja wending their way from the port of Basra, perhaps through Baghdad, toward Samarra, and eventually the central zone of the palace, with the Beja chief, ʿAlī Baba, elaborately dressed in brocade robes and finery, followed by his retinue of Beja tribesmen with next to nothing on and, if we are to take Yaʿqūbī's description at face value, with their nipples cut off, followed by seventy of Qummī's men with the severed heads of Beja victims at the end of their spears. It is not difficult to imagine the impact that this event must have had on the psyche of the people present at the time. Add to these astonishing sights the stories that must have circulated, thanks to Qummī's soldiers, about the gold and emerald mines to be found in Beja lands, as well as details about their habits and living conditions. The Beja must have been the subject of many a lively discussion long after they had come and gone. Or, perhaps, as Ibn Ḥawqal suggests, they never really left. His rendition of the same story places the visit of the Beja king to the Abbasid capital at a slightly earlier date (238/852) and incorporates an intriguing deviation. According to Ibn Ḥawqal, it was not only the chief of the Beja who was dragged off to Samarra but the Nubian king as well,[20] where they were chastised and humiliated for their unruly behavior by being sold as slaves for paltry sums.[21]

> Then, as a result of this imprudence, they [the Būja] all perished—trode on by their camels. For them it was death or captivity. ʿAlī Baba was taken prisoner: placed on a hill, after he swore not to move from there unless the hill disintegrated. After his capture, al-Qummī took him [ʿAlī Baba] and his booty to Aswān. There he sold everything for fifty thousand ounces of gold.
>
> [Next] he sent an ultimatum to Yurkī, the king of Nubia, who came to submit himself. He [al-Qummī] took all his men to Baghdad in the year 238 and presented the two princes to the caliph. There they were put up for auction. The king of the Būja was sold for seven dinars and the king of Nubia for nine. They imposed on each one of them as punishment a daily labor equivalent to the sum of their sale. Al-Qummī returned to Aswān after having secured their agreement on this point.[22]

If Ibn Ḥawqal's version of the outcome is not hyperbole, and the Beja chief and his followers stayed on in Iraq as slaves, then the center would have had the Beja in view for much longer than just the period of the procession. We may also have an explanation for why the Beja revolted again so soon after Qummī's departure as well as a significant fictional connection between the Beja chief and one of the most famous characters in medieval Arabic literature: the poverty-stricken ʿAlī Baba, of the *One Thousand and One Nights*, the famous deceiver of the forty thieves.[23]

The relationship between the two ʿAlī Babas—real and fictional—would require further investigation to determine whether there is indeed a connection between the two characters.[24] If there is a connection, then we would have a powerful example of the effect that the Beja had on the medieval Islamicate imagination such that, via ʿAlī Baba, they also found a permanent berth in the fiction of the period.

Ricocheting in a Husserlian way, the story speaks volumes of the long-term effects of the bizarre on the imagination, such that retentions and reproductions proliferate in the cultural products. Particularly striking is the way in which accounts of the same event vary within the span of a century: from Ṭabarī, writing in the late ninth/early tenth century, to Ibn Ḥawqal, writing in the late tenth century. In the late ninth century, the Beja were being discovered by the center for the first time in a dramatic way. By the late tenth century, they went from being different and *other* to being the *same*. Thus with Ibn Ḥawqal, we get intimate local accounts of the events in the area leading up to the visit to the Abbasid capital. We also get a sense that these were not just "wild people" *out there somewhere*, rather that they were closely linked to the Arabian mainland through the ebb and flow of immigrants. The way in which the details of the story of Qummī's battle vary presents another way in which retentions morph from the origin, influenced by reproductions from the present "now" moment.

Similarly, the maps and the way in which the nomination of the Beja varies on them can be viewed through a Husserlian lens. Fluctuations in spatial allocations are the product of time warps and resonances (retentions and reproductions) that can be viewed and understood through a spatiotemporal lens. On the surface the medieval Islamic world map matrix appears to be frozen in time, but close examination of the mutations of one site reveals that history in fact resonates there.

Representational Variations and Nuances

The fifteenth-century Ottoman map with which I opened chapter 8 shows the Beja as occupying two zones: a square between Abyssinia and the Nile assigned to their tribes, and a larger triangular zone between the Red Sea and the Nile for their deserts (Mafāza al-Būja), including the gold and emerald mines (see fig. 8.1). Other world maps from this Ottoman cluster present a similar view.[25] The space assigned to them is tightly constrained along the shore of the Red Sea, intimately connected with Abyssinia and the Nile. The fifteenth-century Aqqoyunlu manuscript (TSMK A2830) that spawned the Ottoman cluster, discussed in detail in chapter 12, presents a similar double-zone representation of the Beja (see fig. 8.2).[26]

Stepping backward in time from the Ottoman and Aqqoyunlu maps to examine the demarcation of the Beja on KMMS maps from the thirteenth century, we discover surprising variations. The Beja are still marked on these maps but the deserts (mafāza) are no longer indicated as belonging to them. Two examples of this come from the world maps of Ouseley 373 (at the Bodleian) dated to 696/1297 (the earliest extant Persian translation of Iṣṭakhrī's manuscript)[27] and Ahmet 3348 (at TSMK) dated to 684/1285 (produced in Cairo) (see figs. 9.1 and 9.2 for close-up views; for full, color images, see figs. 4.11 and 4.10).

On these maps the Beja are relegated to a rectangular strip of land between Abyssinia and the Nile, roughly comparable in size to the amount of land assigned to Abyssinia. This suggests that the importance of the Beja—at least in the thirteenth-century Egyptian and Iranian imagination—was equivalent to Abyssinia.[28] The amount of space assigned to the former Beja deserts on these maps suggests that the Muslim cartographers still considered these deserts to be significant even though they stripped

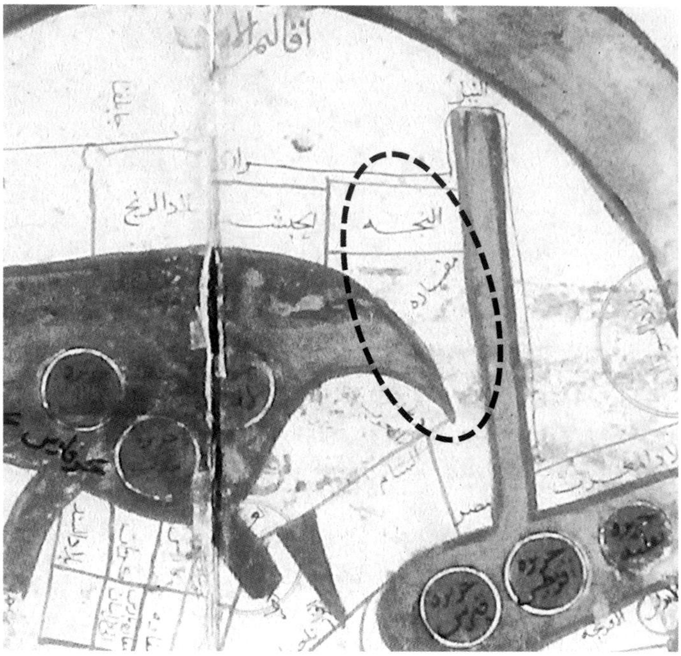

Fig. 9.1. Close-up of Beja lands on Bodleian KMMS world map. Shows that the *mafāza* (deserts) are not assigned to the Beja. 696/1297. Unusual pale-lilac-colored gouache along with red and black ink on paper. Courtesy: The Bodleian Library, University of Oxford. Ms. Ouseley 373, fols. 3b–4a. (For a color version, see fig. 4.11.)

the Beja of ownership of these lucrative gold- and emerald-mining regions. We can read into this cartographic evolution historical confirmation of Muslim control over this space.

This fits with the information on the rich mines in the Beja deserts which, as indicated earlier in this chapter, surfaces through the textual accounts from the ninth century onward. One of the treaties negotiated with the Abbasid caliph in the mid-ninth century specifically states that "Beja country from Aswān to the border between Bādī and Massawa becomes the property of the Caliph" and will be ruled by the Beja chief only in the caliph's name.[29] The unassigned status of these deserts, located on the other side of the Red Sea opposite Arabia, fits with the theory that there was an active immigration of tribes from southern Arabia into the area. The construction of mosques in the region from the ninth and tenth centuries onward, along with Arab familiarity with the Beja language (Bujāwī), can be taken as additional verification of this pattern.[30]

By the tenth and eleventh centuries, the gold trade with West Africa was growing and Muslim penetration of Nubia and the Nubian minefields was on the increase. By the fourteenth century, the gold and emerald mines of the *mafāza* area were all but abandoned because of the richer sources elsewhere in Africa and the

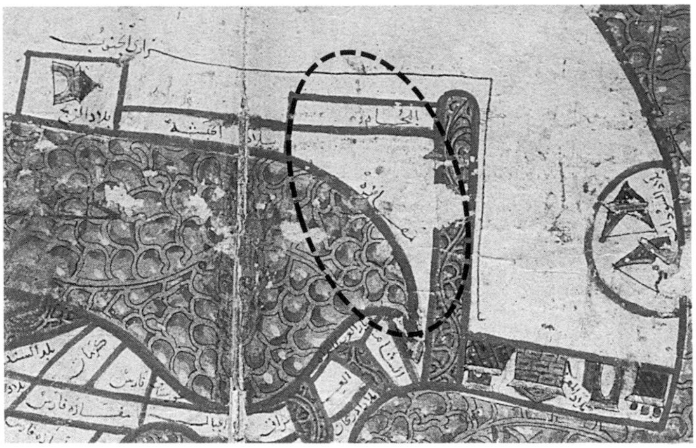

Fig. 9.2. Close-up of Beja lands on KMMS world map from Ahmet 3348. Like Ouseley 373, this map also does not assign the *mafāza* (deserts) to the Beja. 684/1285. Shades of blue, green, red, and brown gouache along with red and black ink on paper. Courtesy: Topkapı Saray Museum, Istanbul. Ahmet 3348, fols. 2b–3a. Photo: Karen Pinto. (For a color version, see fig. 4.10.)

increased difficulty in accessing what little gold remained in the ʿAllāqī mines. One would expect this waning interest to have been reflected on the KMMS maps, but it is not. This anomaly can be explained through the application of the Husserlian model of retentions. Even though the "now" moment of active interest in the Beja was no longer as relevant, it still echoed in the sources as well as the maps as retentions of the past. Another way to view the large spatial allocation to the deserts in these thirteenth-century maps is as a "highway" astride the nexus between southern Arabia and the rest of Africa lying beyond the Nile. "A strip of land between the sea and the Nile; a crossroad between Upper Egypt, Nubia, and Abyssinia," as André Miquel puts it.[31] In addition, Yūsuf Fadl Hasan reminds us that by the twelfth century the port of ʿAydhāb had become a major trading entrepôt on the Red Sea: "Large caravans [criss-]crossed the eastern desert between ʿAydhāb and Qūṣ, transporting merchandise and carrying provisions to miners. It was not long before Suakin joined ʿAydhāb as a trading port from which Muslims penetrated into the interior. Also for more than two hundred years, from 1058 to 1261 [i.e., until the Crusades disrupted the trade], pilgrims from Egypt and North-West Africa on their way to Mecca used the same route. This intensive traffic exposed the Beja tribes to further Muslim influence."[32]

We would expect to find the same emphasis, if not more, on the *mafāza* of the Beja in twelfth-century world maps, but this is not the case. Examining the world map from the Leiden manuscript[33] we note a subtle but significant difference from KMMS world maps of the thirteenth century. As the map in figure 9.3 shows, the trend of not assigning the border to the Beja continues, but with a noticeable twist.

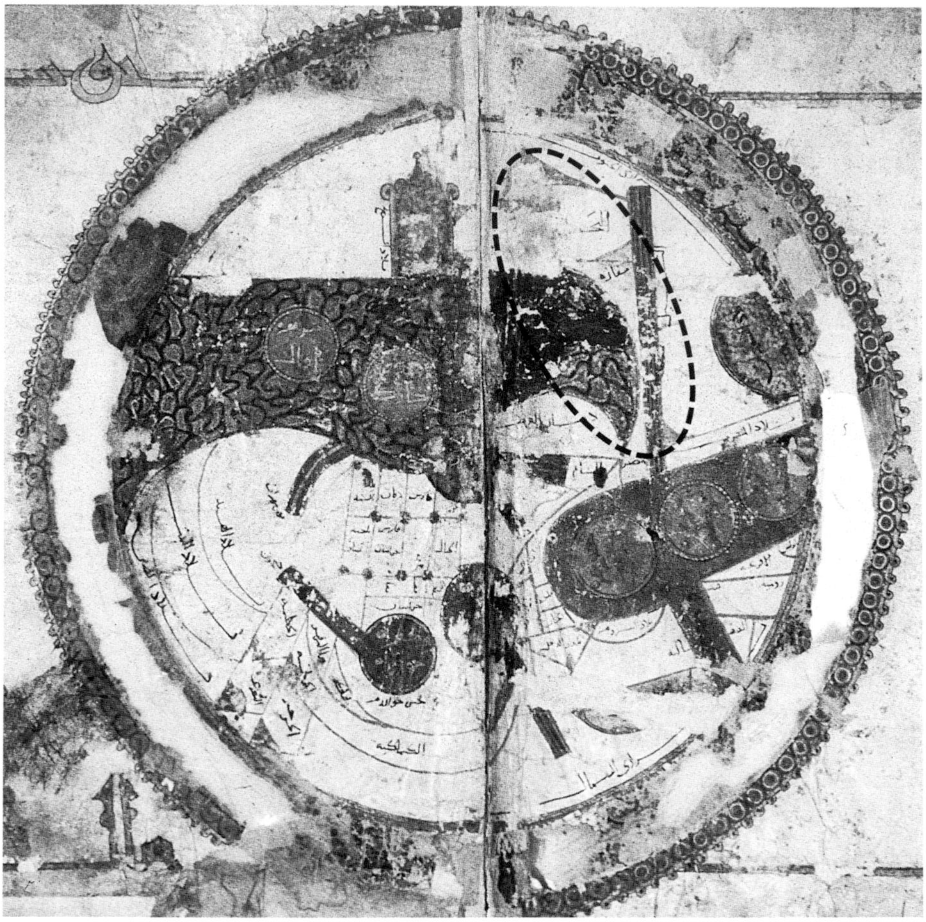

Fig. 9.3. Beja lands marked on Leiden KMMS world map. Beja are assigned more land and their deserts less. 589/1193. Mediterranean. Gouache and ink on paper. Diameter 37.5 cm. Courtesy: Leiden University Libraries. Cod. Or. 3101, fols. 4b–5a. (For a color version, see fig. 1.1.)

Of the two Beja territories, the space assigned to the tribe is larger, but its desert region (*mafāza*) between the Red Sea and the Nile is reduced. We can presume from this distorted representation of the lands of the Beja that their *otherness* was still capturing the Arab imagination even though (in reality) the significance of their deserts had decreased greatly from the ninth century onward. The Islamic maps thus consistently present a conception of the world that is about a century or two behind the prevailing "now" moment.

Moving to the earliest extant map manuscript of 479/1086, TSMK Ahmet 3346, in

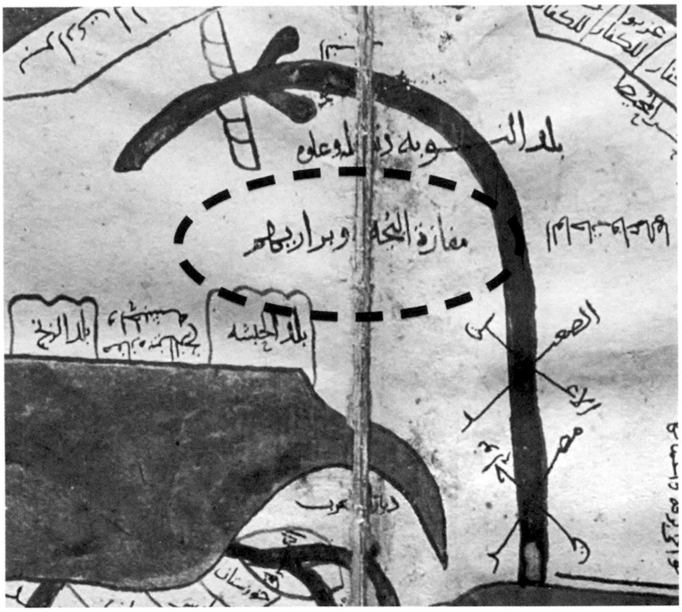

Fig. 9.4. Close-up of Beja lands on earliest extant KMMS world map from Ibn Ḥawqal's *Kitāb ṣūrat al-arḍ* (Book of the Image of the World). 479/1086. Iraq(?). Dark-forest-green and dark-blue and red and black ink on paper. Diameter 36.5 cm. Courtesy: Topkapı Saray Museum, Istanbul. Ahmet 3346, fols. 3b–4a. Photo: Karen Pinto. (For a color version, see fig. 4.7.)

Ibn Ḥawqal's *Ṣūrat al-Arḍ*, we find another variation in the representation of the Beja (fig. 9.4). The Mafāza al-Būja wa barārīhim (Deserts of the Beja and Their Steppes) reign mighty above the lands of Abyssinia and the Zanj and the upper reaches of the Nile. Notably the lands of the Beja are not confined by boundaries. Instead, the maker of this map suggests that the Beja occupy all the open ground between the coast and the upper Nile. This map, produced barely a century after Ibn Ḥawqal's death, demonstrates the manner in which the Beja reigned supreme in the medieval Muslim KMMS cartographer's imagination for at least two centuries following the dramatic Beja visit to the Abbasid center. In this rendition, the location of Beja lands suggests that they are also seen as a crucial crossing point across sub-Saharan Africa for the control of gold-rich West Africa.

There are a few world maps from the fourteenth century onward that replicate the Ibn Ḥawqal world map of the earliest extant KMMS manuscript (Ahmet 3346) instead of the one from Iṣṭakhrī. In one such copy, housed in the Sülemaniye library (Aya Sofya 2577), which contains numerous scribal errors, the Beja are incoherently indicated as "Mafāza al-bahr wa narārīhim" (instead of the customary Mafāza al-Būja wa barārīhim), suggesting that by this point the copyist was unfamiliar with these East African tribes (fig. 9.5).[34]

In the late fifteenth-, sixteenth-, and seventeenth-century Persian renditions, the deserts between the Nile and the Red Sea are marked in the corner between Abyssinia and the Nile as "bīyābān" (the Persian rendition of *mafāza*) (fig. 9.6). They

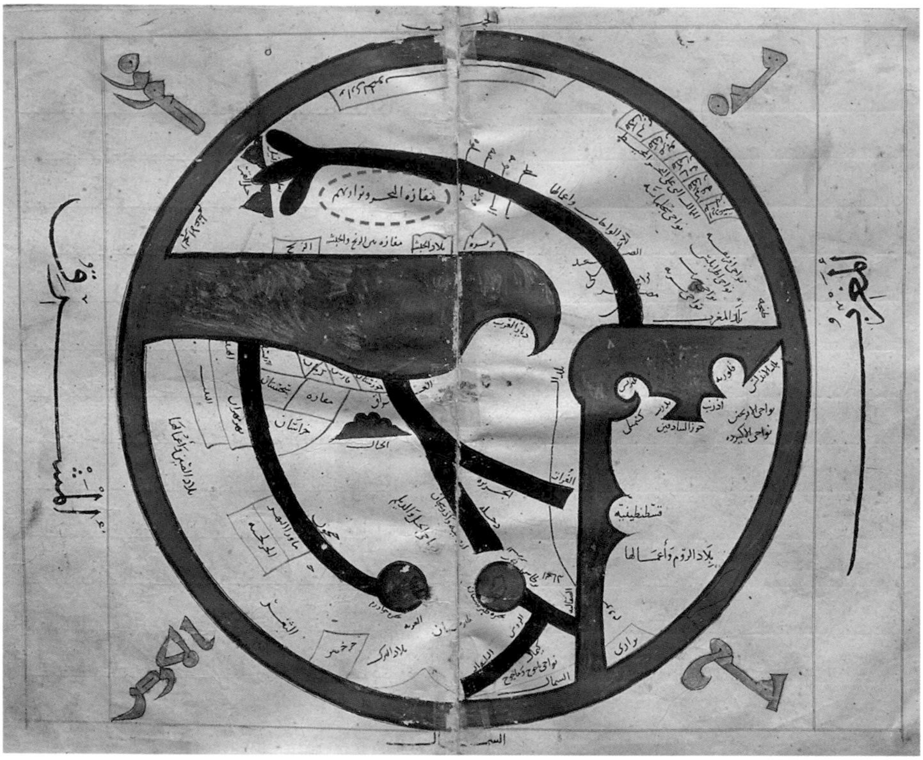

Fig. 9.5. KMMS world map based on the Ibn Ḥawqal model showing the location of the Beja (compare with figs. 4.7 and 9.4). In this copy, the Beja are incoherently labeled as "Mafāza al-bahr wa narārīhim." Listed as fourteenth century but of questionable provenance. Gouache and red and black ink on paper. Diameter 30.4 cm. Courtesy: Sülemaniye Camii Kütüphanesi, Istanbul. Aya Sofya 2577, fols. 3b–4a. Photo: Karen Pinto.

are not even assigned to the Beja. Thus the pattern of representing the Beja in a rectangle above the deserts continues but with significant orthographic variations. For instance, on the world map of Bağdat 334 dated to circa 827/1424, the Beja are marked as "Beje" (or "Beche"), suggesting that the Persian copyist had no idea who the Beja were and, instead of leaving them out and disrupting the established KMMS place-name grid, inserted his best guess. By the seventeenth-century Persian world map of Revan 1646 (fig. 9.7), however, the Beja are marked as "Wilāyāt Beja" (Dominion/Country of the Beja), indicating an information vacuum in which the Iranian Safavid cartographers were under the impression that the Beja still dominated the area eight centuries after their control had faded.[35] This informational lacuna is

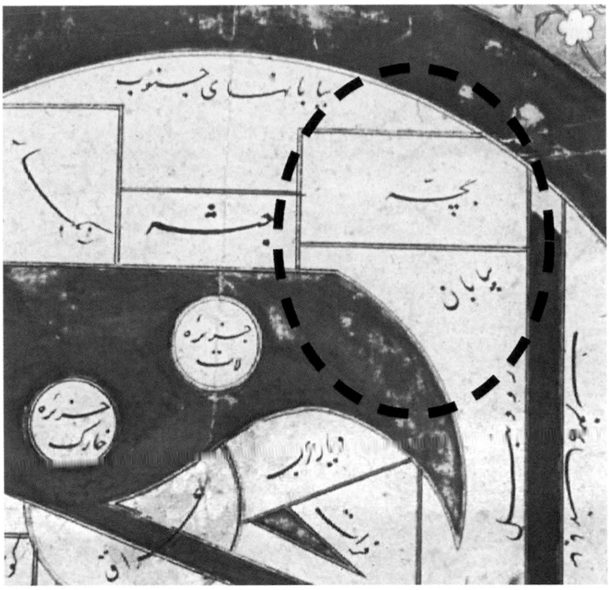

Fig. 9.6. Close-up of Timurid KMMS world map indicating the Beja as "Beje" (or "Beche"). The deserts are marked as *bīyābān* (Persian for *mafāza*) but are not assigned to the Beja. 827/1424. Courtesy: Topkapı Saray Museum, Istanbul. Bağdat 334, fol. 2b. Photo: Karen Pinto. (For a color version, see fig. 4.1.)

further highlighted in another Persian rendition from a century earlier. In the world map of Sülemaniye Aya Sofya 3156, shown in figure 9.8 and dateable on the basis of illumination style to the late fifteenth/early sixteenth century, the name of the Beja is flipped and faces south instead of north as in the other KMMS maps.

Especially intriguing is the fact that the Beja are also located on Idrīsī world maps (see fig. 3.1) along the western littoral of the Red Sea in the approximate location that they are still to be found today (see fig. 8.4). In fact, they are even to be found around the same location on the Idrīsī map in the *Book of Curiosities* (fig. 9.9).[36]

Examination of the Beja in Ibn al-Wardī–type KMMS world maps reveals surprising results.[37] The Beja are marked on all the Ibn al-Wardī maps. The manner of their delineation varies greatly from map to map. Figures 9.10 and 9.11 provide two examples. One map is from the mid-fifteenth century and the other from the late seventeenth century. In the first one, Ahmet 3012, reproduced in figure 9.10, the lands of Nubia, Abyssinia, and the Beja are cordoned off and isolated in small rectangles in the upper bend of the yellow Nile. In a distinct break from the other world maps, the deserts are labeled as belonging to the Ḥabasha (Abyssinians), not the Beja.

The second map, shown in figure 9.11, is taken from an Ottoman Turkish translation (TSMK Bağdat 179) of an Ibn al-Wardī manuscript. By contrast, in this rendition, the Beja and their lands ("Arḍ al-Bujīya," Land of the Beja) are marked prominently within a distinctive double-ringed circle. This dramatic marking on a late seven-

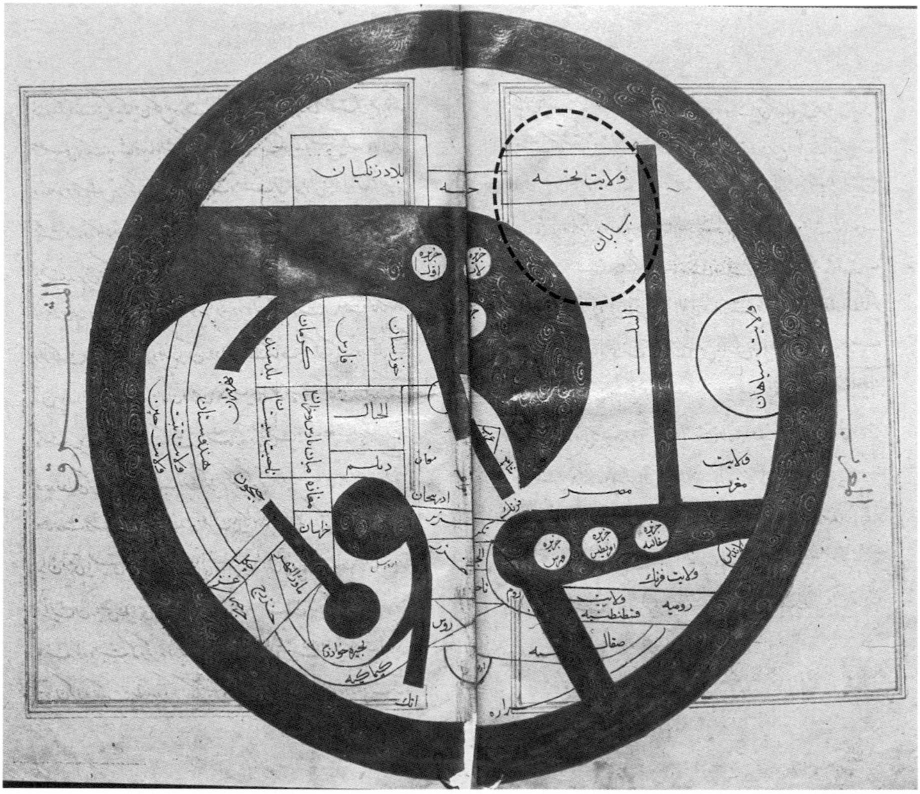

Fig. 9.7. Safavid KMMS world map with the Beja highlighted. They are marked on the map as "Wilāyāt Beja" (Dominion/Country of the Beja), which continues the Persian tradition of not assigning the deserts (*bīyābān*) to them. 1075/1664. Courtesy: Topkapı Saray Museum, Istanbul. Revan 1646, fols. 1b–2a. Photo: Karen Pinto. (For a color version, see fig. 4.14.)

teenth-century Ibn al-Wardī map can be read as an indirect reference to the Funj Empire that developed in southern Nubia in the early sixteenth century.[38]

While it is clear that the Beja must be acknowledged in the annals of Islamic history as occupying a space of significance in the medieval Islamic cartographic imagination, it is also clear that they are not represented in a spatially uniform manner. In the early KMMS maps they dominate the Muslim imagination of East Africa, but they are simultaneously denied full suzerainty over their own lands. The deserts in which they reside are not listed with their name. As the mines in their territory diminish in importance the maps are affected in two ways: either their name is garbled or their lands are assigned to some other group or no group at all, signifying Muslim control.

Fig. 9.8. Beja on Timurid KMMS map. Unusually, the name of the Beja (dashed-line box) is flipped and faces south instead of north. 15th century. Courtesy: Sülemaniye Camii Kütüphanesi, Istanbul. Aya Sofya 3156, fol. 2b–3a. Photo: Karen Pinto. (For a color version, see fig. 4.12.)

From the late fourteenth century onward, their lands are accorded more space, suggesting that their deserts were no longer of any great value or that they had fallen out of Muslim control. It is in this way that we can explain the curious phenomenon of having two prominent spaces assigned to them in the mid-fifteenth-century KMMS world maps—not as markers of importance (as would seem logical upon first read-

Fig. 9.9. Beja marked on *Book of Curiosities* al-Idrīsī map of the world. Unnoticed so far in al-Idrīsī maps are the Beja located along the western littoral of the Red Sea in the approximate location that they are still to be found today. This holds true even for the mid-sixteenth-century copy shown in fig. 3.1. Circa 12th–13th century. Gouache and ink on paper. Courtesy: The Bodleian Library, University of Oxford. Ms. Arab. c. 90, fols. 27b–28a.

ing), but as markers of insignificance. This phenomenon can be read as the cartographic equivalent of saying "Now that your deserts yield neither gold nor emeralds nor virgin territory nor pathways for caravan traffic, you may have them back."

The act of naming, of cordoning off and assigning space, as Christian Jacob puts it, is an essential act of exorcising the ghost of the unknown, of making virgin lands conquerable by providing a memory and a link in the middle of an ocean of unknown territory: "To name is a matter of real exorcism that introduces memory, knowledge, a stable mark; it is an islet of security in the ocean of what is undifferentiated because it is unnamed. Nomination is a mode of symbolic appropriation that gives a memory to virgin lands, a grid that dispossesses the space of its alterity and makes it an object of its discourse, subjected to the constraints of the linguistic reference that wants for every identified place a corresponding name."[39]

Nowhere is this more appropriate than in the case of

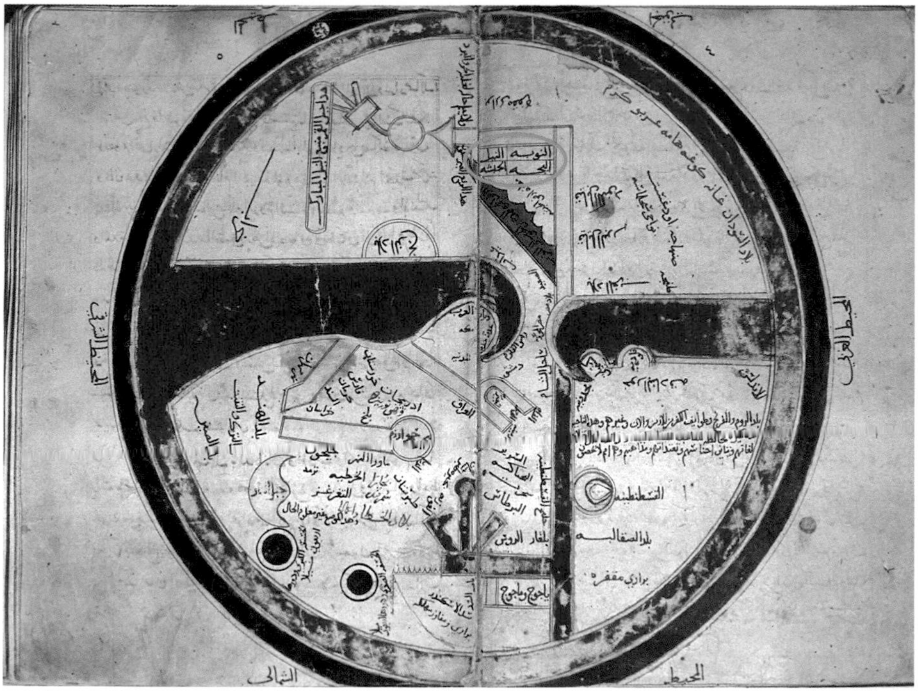

Fig. 9.10. KMMS world map in the style of Ibn al-Wardī (see figs. 9.11, 3.8, and 7.4) with the Beja site indicated by a green circle. In this map the Beja are neatly boxed and constrained with Abyssinia (al-Ḥabasha), Nubia (al-Nūba), and a space giving the name of the river Nile (al-Nīl) under the upper arm below which the boxes are nestled. 867/1462. Gouache and red and black ink on paper. Diameter 28.9 cm. Courtesy: Topkapı Saray Museum, Istanbul. Ahmet 3012, fols. 2b–3a. Photo: Karen Pinto.

Africa, which looms large and menacing on the Islamic horizon. More often than not the large empty space of the most southerly regions stands in stark contrast to the sparse scattering of identified places. Within this sea of the unknown, on its periphery, the Beja reign supreme—not only as manifestations of that *other* world out there in the wilds but also, paradoxically, as a manifestation of *self*. The Beja are, in eastern Africa at the height of the medieval period, one of the final frontiers of Islam where pagan rituals mix with Islamic practices. It is here and not in Egypt that the true mixing of boundaries between Islam, Christianity, and paganism is found. We know that Islam eventually wins out. Conversion accelerates and moves through the region past Nubia and down to the Horn of Africa. But the stepping-stone for East Africa was the Beja lands, and we need to reremember this in our history books.

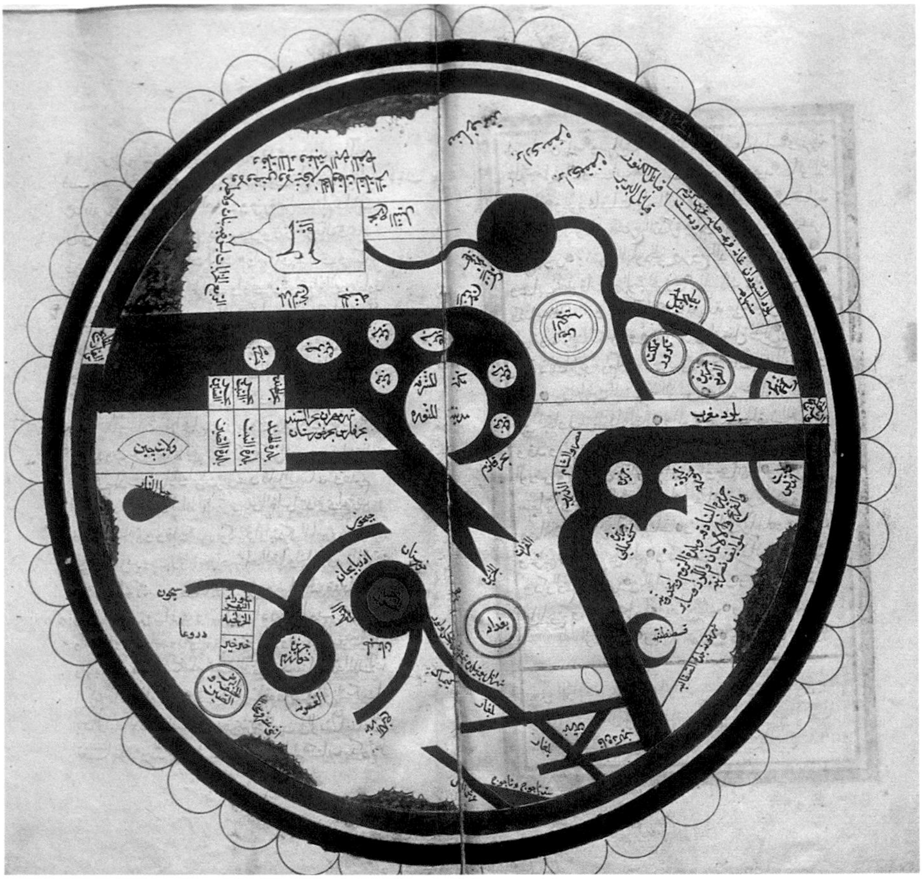

Fig. 9.11. Lands of the Beja (Arḍ al-Bujīya) marked prominently with a double-ringed circle on Ottoman copy of Ibn al-Wardī's *Kharīdat al-ʿajāʾib*. By the seventeenth century this could be an indirect reference to the Funj Empire. As in the case of the Timurid KMMS example (fig. 9.8), the name of the Beja (highlighted with a green circle) is flipped and faces south instead of north. 1092/1681. Turkey. Gouache and red and black ink on paper. 9.7 cm. Courtesy: Topkapı Saray Museum, Istanbul. Bağdat 179, fol. 52b. Photo: Karen Pinto.

In this and the previous chapter, I've shown how close study of the context of a single place-name on medieval Islamic world maps brings to light a people and place left behind in the vast sweep of time. This concludes the second of the three methodologies I employ in the study of the KMMS tradition. In the next chapter, I launch the third and final analysis in this book, a detailed look at the role of patronage in Islamic mapmaking.

10

Meḥmed II and Map Patronage

Deliberate distortions of map content for political purposes can be traced throughout the history of maps, and the cartographer has never been an independent artist, craftsman, or technician. Behind the map-maker lies a set of power relations creating its own specification. Whether imposed by an individual patron, by state bureaucracy, or the market, these rules can be reconstructed both from the content of maps and from the mode of cartographic representation. By adapting individual projections, by manipulating scale, by over-enlarging or moving signs or typography, or by using emotive colours, makers of propaganda maps have generally been the advocates of a one-sided view of geopolitical relationships.

J. B. HARLEY, "Maps, Knowledge, Power" (p. 287)

With these words J. B. Harley set in motion a new era in the critical investigation of maps. Inquiries into the intersection of power and patronage have since been the desideratum of map scholars. Patronage is a tough nut to crack. Specific information regarding sponsorship of cartographers is often difficult to determine. Because of the surreptitious nature of political propaganda, official patronage is frequently concealed from public view and not documented. Rarely do cartographers sign their maps with a dedication to their patrons.

Focusing on the sixteenth century onward, studies such as *Monarchs, Ministers, and Maps* provide inventories of kings, queens, and statesmen whose interest in maps was piqued by the copies of Ptolemy's *Geography* that were coming off the presses in Europe and by the steady trickle of knowledge of discoveries in America, Africa, and India. Decrying "maps of the medieval manner," Peter Barber cites the reign of Henry VIII as "a watershed in the history of map consciousness and map use by king and government," continuing through the reign of Edward VI. A variety of maps were commissioned for military and administrative purposes and the landed gentry began to distinguish themselves with their cartographic skills.[1] Barber asserts that it wasn't until the late sixteenth century and the reign of Elizabeth I that England saw the overt use of cartography as propaganda: "Under Elizabeth, however, cartography came most characteristically to be utilized, in a symbolic manner, in the creation of a personal, imperial imagery that was particularly associated with the queen herself, though it derived from the imagery associated with Charles V and, beyond him, with rulers in medieval Burgundy and ancient Rome."[2]

Catherine Delano-Smith and Robert Kain expand the field with their book *English Maps*. Starting with the premise that "maps are children of their times," the authors lay bare the corpus of English maps, medieval to modern. They are to be lauded for conceiving of all English maps as bound together in a continuum. Their in-depth analysis of the maps is thought provoking but shies away from solving the riddle of the patronage of English maps in the medieval period.[3]

Across the channel, Christine Marie Petto studies the French equivalent of Queen Elizabeth—Louis XIV (the Sun King)—and his patronage of maps in *When France Was King of Cartography*. Petto demonstrates how the "cult of the image" promoted by the Sun King had an impact on the demand for image-based material throughout France, especially maps. To encourage cartographers to make maps depicting provinces of France stamped with his image, Louis XIV granted the title of *géographe du roi* to many of them. This initiated a process that accelerated through the seventeenth and eighteenth centuries into a formal patronage system in which mapmakers attached themselves to the political and religious aristocracy and declared allegiance to their patrons.

Commercially based patronage is easier to document. For this reason, studies examining patronage have tended to focus on the eighteenth century onward when public demand rose and map subscriptions began. In *The Commerce of Cartography*, Mary Pedley shows how the rich and powerful of eighteenth-century France and England orchestrated the use of maps to indoctrinate young minds about the benefits of overseas colonies.[4] Matthew Edney goes a step further in an in-depth study of the British cartographical construction of India and documents the intersection of maps and conquest in *Mapping an Empire*. Edney highlights the close alliance between state and corporate patronage in the imperial project of control. Maps for the British colonists served as a portable panopticon of imperial space.[5] These studies bear out the Harley dictum that "maps anticipated empire."[6]

The increased wealth of archival material from the early modern period enables scholars like Edney, Pedley, and Petto to present statistical details of the patronage of large numbers of maps. This is not the case with studies of mapping in the Renaissance and medieval periods. Emblematic of these fields is the scarcity of documentation pertaining to patronage. We have to glean information on patronage based on marginalia notes in manuscripts, telling stamps, and serendipitous illumination connections. As a result, studies of patronage during the premodern period are few and far between.

Occasionally we have the advantage of the placement of maps in commissioned architectural spaces that are dated and attributed to verifiable patrons. Such is the case with the palaces of Renaissance Italy, in particular the map murals commissioned by Cosimo I and Gregory XIII to adorn the walls of the Guardaroba Nuova and the Vatican. In the *Marvel of Maps*, Francesca Fiorani situates these unique mapping projects in their sociocultural milieu and examines them from a three-dimensional perspective. In doing so, Fiorani seeks to understand these map cycles from the view of the patrons who commissioned them. She provides a convincing narrative for their raison d'être and provides insights into the intentions of the patrons. Fiorani's focus on map milieu resembles the kind of investigation that I undertake in this and the next two chapters regarding the Ottoman sultan Meḥmed II and the set of KMMS copies commissioned during his reign in Istanbul.

The other close parallel to my work is the recent book *The King's Two Maps* in which Daniel Birkholz puts together an exciting historical "whodunit" about the form and identity of two missing and much puzzled-over medieval maps: the mural map commissioned for the Painted Chamber at Westminster Palace in 1236 by King Henry III and King Edward I's painted cloth map (*pannus depictus*) referenced in the Wardrobe Accounts of Westminster. Birkholz takes his reader through a series of bold interpretations that elucidate Henry's and Edward's developing views on the nature of English kingship and the way medieval English maps changed to accommodate the shifting needs of the ruling class.

On the front line of medieval Middle Eastern studies, works on patronage are limited. What little exists is written in an "inventory" style of naming scholars and their sponsors. The works of Ottoman scholars, such as Adnan Adıvar's *Osmanlı Türklerinde İlim*, fall into this category. Enormously useful as references, these works resemble medieval Arabic and Persian biographical dictionaries (ṭabaqāt) or a modern-day "who's who," and do not aim to analyze and understand the nature of the patron-client relationship in depth.[7]

In *Loyalty and Leadership in Early Islamic Society*, Roy Mottahedeh shows how patronage was the glue that held Buyid society together after the Abbasid heyday. In particular, he shows how the personal vow of gratitude for benefit (niʿmah) and patronage (iṣṭināʿ and walāʾ) operated to enforce loyalties in all levels of Buyid society from the overlordship of *ghulām* (Turkish slave / mercenary) armies to clerks, viziers, guilds, and scholars.

In the context of the translation movement of early Abbasid (eighth- to tenth-century) Baghdad, Dmitri Gutas discusses the patronage of the translation movement, first by the early Abbasid caliphs and their vizier families (in particular the Barmakids), and later by independent dynastic rulers, such as the Tahirids, and scholars (e.g., the Banū Musa brothers). This was the impetus of the Greco-Arabic translation movement that grew into an elaborate system of patronage of translators. Gutas shows how the Abbasids exploited this system to create their distinct hybrid Sassanian-Greek-Islamic imperial ideology that set them apart from the Arab rulers. Gutas's work focuses on the "hard sciences" (mathematics, physics, astronomy, optics, etc.); nowhere does he discuss patronage of the carto-geographical tradition.[8]

Sonja Brentjes has picked up the mantle of elucidating the question of patronage of Islamic scientific manuscripts from the medieval to early modern period. In "Courtly Patronage of the Ancient Sciences in Post-Classical Islamic Societies," and in "Euclid's *Elements*, Courtly Patronage and Princely Education," Brentjes explores the role Islamic courts played in copying and translating scientific works such as Euclid's *Elements*. Brentjes discusses various forms that courtly patronage took, including "enforced patronage" that was the hallmark of some of the Turkic dynasties such as the Ghaznavids, Khwarazmshah, and later the Timurids. Well-known scientists were actively sought out and sometimes forced into service. Due to the instability of the latter part of the medieval period in the Middle East, scholars were forced to switch patrons frequently to secure their own safety. This resulted in scholars often dedicating and rededicating their works, showing how they were at the mercy of the rapid rise and fall of rulers. We'll see in the next chapter that Meḥmed II's court was no exception to this practice. Brentjes devotes only a few lines to two carto-geographical manuscripts. The survey nature of her investigation means that while she does an admirable job of discussing patronage of the sciences in the medieval Islamic world, she does not address one patron in depth. Nor does she look at the sociocultural impact of this patronage.[9]

No work has been done on the question of patronage of Islamic cartography. The closest parallel analysis to the one that I present in this set of chapters is by Sumathi Ramaswamy in her article "Conceit of the Globe in Mughal Visual Practice."[10] In this article, Ramaswamy analyzes representations of the Mughal emperor Jahangir and the globes that he commissioned. These images show Jahangir standing on these globes in different poses, sometimes with rival rulers. The most famous of these is one painted by his favorite painter, Abū'l Ḥasan, showing Jahangir towering over his rival the Safavid shah ʿAbbās. Ramaswamy sees the imperial Mughal penchant for depicting themselves centered on the globe as a Mughalī retort to Europe, in particular to European cartographic portrayals that diminished the importance of their lands. It was Jahangir's way of saying he was "the world-king." In this sense, Jahangir's endeavor to present himself as a caesar resembles Meḥmed's exploitation and commissioning of a set of KMMS world maps.

Meḥmed II's Love of Maps

The tastes and inclinations of Meḥmed II (r. 1444–1446 and 1451–1481), the Ottoman sultan and famed conqueror (fātiḥ) of Constantinople, are hidden behind a cloud of controversy, stemming partly from the dearth of extant material, and partly from the lens through which we view him (fig. 10.1).[11] Opinion has swung widely from Meḥmed "the earliest Ottoman Sultan of decidedly westerly leanings," to Meḥmed "the Persiante," interested in things Islamic and Middle Eastern, promoter of Persian and an avid fan of Jamīʿ and Mīr ʿAlī Shīr Nawāʾī.[12]

While both sides have received attention,[13] it is the Meḥmed of Western taste and inclination that has dominated scholarship, at the expense of his Eastern, Muslim side.[14] Nowhere is this emphasis more apparent than in the case of maps and Meḥmed's patronage of them. Meḥmed's interest in geography and cartography is well known, but focus thus far has been exclusively on his interest in and demand for European maps—particularly those of Italian origin—with no mention of his patronage of traditional Islamic geographic material.[15]

Can a series of classical medieval Islamic map manuscripts copied in Istanbul from the 1470s onward be used to shed new light on this matter? I intend to reevaluate Meḥmed's cartographical interests by inserting into the historical picture a set of Iṣṭakhrī's *Kitāb al-masālik wa-al-mamālik*, accompanied by maps, that were copied in Istanbul[16] during the latter half of Meḥmed's reign. I will lay out the possible circumstances under which the map manuscripts were copied and suggest ways in which the maps can be interpreted. I contend that this hitherto unknown KMMS Ottoman cluster reveals surprising insights into another dimension of Meḥmed: his use of propaganda.[17]

The extant record tells us that Meḥmed was particularly fond of pictures, especially portraits of himself (see fig. 10.1), and maps.[18] It has been presumed that he preferred European, particularly Italian, products.[19] Many accounts attest to his interest in Italian portrait painters,[20] the most famous of these being the Venetian Gentile Bellini who spent two years in Istanbul (1479–1481) and to whom is attributed a number

Fig. 10.1. Portrait of Ottoman sultan Meḥmed II (1430–1481) smelling a rose. Surprising depiction of the famed conqueror of Constantinople in a decidedly romantic pose appreciating the aroma of a flower—perhaps from his palace's inner garden of which he was particularly fond. Meḥmed was a fan of pictures, especially portraits of himself. Painted either by an Ottoman court painter, possibly Sinān Beg, or a pupil of Gentile Bellini. Circa 1470–1480. Gouache on paper. 39 × 27 cm. Courtesy: Topkapı Saray Museum, Istanbul. Hazine 2153, fol. 10a.

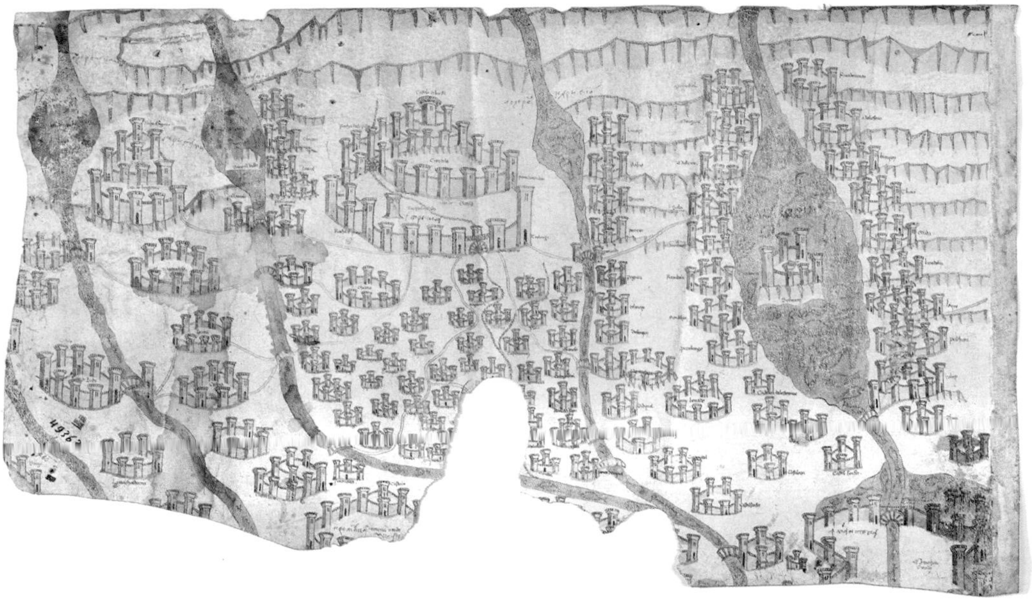

Fig. 10.2. Anonymous *Terraferma* fragment of Venice. Meḥmed II asked Bellini to prepare a map of Venice. This could be the end result. Circa late 15th century. Green, brown, and beige gouache on parchment. 39 × 67 cm. Courtesy: Topkapı Saray Museum, Istanbul. Hazine 1830. Photo: Bahadır Taşkın.

of portraits of Meḥmed and possibly a map of Venice made on the sultan's request.[21] It is reported that Meḥmed asked Bellini to prepare a map of Venice, and Ahmet Karamustafa postulates that the anonymous *Terraferma* map (fig. 10.2) showing Venice and its hinterland, dated to the fifteenth century and housed in the Topkapı Saray library, could be the final product of this request.[22]

Meḥmed's keen interest in acquiring maps of European provenance was common knowledge among his enemies and the cause of much concern, particularly among the Venetians.[23] Giàcomo de' Languschi is quoted in Zorzo Dolfin's *Cronaca* as saying about the young sultan soon after the conquest: "He possesses a map of Europe with the countries and provinces. He learns of nothing with greater interest and enthusiasm than the geography of the world and military affairs; he burns with desire to dominate; he is a shrewd investigator of conditions. It is with such a man that we Christians have to deal."[24]

Was Languschi referring to the map of the Balkans stuck into the middle of a manuscript presently housed at the Bibliothèque nationale de France in Paris (fig. 10.3)? Located on

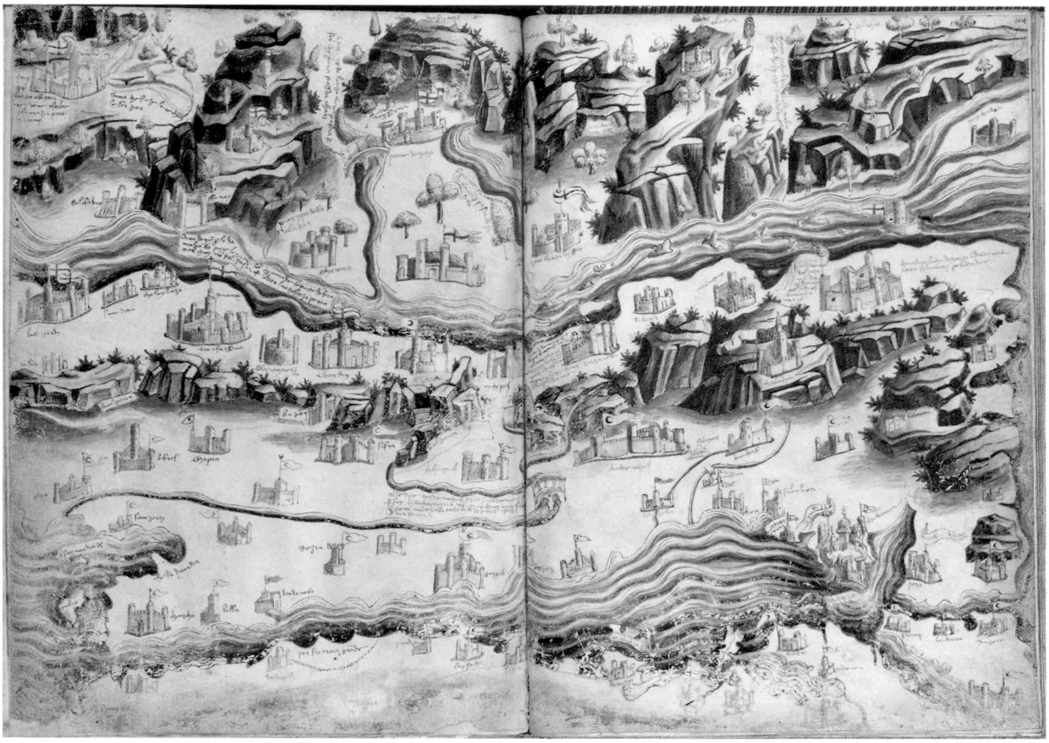

Fig. 10.3. Detail from Paolo Santinida Duccio's *Tractus de re militari et machinis bellicis*. Map facing north shows the Balkans up to the Bosphorus, including the completed fortresses of Anadolu Ḥiṣārı and Rumeli Ḥiṣārı flying Ottoman flags. Circa mid-15th century. Shades of green, brown, and beige gouache on parchment. 32 × 37 cm. Courtesy: Bibliothèque nationale de France, Paris. Cod. Lat. 7239, fols. 113b–114a.

folios 113b and 114a, it follows an elaborately illustrated military treatise (around 400 miniatures), Paolo Santini da Duccio's *Tractus de re militari et machinis bellicis* of 110 folios.[25] In a bold reevaluation, Babinger argues that the map was owned by Meḥmed and that it shows, along with the Bosphorus and parts of the Balkans, the completed fortresses of Anadolu Ḥiṣārı and Rumeli Ḥiṣārı flying Ottoman flags.[26] The manuscript contains the telltale stamps of Meḥmed's son Bāyezīd II, confirming that it was in the possession of the Ottoman sultans at least until the early sixteenth century.[27] Perhaps this is the manuscript that the fifteenth-century Italian chronicler Ducas was referring to when he noted that prior to the conquest of Constantinople, Meḥmed "immersed himself in illustrated Western works on fortifications and siege engines."[28]

In the autumn of 1472, Jakob Unrest, the Carinthian ecclesiastic, reported in his *Chronicon Austriacum* about the Ger-

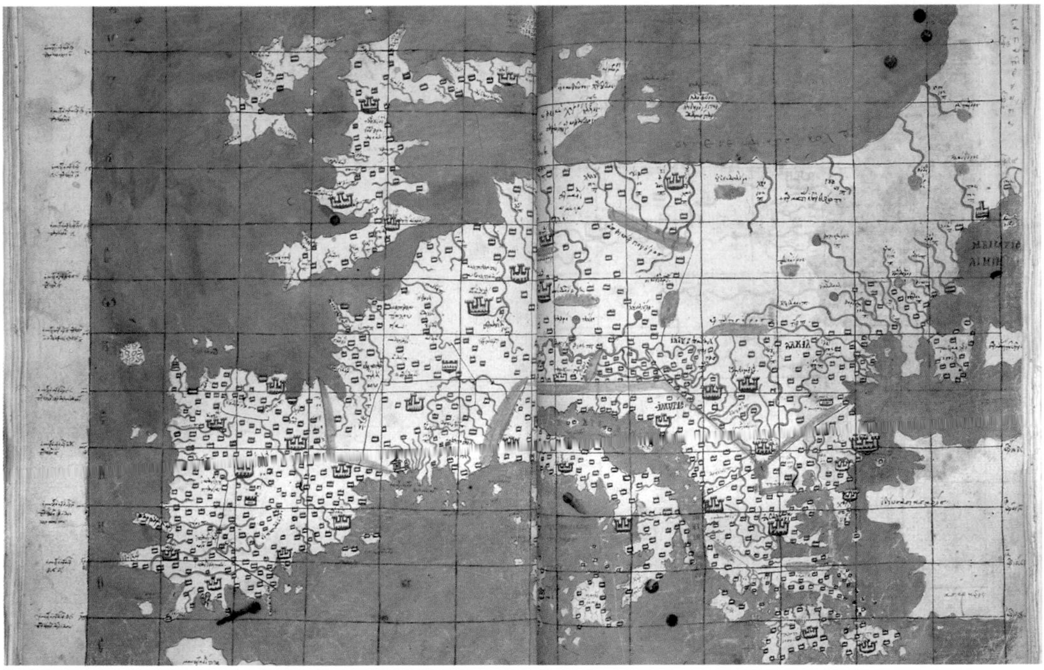

Fig. 10.4. Map of Europe from the manuscript of Ptolemy's *Geographike hyphegesis* probably consulted by Meḥmed II during the summer and autumn of 1465. It does not contain a colophon but has been identified as late fourteenth or early fifteenth century with the end date supplied by the signature of Joseph Bryennios and a 1421 date of deposit. Light-green-blue and pale-yellow and red and black ink on paper. 47 × 56.3 cm. Courtesy: Topkapı Saray Museum, Istanbul. Gayri İslam 27, fols. 83b–84a.

man provinces that "the Turkish emperor has had all the cities in these countries indicated on his map, and has received information from an exiled priest and two prelates."[29] By this time, Unrest may have been referring to the maps in one of the two Ptolemy manuscripts that Meḥmed inherited from the Byzantine imperial collection following the conquest.

Meḥmed is reported to have taken an active interest in one of the two Ptolemy-attributed *Geography* manuscripts still to be found in the Istanbul libraries.[30] According to Kritobolus, during the summer and autumn of 1465,[31] while resting his battle-weary forces, the sultan examined a Ptolemy manuscript at length.[32] Meḥmed must have been taken by what he saw because he spent months poring over the manuscript with the assistance of his polyglot friend George Amirutzes of Trebizond[33] from whom he commissioned an Arabic translation,[34] along with a single amalgamated world map that was based on the numerous regional maps contained within the Ptolemy manuscript.[35]

The map prepared by Amirutzes is no longer extant, so we can only guess at what it looked like.[36] There are, however, two Arabic translations of Ptolemy's text housed in the Sülemaniye library in Istanbul, which, although unsigned and undated, may be the end product of Meḥmed's request of the Amirutzes family (fig. 10.4).[37] A key piece of information, corroborating Kritobolus's account of Meḥmed's reading of Ptolemy's *Geography*, comes from a letter sent by George Amirutzes to Meḥmed on 25 February 1466. In it we learn that Amirutzes was planning to send the sultan a copy of his own Latin translation of Ptolemy's *Almagest* (Comparationes philosophorum Aristotelis et Platonis) with an introduction in Greek. This news, coming so soon after Meḥmed's extensive examination of Ptolemy's *Geography*, suggests both that the reading happened and that it had enough of an impact on the sultan that he desired a copy of Ptolemy's other work as well.[38]

The library of Topkapı Saray houses two additional intriguing map manuscripts. One is the controversial Benincasa Atlas, dated by some map scholars to the mid-fifteenth century, and by others to the mid-sixteenth. Marcel Destombes quarrels with Marina Emiliani Salinari's Benincasa attribution, while Leo Bagrow goes so far as to suggest that the world map in the atlas may be the missing Amirutzes world map.[39]

The other is an exquisite lapis lazuli and gold Berlinghieri atlas dated to circa 1480 (fig. 10.5). What makes this atlas so unique is that it carries on the opening folio a personal dedication from the famous Florentine cartographer Francesco Berlinghieri to Meḥmed. This is then scratched out and the atlas is rededicated to Bāyezīd II, suggesting that it was sent to Istanbul shortly after the sultan's death in May 1481.[40] That such an elaborate atlas was intended as a gift for Meḥmed would seem to confirm both Meḥmed's interest in maps, and the awareness of European map artists of this cartographic passion for which Meḥmed must have been willing to pay handsomely—hence the rich pigments.

Aside from these individual references, a large number of European maps and map manuscripts are preserved in the non-Islamic (*Gayrı İslam* or G.İ.) section of the library of Topkapı Saray. Ahmet Karamustafa notes that "even a partial listing of the non-Ottoman maps preserved in the Topkapı Sarayı Müzesi in Istanbul . . . would be sufficient to demonstrate that the Ottomans came into close contact with contemporary cartographic practice of the Latin cultural areas of the Mediterranean." Which of these hail from Meḥmed's collection and which ones he saw is a matter of dispute.[41]

That Meḥmed had a need for and a fascination with naturalistic Western representation, as manifested in maps and other "arty"-facts, cannot be denied. In this sense, one can only concur wholeheartedly with Julian Raby when he says, "The Sultan's driving imperial ambition subsumed a fascination with the art of war, with contemporary politics, with history and geography."[42] To Raby's statement I would

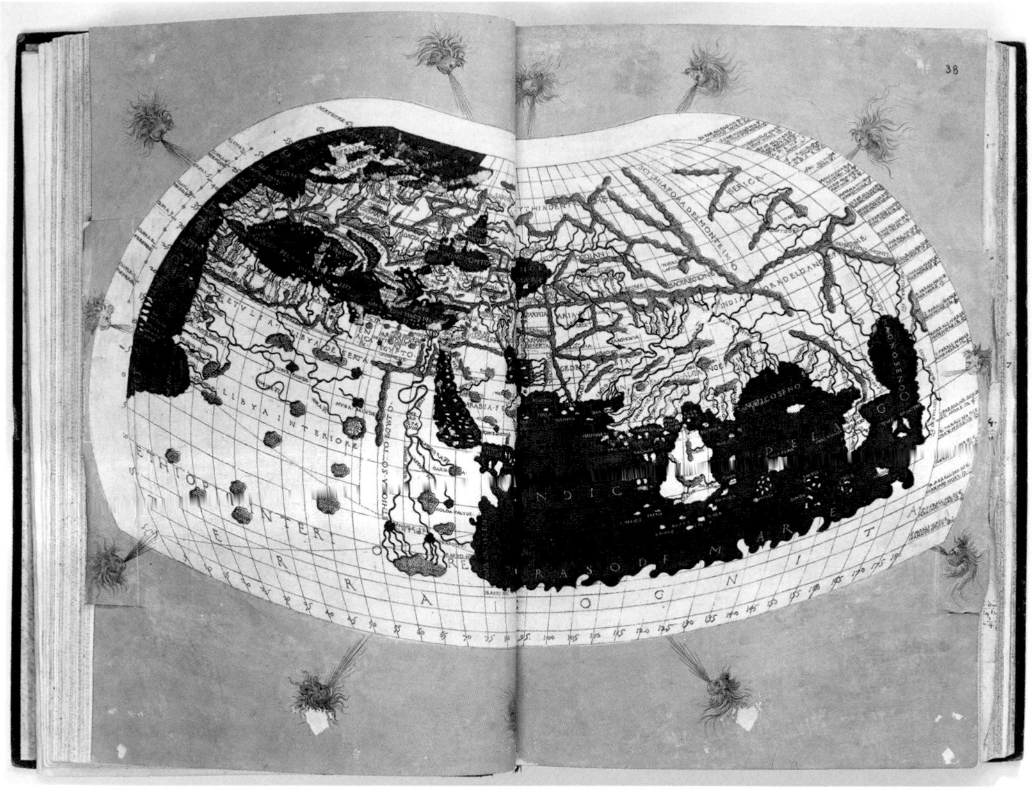

Fig. 10.5. World map from a Berlinghieri atlas. Written in the *terza rima* form, this is an early example of a hand-painted incunabula map by Francesco Berlinghieri. Circa 1480–1481. Ultramarine and gold pigments on copper engraved print. 42 × 51 cm. Courtesy: Topkapı Saray Museum, Istanbul. Gayri İslam 84, 37b–38a. Photo: Bahadır Taşkın.

add "maps" as the intersection of two of the sultan's most passionate interests: war and art.

Not only was Meḥmed an avid collector of Italian art and European maps, but as an extant sketchbook from his childhood days indicates, he dabbled in "European-style" portrait sketches himself (fig. 10.6).[43] This proves that Meḥmed was exposed to European portraiture technique as a child, and from this early age, he indulged in and expressed a passion for drawing pictures and sketching. Thus, from the extant visual record, one could argue that Fâtih was a very visually oriented person and that through maps he brought his fondness for pictures to bear on his obsession with conquest.

In the days immediately preceding the conquest of Constantinople, Ducas tells us, the sultan busied himself preparing plans and sketches of the city, the Bosphorus, and its fortifications.

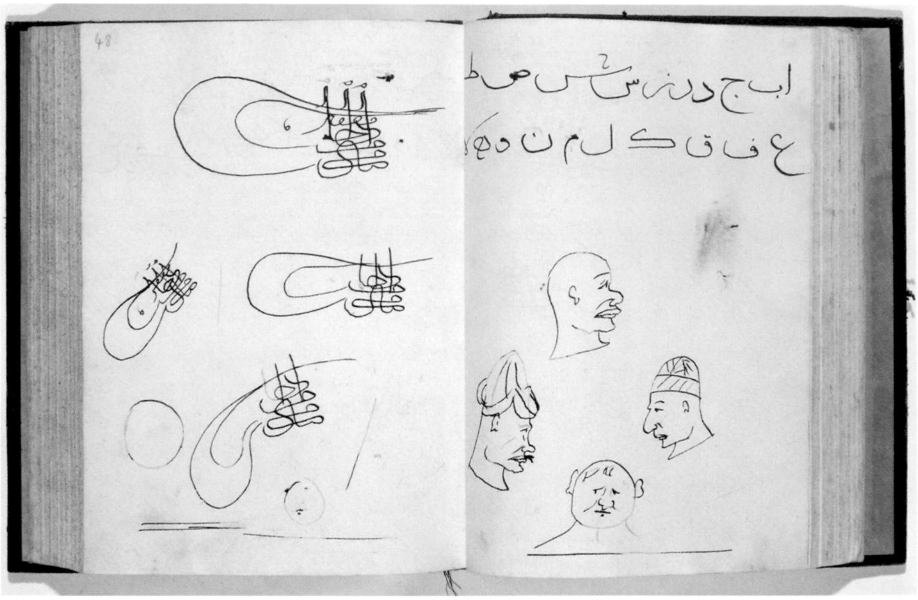

Fig. 10.6. Sketches from Meḥmed's schoolbook. Circa 1440–1445. Ink on paper. 21.5 × 38.5 cm. Courtesy: Topkapı Saray Museum, Istanbul. Hazine 2324, fols. 47b–48a.

> War was constantly on Mehmed's thoughts. Often at night he strolled through the city incognito, accompanied only by two intimates, to inform himself of the state of mind among the population and in the army. . . . He himself spent the remainder of the night and many more nights working feverishly on his plans. He sketched the walls of the city, the battle lines and outposts, the positions of the siege machines, batteries, and mines; he consulted persons familiar with the situation in Constantinople and the state of the fortifications. At this time, and perhaps earlier, *Mehmed appears to have immersed himself in illustrated Western works on fortifications and siege engines.*[44]

Meḥmed's hand in designing the two Bosphorus fortresses (Rūmeli Ḥiṣārı and Anadolu Ḥiṣārı), through which he launched the conquest, has long been posited. Corroboration for Meḥmed as architect comes from Gülru Necipoğlu's *Architecture, Ceremonial, and Power: The Topkapı Palace in the Fifteenth and Sixteenth Centuries*, through which we learn that Meḥmed not only fancied himself an artist but an architect as well.[45] The haphazard layout of Topkapı Saray is the product of the fact that Meḥmed designed his own palace atop the Golden Horn.

Support of art and architecture in the Muslim world by the ruling circles is not unusual and dates back to the late Umayyad and early Abbasid period. The Timurid rulers and amirs are particularly well known for their interest in and active support of the visual arts. Where Meḥmed stands out in Middle Eastern history is in bringing maps to bear on military strategy and in making his own maplike sketches. Other than the Umayyad general Ḥajjāj ibn Yūsuf (d. 714), governor of Khurāsān and the eastern part of the Umayyad Empire, who is said to have commissioned maps of the Daylam region (ca. 702) and Bukhara (707) for strategic purposes, we have no other references to the use of maps by Muslim conquerors for strategic purposes prior to Meḥmed. The only other reference to the use of maps by Muslim generals for strategic purposes comes from early eighth-century sources and the maps are no longer extant.[46]

Given Meḥmed's well-demonstrated interest in and possession of Western cartographic models, why do a significant series of medieval Arabic geographical treatises with stylized maps herald from Meḥmed's time? Did he sponsor the copying of this cluster? If so, why? Did he see them as a beautiful art form to be emulated? Did he shun the more mimetic European cartographic examples in his collection in favor of the nonmimetic image conveyed through the classical KMMS maps? Or did he deem public sponsorship of these manuscripts appropriate or expedient because of their stature as Islamic geographical classics? Why did he not commission copies of Western geographical material for his public libraries instead?

In order to frame answers to these questions, it is first necessary to familiarize ourselves with the hallmarks of the KMMS Ottoman cluster. In the next chapter, I look at the defining physical and artistic characteristics of this Ottoman cluster, followed by a consideration of the originating impulse that produced this unique cluster of documents.

11

The KMMS Ottoman Cluster

There are six copies of Iṣṭakhrī's *Kitāb al-masālik wa-al-mamālik* that, on the basis of their maps, illustration technique, paper, pigments, style of script, number of lines, and size, can be identified as a set originating in Istanbul circa 1474 during the period of Meḥmed II (r. 1444–1446 and 1451–1481), stretching into the reign of his grandson Selim I (r. 1512–1520). One may well ask why six copies of a manuscript copied in Istanbul should be so significant. To date we know of only thirty-five extant copies of Iṣṭakhrī's KMMS worldwide. Six out of thirty-five amounts to 17 percent of the total. That 17 percent of all the known KMMS copies were made in the same place within a fifty-year period is significant. In fact, these illustrated manuscripts are doubly significant. They also represent the only known illustrated manuscripts that are definitively dateable—on the basis of their colophons—to Constantinople/Istanbul following Meḥmed's conquest.[1]

Listed by order of date, the KMMS Ottoman cluster world maps are as follows: at the Sülemaniye library, Aya Sofya 2971a (see fig. 4.5; and for an English translation, see fig. 4.6) and Aya

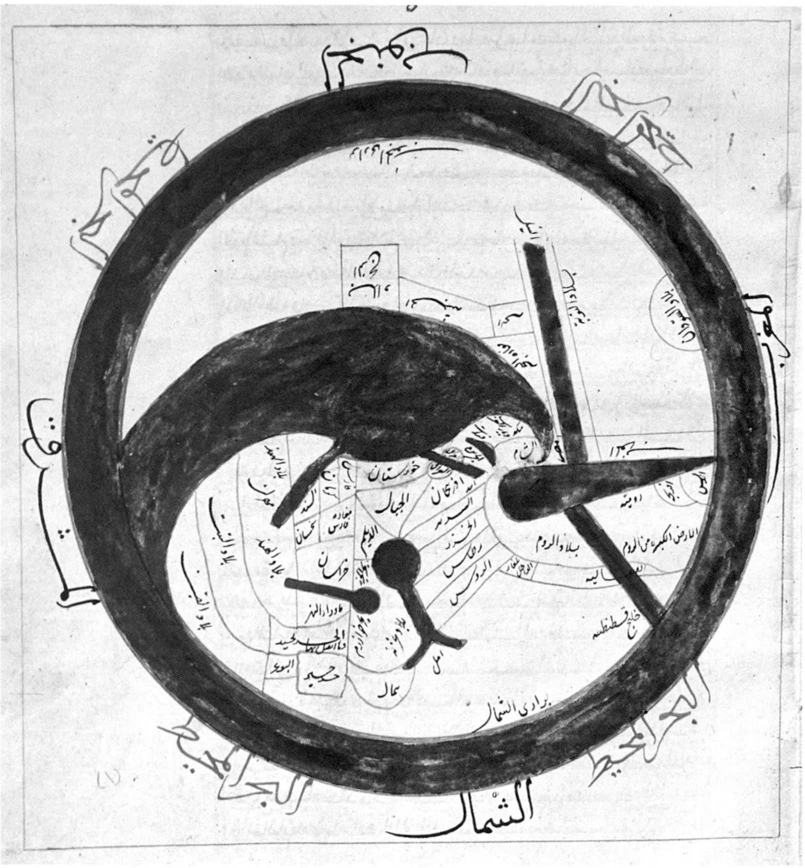

Fig. 11.1. KMMS world map of the second manuscript in the Ottoman cluster. This map has been traced from the first world map in the Ottoman cluster series: Aya Sofya 2971a (see fig. 4.5). As with most KMMS world maps, south is on top. 878/1473. Ottoman. Blue gouache and red and black ink on paper. Diameter 19.6 cm. Courtesy: Sülemaniye Camii Kütüphanesi, Istanbul. Aya Sofya 2613, fol. 3a. Photo: Karen Pinto. (For a color version, see fig. 8.1.)

Sofya 2613 (fig. 11.1); at the Topkapı Sarayı Müzesi Kütüphanesi, Ahmet 3349 (fig. 11.2); at the Dār al-Kutub, Jughrāfiyā 256 and 257;[2] and at the British Library, Or. 5305 (fig. 11.3).[3]

Four of the copies, namely, Aya Sofya 2613, Ahmet 3349, Jughrāfiyā 257, and Or. 5305, contain dated colophons,[4] which state that they were copied by the *faqīr* (poor man) Ibrāhīm ibn Aḥmad al-Sinābbī (i.e., from Sinop) in the "Bilād Qusṭanṭīniyya" ("Land of Constantinople," the name that was used by the Ottomans until "Istanbul" came into regular use), on the fifth day of the start of Shawwāl 878 (22 February 1474). Ahmet 3349 and Or. 5305 carry the date of copy (i.e., AH 878/1474 CE) but not the place.[5]

The paper of Or. 5305 carries a telltale watermark that places it in the vicinity of 1520. It provides a terminus date for the cluster and indicates that KMMS map manuscripts of this type were being copied in Istanbul between 1474 and 1520. Al-

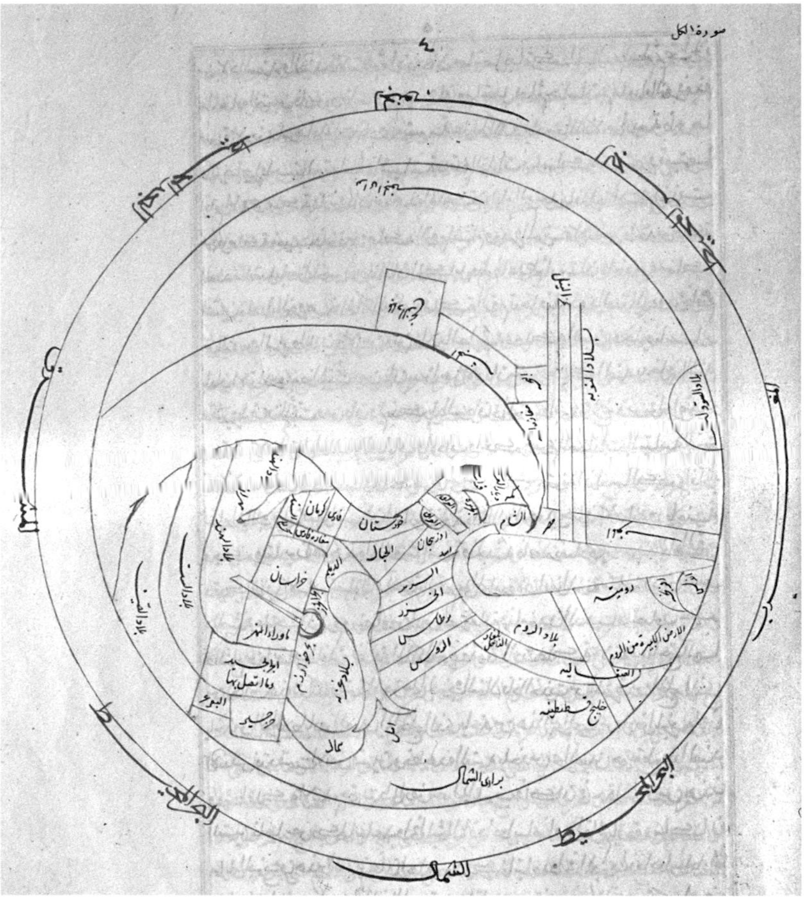

Fig. 11.2. World map from a KMMS Ottoman cluster manuscript. The maps in this manuscript are not colored. It is as if the artist never finished the project. Circa late 15th / early 16th century. Red and black ink on paper. Diameter 19.3 cm. Courtesy: Topkapı Saray Museum, Istanbul. Ahmet 3349, fol. 3a. Photo: Karen Pinto.

though Aya Sofya 2971a does not carry a colophon, I have identified it on the basis of the execution of the maps as the first manuscript in the Ottoman cluster series. The maps of Aya Sofya 2971a are freehand copies of maps in another KMMS manuscript, Ahmet 2830, which came to Topkapı Saray as part of a ransom payment that, as I argue below, ended up inspiring the Ottoman cluster series. Just like Aya Sofya 2971a, Ahmet 2830 does not contain a colophon.[6] The maps of the rest of the Ottoman cluster manuscripts appear to be tracings from the original master copy, Aya Sofya 2971a.

All the Ottoman cluster manuscripts are on thin, highly polished paper in tight, late *naskhī* script with few diacritical marks. They are strikingly similar in other respects too. Furthermore, the manuscripts of Aya Sofya 2971a, Aya Sofya 2613, Ahmet 3349, Jughrāfiyā 257, and Or. 5305 average the same number of lines per page: twenty-five. Rubricated words are exactly the same, and all the manuscripts are the same size,

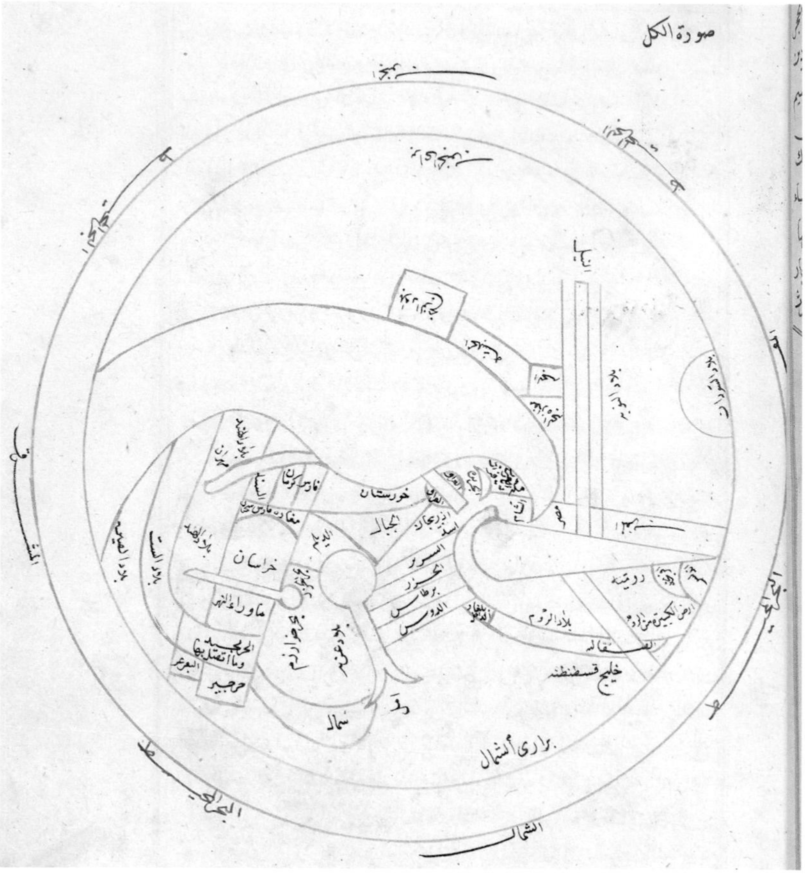

Fig. 11.3. Last KMMS world map in the Ottoman cluster series. Delicate outlines painted in gold on egg-white polished European paper. The grape-with-crown watermark (4.65 × 2.8 cm) suggests a date for the map of circa 1520. The coloring does not conform to the typically strict color-coding of other Ottoman cluster maps. The use of gold ink suggests that this final version of the Ottoman cluster was completed for a high-ranking dignitary—possibly the then Ottoman sultan Selim I (Meḥmed's grandson) who defeated the Mamluks in 1517. Gold and black ink on paper. Diameter 19.2 cm. Courtesy: The British Library Board, London. Or. 5305, fol. 3a. © The British Library Board.

approximately 32 × 22 cm. As if traced, the maps are almost identical in measurement. The world maps consistently have a diameter of approximately 19–20 cm. The same holds true for the other maps, such as the one for the Persian Gulf, which is usually 24 × 17 cm. Pigments tend to be the same: dull-blue washes for the seas, reddish browns for the mountains, and pale pinks or oxidized copper greens for the deserts, with red ink as the preferred color for the outlines of the land masses and the territorial demarcations. Ahmet 3349 and Or. 5305, the final manuscript in the series, represent a departure from the strict color code. None of the maps are painted. Instead all are outlined in either red or gold respectively. The Ottoman cluster characteristically contains single-page world maps only, in contradistinction to the double-folio world maps of other KMMS manuscripts.[7]

Aya Sofya 2971 and 2613, Jughrāfiyā 257, and Ahmet 3349 carry the seal of the Ottoman sultan Maḥmūd I (r. 1730–1754) on the opening pre-title folio, along with a laudatory dedication to the sultan and a large, thick-nibbed, black ink signature. The Maḥmūd I stamps and dedications are not significant because they are a vestige of a comprehensive inventory of manuscripts ordered by Maḥmūd I in the mid-eighteenth century. Many other manuscripts in the Istanbul collections contain this stamp.[8] On the other hand, all the Ottoman cluster manuscripts, with the exception of the Cairo manuscript (Jughrāfiyā 257),[9] carry Bāyezīd II seals on their opening and closing folios.[10] This is significant, as it confirms the dating of this cluster and suggests that Bāyezīd II (Meḥmed's son and successor who ruled from 1481 to 1512) perused these manuscripts and was familiar with the form and content of the maps. Meḥmed, unlike his successors, was not in the habit of stamping his collection. This we know from the rarity of the occurrence of his stamp. His son Bāyezīd II, however, seems to have stamped everything that came his way.

All these early Ottoman KMMS copies also contain one curious and striking "textual" anomaly, which sets them apart from other KMMS Iṣṭakhrī manuscripts. After the opening introductory paragraph, which is identical to other KMMS copies, a major relocation of text takes place: the section discussing the Arabian Peninsula is moved forward, approximately four folios.[11] It precedes the usual opening segment describing the world and its four "pillar-like" empires, of which Iranshahr, lying in the ideal fourth clime, is the mightiest. This is a significant relocation because the mother manuscript from which the text has been taken does not contain this alteration. Therefore this text relocation must have been deliberate. The creator of the Ottoman copies consciously privileged the Arabian Peninsula above Iran. One would think that there is nothing unusual in privileging the Arabian Peninsula since it is the site of the birth of Islam. However, alteration of a classical Arabic manuscript is highly unusual and an unlikely innovation of a low-ranking copyist. Only someone of high rank could have authorized it. Given the active interest that chroniclers report Meḥmed took in the day-to-day administration of his palace, and geographical texts in particular, it is likely that the order came from him. The shift in the priority of the text of the Arabian Peninsula over that of Iran is in keeping with Ottoman political ambitions in the Arab world and, especially, with their interest to take over as the caretakers of the holy cities of Mecca and Medina—which Selim I (Meḥmed's grandson) accomplished with the defeat of the Mamluks in 1517.

The Maps of the KMMS Ottoman Cluster

It is through the distinctive delineations of the maps that one can most easily identify these KMMS copies as part of a cluster. The maps of the Arabian Peninsula (fig. 11.4), for instance, have a unique overelongated shape, and the shape of the sea in the Mediterranean maps (fig. 11.6) is distinctively oblong.[12] The hallmarks of the Ottoman cluster maps are apparent when compared to other maps of the KMMS tradition (compare fig. 11.4 with fig. 11.5, and fig. 11.6 with fig.7.2).

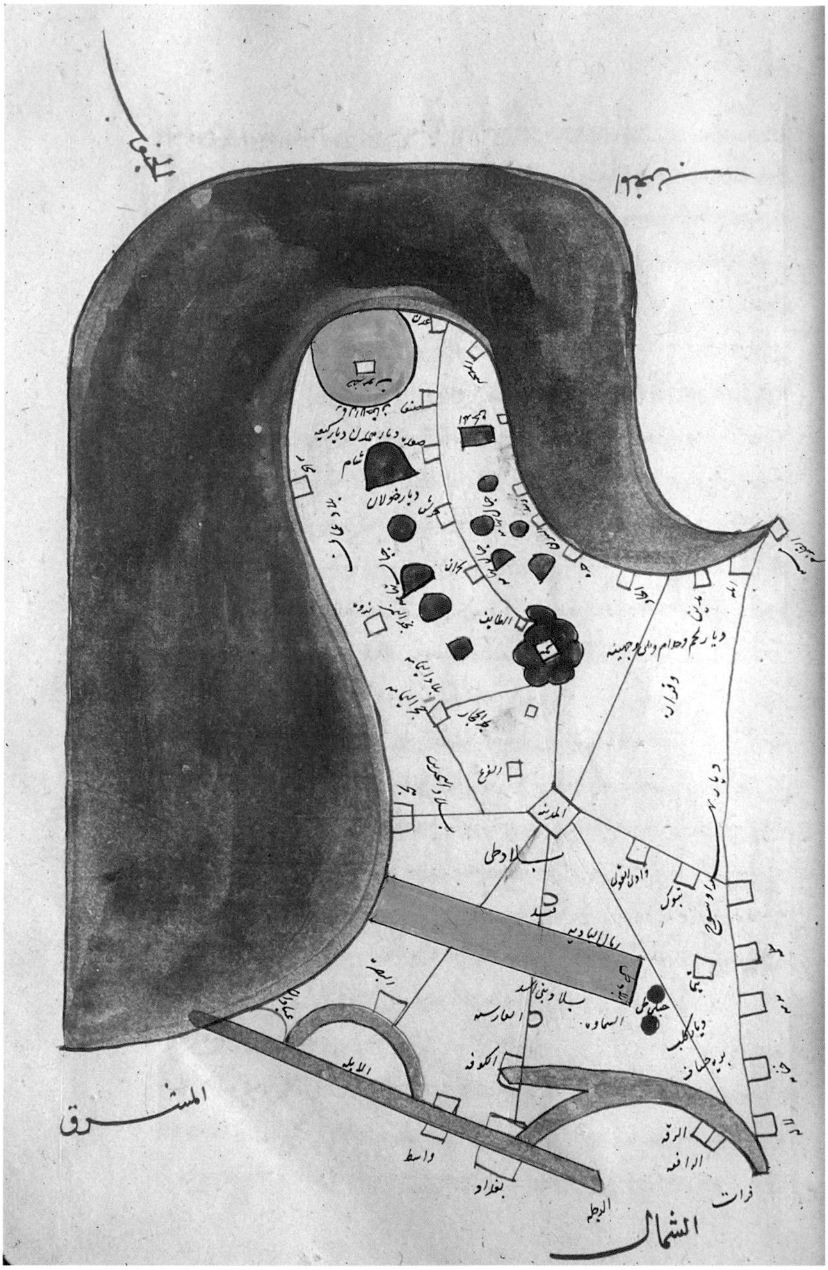

A

Fig. 11.4. Comparison of KMMS Arabian Peninsula maps in the Ottoman cluster. Version A with blue gouache comes from the earliest Ottoman cluster copy from 1473 (Aya Sofya 2971a, fol. 6a) and version B is taken from another copy in the Ottoman cluster series (Aya Sofya 2613, fol. 6a). The parallels between these two versions are immediately visible: for instance,

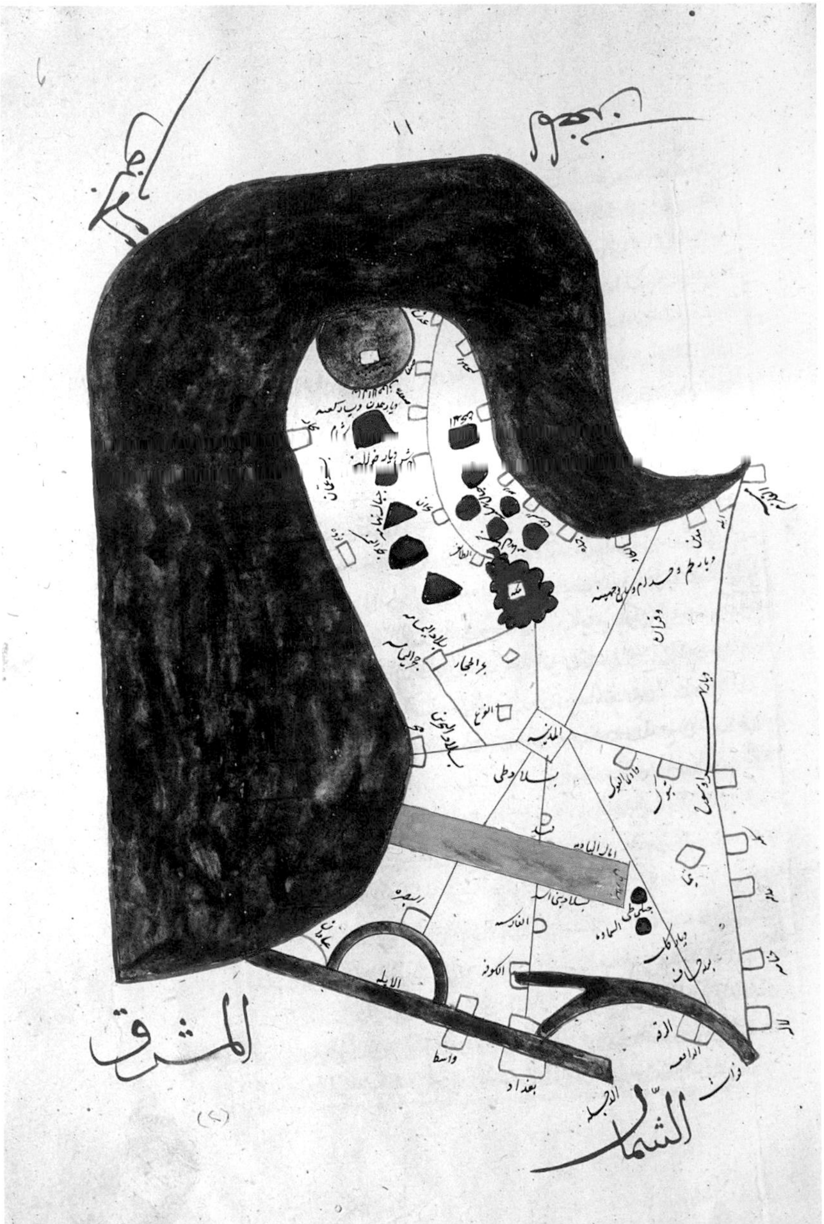

B

the similarity in the decoration around the marker for Medina, the way the lines radiate out from Medina, and so on. Only the color coding is different in places. The map from Aya Sofya 2613 is clearly a copy of the map in Aya Sofya 2971a. Courtesy: Sülemaniye Camii Kütüphanesi, Istanbul. Aya Sofya 2971a, fol. 6a, and Aya Sofya 2613, fol. 6a. Photos: Karen Pinto. (Compare with fig. 12.2a.)

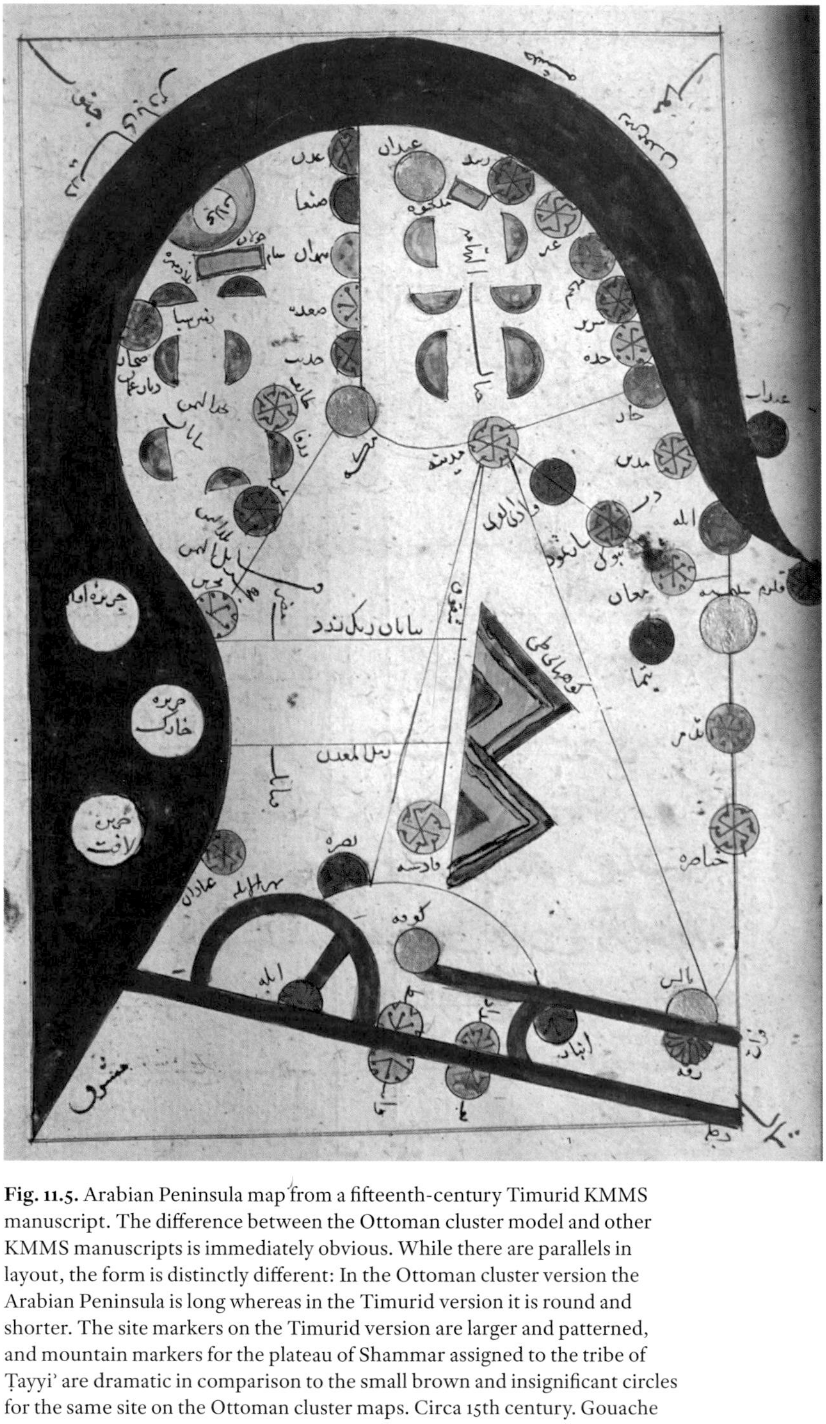

Fig. 11.5. Arabian Peninsula map from a fifteenth-century Timurid KMMS manuscript. The difference between the Ottoman cluster model and other KMMS manuscripts is immediately obvious. While there are parallels in layout, the form is distinctly different: In the Ottoman cluster version the Arabian Peninsula is long whereas in the Timurid version it is round and shorter. The site markers on the Timurid version are larger and patterned, and mountain markers for the plateau of Shammar assigned to the tribe of Ṭayyiʾ are dramatic in comparison to the small brown and insignificant circles for the same site on the Ottoman cluster maps. Circa 15th century. Gouache and black ink on paper. 19.5 × 14.5 cm. Courtesy: Sülemaniye Camii Kütüphanesi, Istanbul. Aya Sofya 3156, fol. 8a. Photo: Karen Pinto.

It is through the world maps of this cluster that one can best understand the intentions of the patron of this manuscript series. In keeping with the thrust of this book this chapter focuses on an analysis of the world maps of the Ottoman cluster. All the seas and rivers on these world maps have distinctive trademark shapes (see figs. 11.1, 11.2, and 11.3, and 4.5 and 4.6).[13] In the northwestern quadrant of the image (i.e., the lower right-hand side), one sees the elongated, tear-shaped Mediterranean, with two outstretched arms representing the Nile, perpendicular on one side, and the Bosphorus, aligned at a forty-five-degree angle, on the opposite shore. Together they give the Mediterranean a bulging cross-like appearance. From the eastern end of the map, the combined Persian Gulf/Indian Ocean sweeps in from the left of the image, threatening to hook onto the Mediterranean. One of the most distinctive features of the world maps in this Ottoman series of KMMS is that the Persian Gulf appears almost on the verge of breaking through.

As with other KMMS-type world maps, the landmasses are left uncolored, save for the red ink outlining. Consequently, it is the blueness of the water surrounding the three continents that causes the landmasses to emerge from the page. The stark whiteness of the landmasses also serves to heighten the visual conflict between the threatening Persian Gulf and the placid Mediterranean. The African landmass, which sweeps across the top of KMMS world maps, has a pronounced dagger or crescent shape. Below, as if sheltered by Africa, is the double-humped landmass that represents Asia. Off in the lower right corner of the image is the triangular European landmass marooned between the Mediterranean Sea and the Encircling Ocean. The stark, unadorned simplicity of the maps and the dramatic shapes of the lands and seas are among the most visually striking features that distinguish the world maps of this series.

There is one final but crucial aspect that individualizes all the maps in this Ottoman cluster, namely, the unsophisticated hand and lackluster painting technique. The outlines of the maps contain within them a crudeness of hand that shows through in uneven outlining, smudges, and lines that sometimes run one into another. The paintwork, too, is patchy and unevenly applied, while the colors employed tend to be watered-down and pale. The color palette is limited. In the world maps it is restricted to the blue of the sea, the white of the paper, and the red of the outlines and rubrication. In the regional maps, such as the Arabian Peninsula and Mediterranean maps, the blue and white monotony is broken by an occasional red-brown mountain and pale pink or oxidized-copper-green desert (see figs. 11.4 and 11.6). The painting technique of this cluster of map manuscripts raises a crucial and controversial issue: the question of the artistic caliber of Meḥmed's Istanbul atelier.

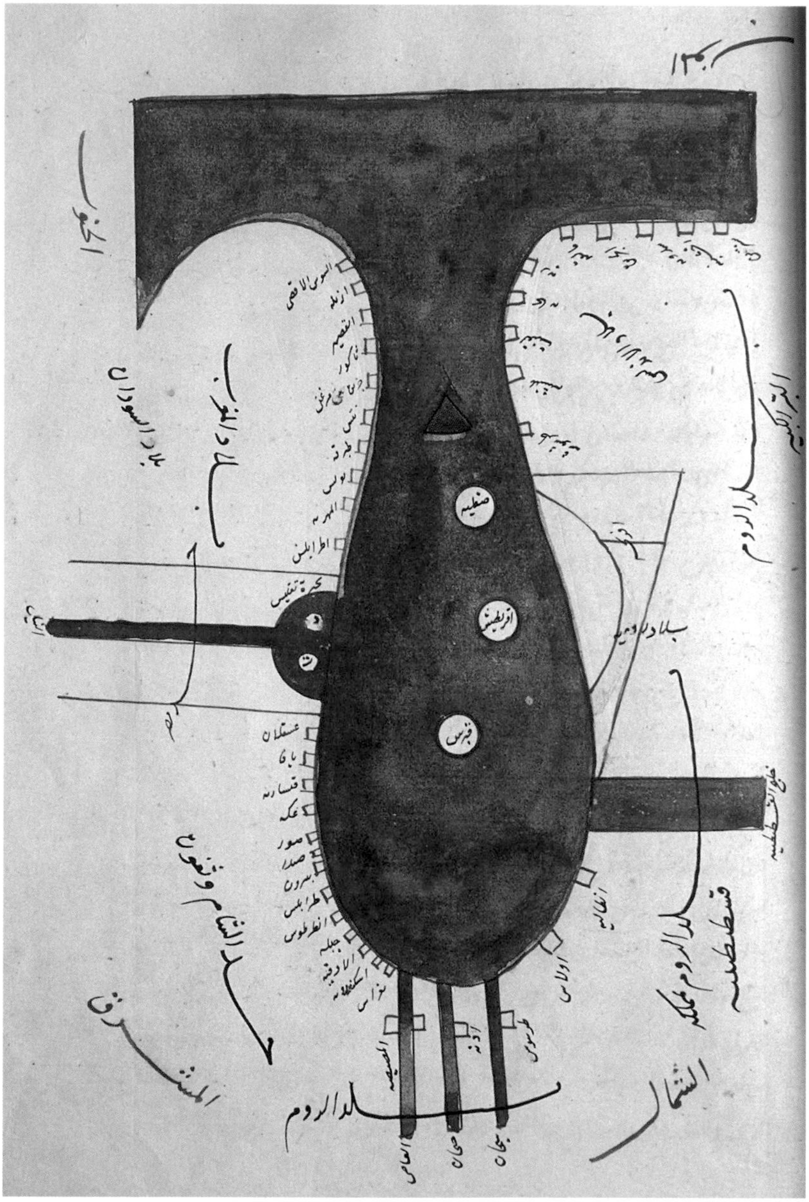

A

Fig. 11.6. Comparison of Mediterranean maps in the Ottoman cluster. That they are part of the same stemma is clear. The Mediterranean has a characteristic elongated shape that is distinctive of the Ottoman cluster with small island markers for Sicily, Crete (always off center), and Cyprus, capped by an almost insignificant triangular marker for the mythical Jabal al-Qilāl that typically guards the mouth of Mediterranean in KMMS maps. (For a dramatic and more common example of Jabal al-Qilāl in KMMS maps, see fig. 7.2) Unlike the hesitant execution of the earliest version from Aya Sofya 2971a (*A*), the version from the last Ottoman cluster copy housed at the British

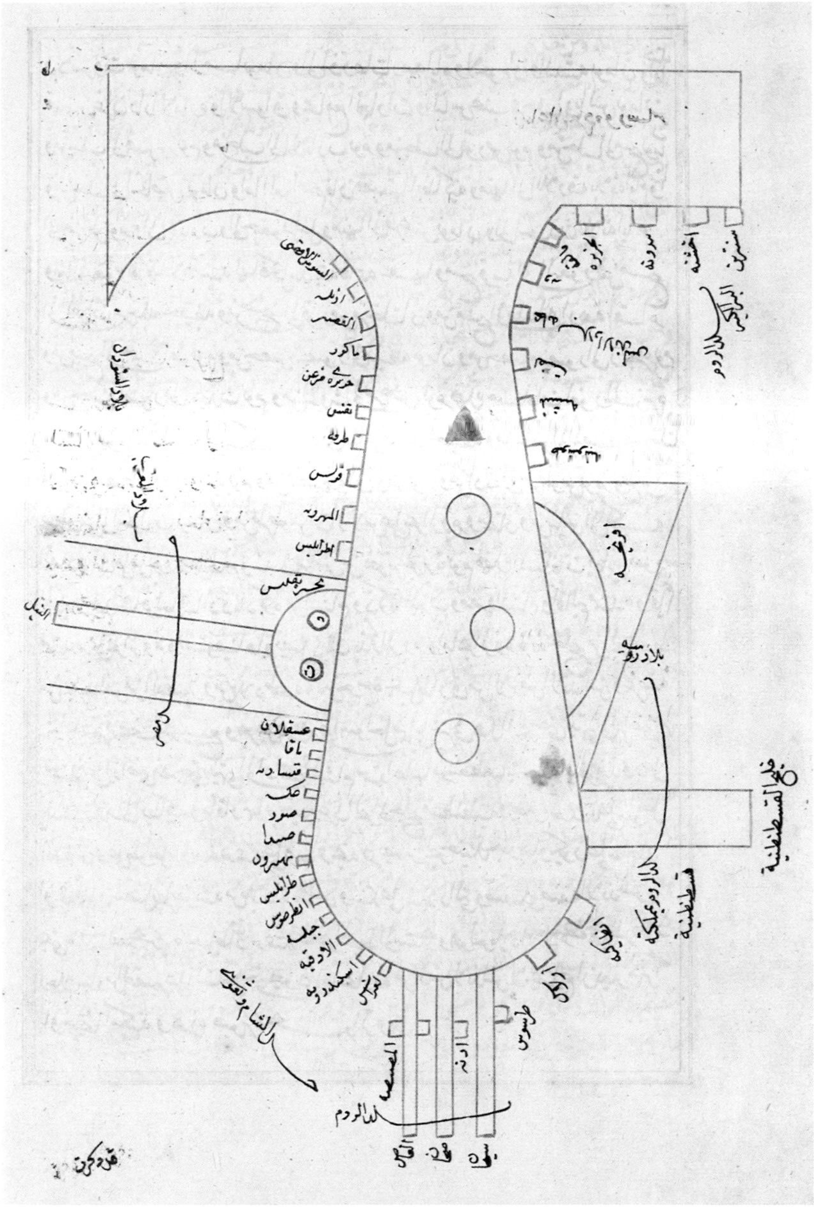

B

Library (Or. 5305) (*B*) displays a confidence in execution that comes with the familiarity of a well-known and often-copied form. *A*, 1473–1520. Blue gouache and red and black ink on paper. 23 × 15 cm. Courtesy: Süleymaniye Camii Kütüphanesi, Istanbul, Aya Sofya 2971a, fol. 24a. *B*, 1473–1520. Gold ink on paper. 23 × 15 cm. Courtesy: The British Library Board, London. Or. 5305, fol. 20b. © The British Library Board. (Compare with fig. 12.2b, the Mediterranean map of the mother manuscript that spawned the Ottoman cluster.)

Meḥmed's Istanbul Atelier

An obvious question is the extent to which the painting style of these Ottoman cluster maps matches the painting style of other illustrated manuscripts from Meḥmed's Istanbul atelier of the mid to late fifteenth century. Here we are faced with a dilemma, because the question of the quality of Islamic painting during the period of Meḥmed II is an ongoing matter of considerable debate. Research on the subject has been hampered by the lack of illustrated manuscripts containing colophons, in particular the large number of lavishly illustrated Turkman Aqqoyunlu manuscripts without colophons that have been erroneously attributed to Meḥmed's Istanbul atelier. I shall return to the discussion of these Aqqoyunlu manuscripts when I discuss one of them, Ahmet 2830, as the source of the Ottoman cluster. For the time being, I remind the reader that these KMMS Ottoman cluster manuscripts are the only illustrated manuscripts with a colophon indicating that they were produced in Istanbul. In this sense, they are intriguing not only from the point of view of cartographic history and what they can tell us about the Ottoman sultan Meḥmed II, but also from the point of view of art history and what they can tell us about the art of the period.

When it comes to placing the Ottoman cluster illustrations in the artistic milieu of the period, only three extant illustrated manuscripts can be relied on. Each of these manuscripts fits with the painterly style of the Ottoman cluster and provides a clear indication of the caliber of the palace copyists and scribes. The relevant manuscripts were produced in several major Ottoman centers. One such place was Amasya, in northeastern Anatolia, to which the early sultans sent their young sons to safeguard them from popular uprisings and troop mutinies in the capital. Murād II, Meḥmed's father was born there, and Meḥmed and his mother went to Amasya in 1434 when he was two years old. Another one of these centers was Edirne (Adrianople), the capital of the Ottoman Empire before Meḥmed conquered Constantinople.

It seems logical to presume that when Meḥmed moved his capital to Istanbul, he also brought the copyists and artists from his palace in Edirne to his new palace in Istanbul. Thus these three manuscripts provide an indication of the caliber of the sultan's atelier. The earliest of these is the *Dilsizname* [Book of the Mute] by Badīʿ al-Dīn al-Tabrīzī, completed in Edirne in 1455–1456.[14] The second is the *Külliyat-i Kātibī* [Complete Works of Kātibī], a manuscript that has been identified by Filiz Çağman as stylistically related to the *Dilsizname*.[15] The third is the only known illustrated copy of the *Cerrāḥiyyetü'l-Ḥāniyye* (Imperial Surgery) by Şerefeddin (Sharaf al-Dīn) Sabuncuoğlu. It is a medical manuscript executed during the author's lifetime and completed, according to the colophon, in Amasya in 1465 (fig. 11.7).[16]

The illustrations in these three manuscripts share the painterly characteristics of the Ottoman cluster maps. All are in a rough, unsophisticated hand, with thick lines and a flat, gouache style of painting in dull colors with a preponderance of geomet-

Fig. 11.7. Dental work on a patient from *Cerrāḥiyyetü'l-Ḥāniyye* (Imperial Surgery) by Şerefeddin (Sharaf al-Dīn) Sabuncuoğlu. Note the similarity of pigments and painting technique between the illustrations of this manuscript and the maps of the Ottoman cluster. This is the only known illustrated copy completed during the author's lifetime. 1465. Amasya, Turkey. Gouache and red and black ink on paper. Courtesy: Bibliothèque nationale de France, Paris. Suppl. Turc. 693, fol. 29a.

rical (particularly hexagonal tile) shapes. The illustrations in the *Cerrāḥiyyetü'l-Ḥāniyye* are striking in the simplicity of their layout and the flat tone and limited color scheme of dull blues, olive greens, pinks, and oranges, which together with the red outlining, closely parallel the pigments used in Ottoman cluster maps. The most distinctive color in the palette of these early Ottoman miniatures appears to have been a "dusty rose" pink, and it is precisely this shade of pink that is employed to demarcate the deserts in the maps of Aya Sofya 2971a. (Compare fig. 11.7 with figs. 11.4 and 4.5.) This "dusty rose" pink features in other manuscript of the same period, such as Topkapı Saray's *Külliyat-i Kātibī*[17] (fig. 11.8) and an illustrated copy of Amīr Khusraw Dihlawī's *Khamsa*, with a colophon indicating that it was made in Istanbul on Ramadan 903 (April 1498) during the period of Meḥmed's son and successor, Bāyezīd II.[18]

Late Byzantine manuscript illumination provides another source of comparison for painting styles. Once again the parallels with the Ottoman cluster are notable: the rough and patchy

Fig. 11.8. The sultan with his retinue at a party from *Külliyat-i Kātibī*. Note the "dusty rose" pigment of the floor in the foreground that is found in other manuscripts of the period and on the maps of the earliest copy of the Ottoman cluster, Aya Sofya 2971a. (Compare with fig. 11.4.) Circa 1480. Gouache on paper. 8 × 6.4 cm. Courtesy: Topkapı Saray Museum, Istanbul. Revan 989, fol. 93a.

Fig. 11.9. "Leo Offering His Bible to the Virgin," frontispiece of the Bible of Leo Sakellarios. Shows the rough and patchy painting technique that set apart late Byzantine manuscript illustration from earlier Byzantine work. Circa 940. Gouache and gold on vellum. 41 × 27 cm. Courtesy: Biblioteca Apostolica Vaticana. Ms. Vat. Reg. gr. I, fol. 2b.

painting technique and the choice of pale, dull pigments that set apart late Byzantine manuscript illustration from earlier Byzantine work (fig. 11.9). Like the maps of Aya Sofya 2971a and Aya Sofya 2613 (compare fig. 11.9 with figs. 4.5, 8.1, 11.4, and 11.6), Byzantine miniaturists employed flat opaque gouache washes in which dull blues and browns abound, with the occasional burst of red and purplish pink, and plenty of gold.[19] In his article on Ottoman painting in the fifteenth century, Ernst Grube notes the structural similarities between early Ottoman miniature painting and Byzantine mural painting:

Yet in Ottoman Anatolia, an immediate contact with Byzantine painting, both in the form of manuscript illustration and of monumental painting, is of course to be expected, and it would appear from Ahmedi's *Iskandar-nama* that Byzantine painting fascinated and inspired early Ottoman painters for their own work. It can be shown, I believe, that the contact with Byzantine painting in Anatolia was one of the major experiences for young Ottoman painters and that many of the elements that went into the making of the early Ottoman style can be traced back to suggestions offered by Byzantine painting, especially of the 14th century.[20]

Much work remains to be done to solve the puzzle of the quality of the palace atelier in Constantinople / Istanbul during Meḥmed's time, and one can only hope that specialists in the field will take into consideration the painting techniques of the maps of the KMMS manuscripts. In keeping with the rigid boundaries of art history, there has been a tendency to ignore those illustrated manuscripts that fall under the rubric of "scientific." As a consequence the illustrations of one of the crucial manuscripts from the period, the *Cerrāḥiyyetü'l-Ḥāniyye,* have yet to be studied in depth. It is not surprising, then, to find that the KMMS map manuscripts have also been overlooked in the general picture of the quality of the early Ottoman Istanbul atelier.

The evidence suggests that there was continuity between the capabilities of the artists of Edirne and Amasya and those in Istanbul. The shift of the capital from Edirne to Istanbul may well have meant the transfer of artisans from one court to another. The colophonless *Külliyat-i Kātibī,* which matches the Edirne painting style, could just as easily have been produced in the early days of Meḥmed's postconquest Istanbul palace atelier. I find Esin Atıl's 1973 evaluation of the matter compelling when she suggests that "the crude and provincial quality seen in the paintings" in the *Dilsizname* and *Cerrāḥiyyetü'l-Ḥāniyye* manuscripts,

> is indicative of the fact that the painters, having no prototypes to follow, had to rely upon their own resources to illustrate contemporary texts. *Regardless of the poor aesthetic quality of the two manuscripts from the reign of Mehmed II,* the artists must be credited with originality and creativity, if not with outstanding performance, for having made the first local attempts at illustrating manuscripts. It is rather significant that these two Ottoman works are not derivative or based on traditional examples. One could speculate that perhaps the imperial libraries at this time were devoid of, or had a limited amount of, illustrated books and the painters could not find the appropriate models to follow. *It is in the reign of the ensuing Sultan, Bayezid II, that we find strong Persian and Turkman influences, which indicate that the royal libraries had begun to expand and to include a variety of works from the East.*[21]

The *Dilsizname,* the *Kulliyat-i Kātibī,* and particularly the *Cerrāḥiyyetü'l-Ḥāniyye* represent not only the illustration style and capabilities of the miniaturists of Edirne

Fig. 11.10. *Taqwīm-i ta'rīkhī* (Historical Calendar) prepared during the reign of Meḥmed's father, Murād II (r.1421–1444 and 1446–1451) and possibly the earliest extant Ottoman world map. The presentation is distinctly different from the KMMS model. 5 Rajab 833/30 March 1430. Gouache and gold, red, and black ink on paper. Diameter 24.5 cm. Courtesy: The Chester Beatty Library, Dublin. Ms T402, fols. 12b–13a. © The Trustees of the Chester Beatty Library.

and Amasya but indirectly those of the Istanbul atelier as well.[22] The painting style depicted by the Ottoman cluster confirms this through its colophon and provides us with a window onto the capabilities of postconquest ateliers, especially that of the court.[23] The Ottoman cluster suggests that the postconquest ateliers were still in their infancy and that they were in no way close to the sophistication of the Timurid or Turkman ateliers of the time, in contradistinction to the prevailing Ottoman art historical scholarly view.[24]

Identification of an Ottoman cluster of KMMS manuscripts raises questions about their origins and sources and also about the extent to which the maps represent a type that was circulating within Ottoman Anatolia prior to the last quarter of the fifteenth century or whether the impetus for their creation came from outside. It could be that KMMS manuscripts were present in Anatolia during the time of Meḥmed's predecessors, but evidence for this is lacking.

A single reference from within Ottoman territory before Meḥmed II concerns a classical Islamic world map in a *taqwīm*, or calendar, made during the reign of Meḥmed's father, Murād II (r. 1421–1444 and 1446–1451), *Taqwīm-i ta'rīkhī* [Historical Calendar] (fig.11.10).[25] This map does not resemble the KMMS world map in any way, and no other references to Islamic map manuscripts in the Ottoman context prior to the Ottoman cluster are known. The silence implies that the KMMS-type manuscript with its complement of twenty-one maps was not known in Ottoman circles prior to Meḥmed II's time. Following extensive KMMS manuscript examination, it is my deduction that the Ottoman cluster was generated by something quite different, a manuscript that came to the Ottoman palace as a gift from the East. This manuscript, Ahmet 2830, is still housed in the Topkapı Saray library in Istanbul.

In the next chapter, I examine this remarkable document and argue that it is not only fit for a sultan but particularly well suited to serve a sultan's imperialistic ambitions and public relations requirements.

12 Source of the Ottoman Cluster

In the early 1470s Meḥmed was presented with an exquisitely illuminated copy of Iṣṭakhrī's *Kitāb al-masālik wa-al-mamālik*. TSMK Ahmet 2830 is a beautiful manuscript with twenty-one maps—expressions of the medieval Islamic KMMS map tradition at an artistic high. The rich lapis lazuli and elaborate gold ornamentation alone bespeak the intention of presentation to a sultan. On highly polished, thin paper, the maps are decorated in a delicate gold filigree pattern of foliated leaves, diaphanous fish, and the occasional snake with the tiniest hint of a bright fuchsia pink (figs. 12.1 and 12.2).[1]

The manuscript opens with an elaborate double folio displaying two *shamsa*s, or medallions, which contain a laudatory dedication to Meḥmed II (fig. 12.3). These very fine opening medallions are followed by two elaborately illuminated opening folios with the name of the book, *Kitāb ṣūrat al-aqālīm al-sab'a* [Picture Book of the Seven Climes], and the author delicately inlaid in white floriated *thulth* script on a foliated backdrop of knotted leaves and flowers (islim-i bargi and band-i rūmī) in a brownish gold, set against a dark

Fig. 12.1. Close-up view of Aqqoyunlu KMMS map of the world from the mother manuscript that spawned the Ottoman cluster. The lavish lapis lazuli and gold pigments are commensurate with an expensive production intended for presentation to Sultan Meḥmed II as the opening medallions confirm (fig. 12.3). Circa mid to late 15th century. Lapis lazuli (ultramarine) gouache with gold and black ink on paper. Diameter 26.4 cm. Courtesy: Topkapı Saray Museum, Istanbul. Ahmet 2830, fol. 4a. Photo: Karen Pinto.

lapis lazuli background with evenly spaced blue sprays projecting into the margins (fig. 12.4).[2]

The artist responsible for this magnificent artwork is unknown. There is little to identify the provenance of TSMK A2830 since a colophon is lacking and the style of the illumination is mixed. Fehmi Edhem Karatay, in his 1966 catalogue of manuscripts in Topkapı Saray library, identifies TSMK A2830 simply as a copy prepared for Meḥmed, without attribution to a particular atelier.[3] Tucked inside the manuscript is a little card, of the kind used in exhibitions, which also mentions the dedication to Meḥmed and goes beyond Karatay to posit a date of 1460–1470 for the manuscript and to say it was executed in the atelier of Topkapı Saray.

Whoever prepared the reference card now in TSMK A2830, however, was making his or her evaluation based solely on the opening double-folio ex libris dedication to Meḥmed. That person was not alone in this approach. Tahsin Öz, for instance, has also attributed a number of manuscripts to Istanbul solely on the basis of their opening dedications to Meḥmed.[4] But one cannot specify the place of origin of a manuscript simply on the basis of the person to whom it is dedicated. There must be confirmation from the colophon or from an established painting style in order to situate a manuscript. Neither of these factors obtains in the case of TSMK A2830.

As already noted, the quality of Islamic miniature painting in postconquest Istanbul remains an open question. There are two firmly dated Ottoman manuscripts with colophons, the *Dilsizname* of Edirne and the *Cerrāḥiyyetü'l-Ḥāniyye* of Amasya, to which we can now add, for Istanbul, the illustrated geographical manuscripts of the Ottoman cluster. These collectively suggest a style that is rough and evidently in an evolutionary phase, a style that is far from the sophistication and maturity of that displayed in TSMK A2830.

The Timurid Tabrīzī-Herātī school stands out as one possible provenance for TSMK A2830. The delicate gold fish pattern is reminiscent of the golden sprays of animals amid flora that adorn many a garment in Herātī miniatures. One of the best examples of this comes from a copy of Niẓāmuddin's *Kalīla wa Dimna* dated Muḥarram 833/October 1429: "Bāysunghur ibn Shāh Rukh Seated in a Garden" (TSMK Revan 1022, fols. 1b–2a), where even the central portion of the rug that Bāysunghur sits on is covered by a delicate foliated repeat pattern in gold of the kind found on the maps of TSMK A2830.[5] Similarly, the clothing of the figures in the famous "Humāy and Humāyūn in a Garden Miniature," which comes from an anthology dated to around 1430, carry similar patterns. A particularly close parallel between a miniature attributed to the Herātī school and the foliated opening folios of TSMK A2830 is found in the gold pattern on the walls behind the central figures in a painting entitled *Tahmīna Enters Rustam's Chamber*, dated to the 1430s (fig. 12.5).[6] The upper left corner of a blue-and-white mural in an illustrated copy of Niẓāmī's *Khamsa*, "Khusraw receives Farhād" (TSMK Hazine 781, fol. 62a), painted by Khwāja 'Ālī al-Tabrīzī,

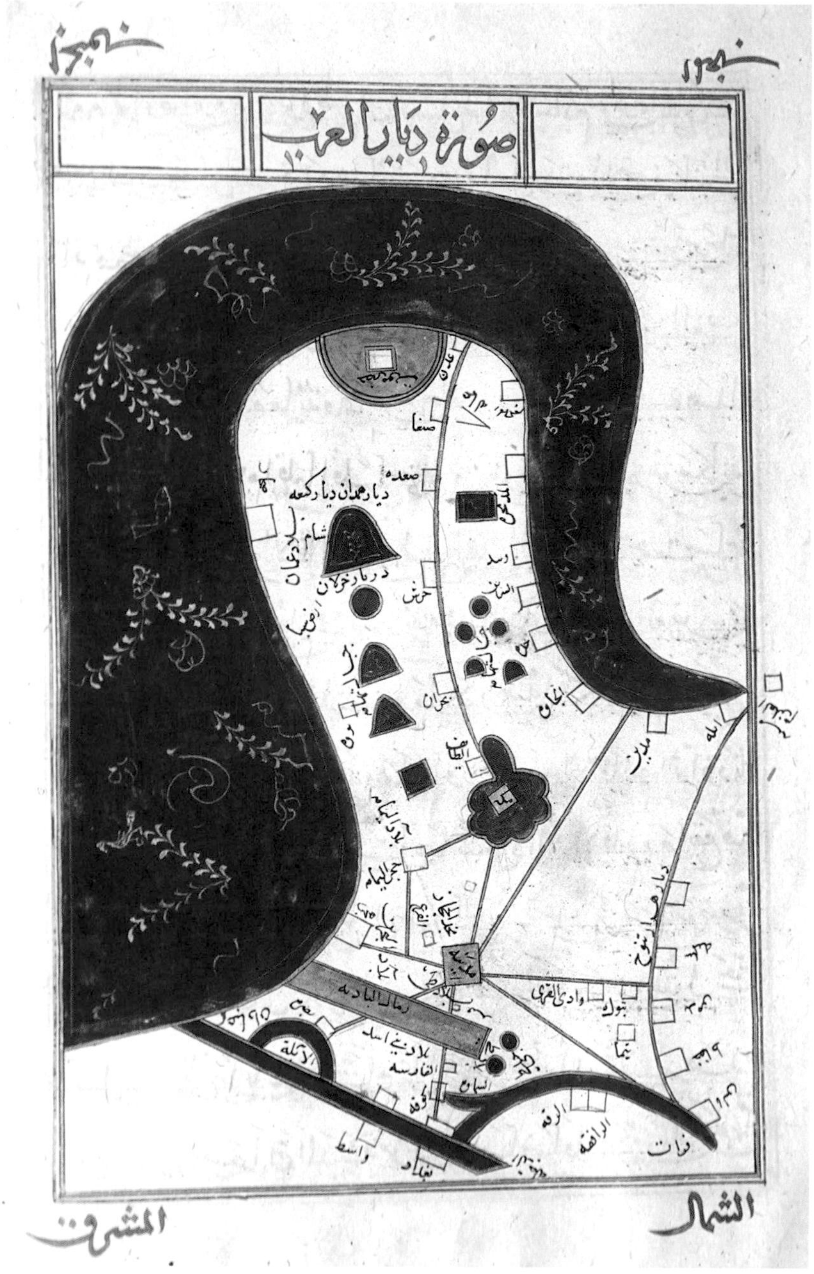

A

Fig. 12.2. *A*, Arabian Peninsula map from the Aqqoyunlu KMMS manuscript Ahmet 2830. Compare this map with Arabian Peninsula maps from the Ottoman cluster (fig. 11.4) to see the parallels in form and design. The copyist of the first Ottoman cluster copy (Aya Sofya 2971a) even tried to match the color code. Circa mid to late 15th century. Lapis lazuli (ultramarine) gouache with gold and black ink on paper. 18.3 × 12.7 cm. Courtesy: Topkapı Saray

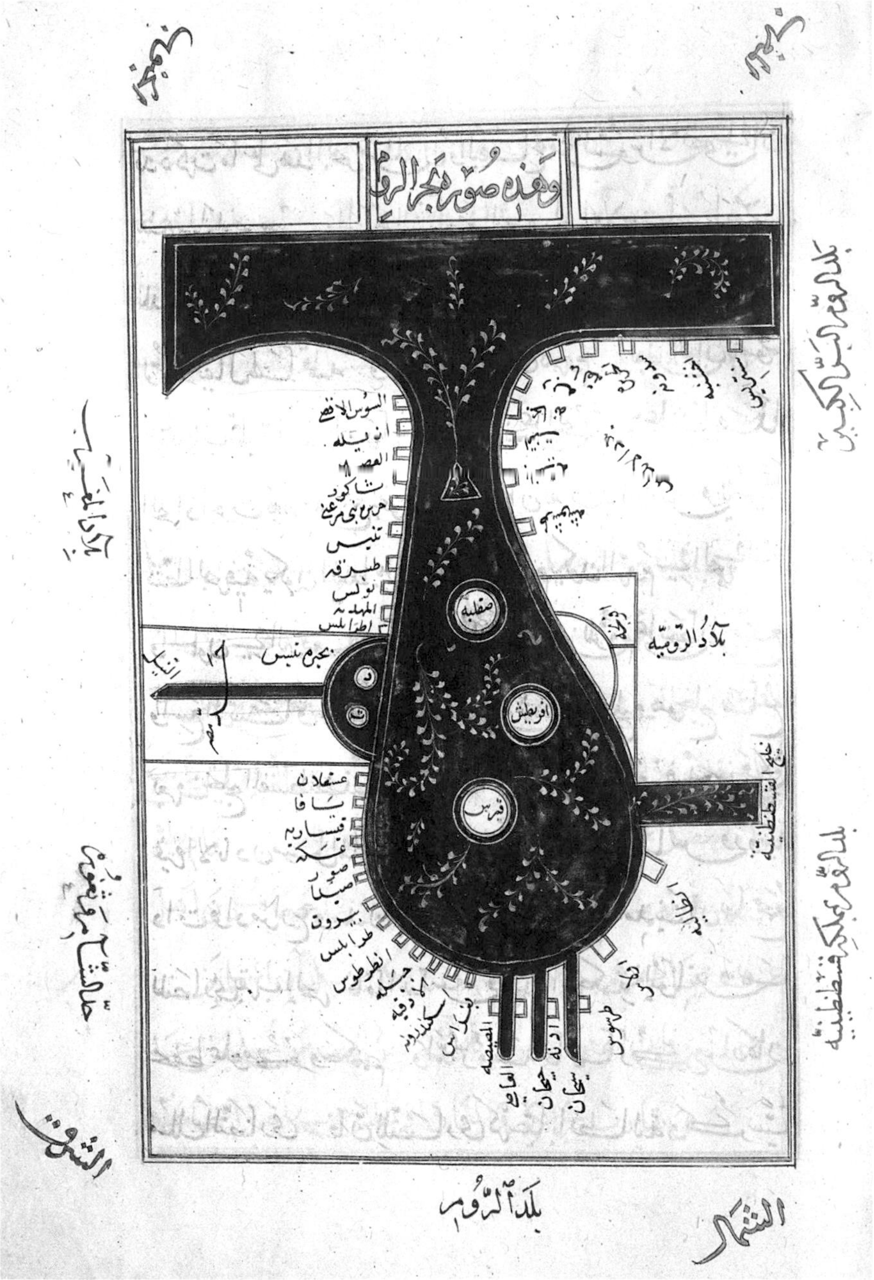

B

Museum, Istanbul. Ahmet 2830, fol. 9b. Photo: Karen Pinto. *B*, Mediterranean map from Aqqoyunlu KMMS manuscript. Compare this map with Mediterranean maps from the Ottoman cluster (fig. 11.6) to see the parallels in form and design. Note, in particular, the elongation of the Mediterranean in the Ottoman cluster versions. Circa mid to late 15th century. Lapis lazuli (ultramarine) gouache with gold and black ink on paper. 18.3 × 12.7 cm. Courtesy: Topkapı Saray Museum, Istanbul. Ahmet 2830, fol. 39a. Photo: Karen Pinto.

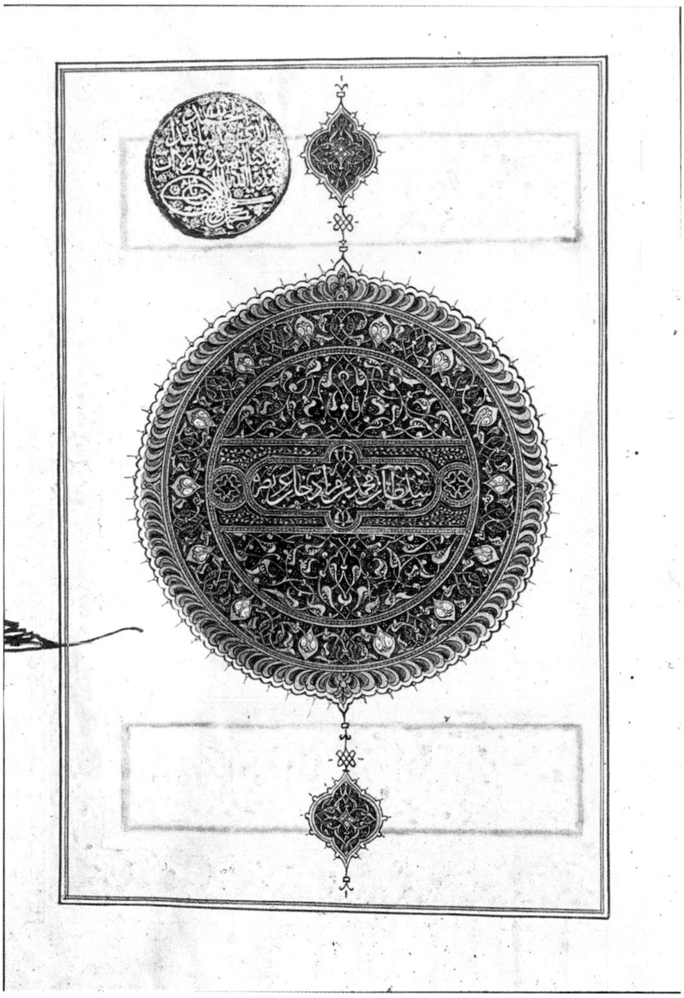

A

Fig. 12.3. *Shamsa* medallions containing laudatory dedication to Meḥmed II. The first medallion (*A*) reads, "Sulṭān Muḥammad son of Murād Khān—May his victory be glorified" (Sulṭān Muḥammad Khan ibn Murād Khān ʿazza naṣruhu).

contains animal and vegetal motifs similar to those used to ornament the seas in the TSMK A2830 world map (fig. 12.6).[7] The style of illumination of the maps, thus, has something in common with a distinctive type of Herātī and Tabrīzī decorative motif.

The illumination of the opening text folios has a curious repeat pattern of three white dots triangularly clustered against the ultramarine background and delicately interspersed in between the pattern of foliated leaves (see fig. 12.4). On the basis a Shīrāzi miniature depicting Tīmûr Lang (1336–1405),[8] founder of the Timurid dynasty in Central Asia and eastern Iran, Yuka Kadoi has brilliantly identified a three-dotted pattern on the parasol shading Tīmûr during a hunt as a Timurid symbol of "the king as cosmic ruler" (fig. 12.7).[9] In keeping with the opening medallions dedicating

The second medallion (*B*) contains the first part of the inscription "Service was rendered to His Highness the just / righteous Sultan" (Khudima li-ḥaḍrat al-Sulṭān al-ʿĀdil). The circular stamp in the top left corner of folio 1a was added in the eighteenth century during a comprehensive inventory of the manuscripts in Topkapı Saray ordered by the Ottoman sultan Maḥmud I (r. 1730–1754). Circa mid to late 15th century. Lapis lazuli (ultramarine) gouache with gold and white writing on paper. Diameter 9.6 cm. Courtesy: Topkapı Saray Museum, Istanbul. Ahmet 2830, fols. 1a–0b. Photo: Karen Pinto.

B

TSMK A2830 to Meḥmed II, this subtle three-dotted pattern can be read as a subliminal message enhancing the Ottoman sultan's glory. It brings Shīrāzi influence to bear on our understanding of the provenance of TSMK A2830, and it confirms my reading that this manuscript cannot be of Ottoman provenance. This three-dotted stamp of Timurid divine kingship is not a visual metaphor used by Ottoman illustrators.

There are additional illustrated features in TSMK A2830 that fit neither the Herātī nor the Tabrīzī style. These are the chain-like bands that gird the text of the opening two folios and the outer band of blue, yellow, and orange scallops, which encircle the ex libris dedication medallions, that do not resemble Timurid work. The chain-like

الارض بل جعلت كل قطعة اقرن بها مفردة مصوّرة تحكي
موضع ذلك الاقليم ثم ذكرت ما يحيط به من الاماكن وما في
اضعافه من المدن والبقاع المشهورة والبحار والانهار
وما يحتاج الى معرفته من جوامع ما يشتمل عليه ذلك الاقليم
من غير ان استقصيت ذلك كراهة الاطالة التي تؤدي الى ملال
من قراه ولانّ الغرض في كتابي هذا تصوير هذه الاقاليم التي
لم يذكرها احد علمته واما ذكر مدنها وجبالها وانهارها وبحارها

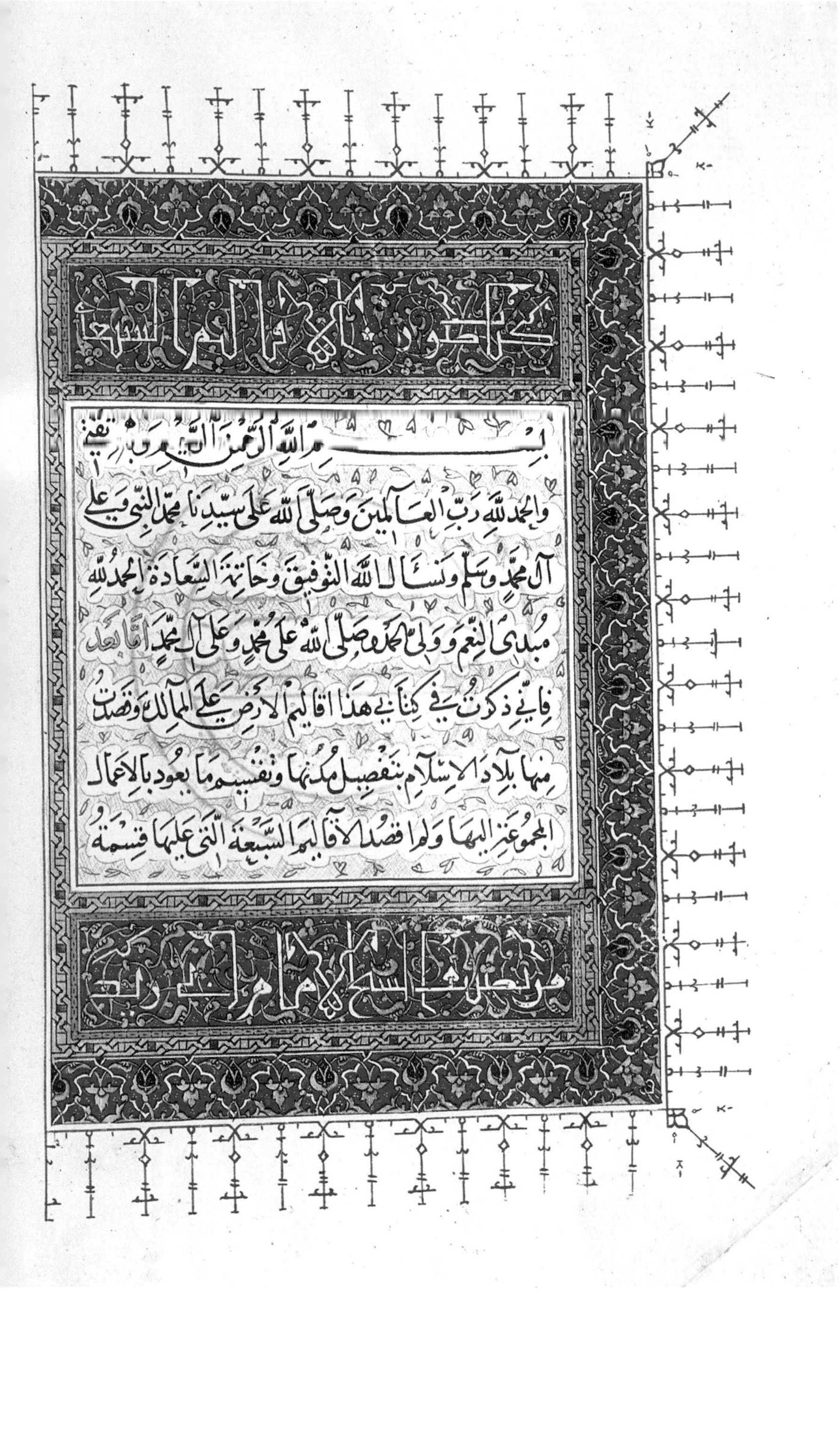
بسم الله الرحمن الرحيم
والحمد لله رب العالمين وصلى الله على سيدنا محمد النبي وعلى
آل محمد وسلم ونسأل الله التوفيق وخاتمة السعادة الحمد لله
مبدي النعم وولي الحمد وصلى الله على محمد وعلى آل محمد أما بعد
فإني ذكرت في كتابي هذا أقاليم الأرض على الممالك وقصدت
منها بلاد الإسلام بتفصيل مدنها وتقسيم ما يعود بالأعمال
المجموعة إليها ولم أقصد الأقاليم السبعة التي عليها قسمة

Fig. 12.4. Elaborately illuminated opening two folios of the Aqqoyunlu KMMS manuscript. Circa mid to late 15th century. Lapis lazuli (ultramarine) gouache with gold and white writing on paper. 31.5 × 21.5 cm (each folio). Courtesy: Topkapı Saray Museum, Istanbul. Ahmet 2830, fols. 2a–1b. Photo: Karen Pinto.

A

Fig. 12.5. *Tahmīna Enters Rustam's Chamber* from a Timurid anthology (*A*). See the medallion in the top left corner and the band above Rustam's bed (*B*) for illumination parallels with the opening folios of TSMK Ahmet 2830 (fig. 12.4). Circa 1434–1440. Herāt. Gouache with gold and ink on paper. 21 × 10.8 cm. Courtesy: Harvard Art Museums / Arthur M. Sackler Museum, Gift of Mrs. Elise Cabot Forbes and Mr. Eric Schroeder and Annie S. Coburn Fund, 1939.255. Photo: Imaging Department © President and Fellows of Harvard College.

Fig. 12.5. (continued)

B

design is reminiscent of Mamluk manuscript illumination where it was one of the most common features in their book decoration vocabulary, while the outer scalloped band of the opening two *shamsa*s—oddly—resemble a late thirteenth-/early fourteenth-century Ilkhanid illumination (fig. 12.8).

The identification of the outer scallop pattern of the medallions proved to be challenging and was only resolved by the *shamsa* medallion dedication on the opening folio of the *Mānafiʿ al-Ḥayawān* manuscript housed at the Morgan Library and Museum in New York (M.500), which provides a close match.[10] But what connection could the illumination of the Morgan's M.500, dated by its colophon to early fourteenth-century Maragheh, Iran, have to the TSMK A2830 made almost a century later?[11]

An answer comes to light when we reframe the question as, which atelier in the decades of the 1460s and the 1470s could have produced a manuscript with Timurid Herātī/Tabrīzī/Shīrāzi, Mamluk, and Ilkhanid miniature influences intertwined? B. W. Robinson's monograph *Fifteenth-Century Persian Painting: Problems and Issues* provides a solution to the enigma of the provenance of TSMK A2830. Robinson dis-

Fig. 12.6. "Khusraw receives Farhād" from Niẓāmī, *Khamsa*, painted by Khwāja ʿĀlī al-Tabrīzī. Upper left corner of a blue-and-white mural (green box) contains animal and vegetal motifs similar to those used to ornament the seas in the TSMK Ahmet 2830 world map. 849/1445–1446. Tabrīz or Herāt. Gouache with gold and ink on paper. 24.1 × 16 cm. Courtesy: Topkapı Saray Museum, Istanbul. Hazine 781, fol. 62a.

cusses how, following the collapse of the Timurids after Shāh Rukh's death in 1447, the Turkmans of eastern Anatolia and western Iran, first the Qaraquyunlu and then the Aqqoyunlu, absorbed miniaturist artists from a wide range of different courts, including Baghdad, resulting in the ultimate hybrid atelier of the Muslim world in the mid-fifteenth century. Robinson points out:

> It seems inevitable to attribute these mixed style manuscripts to Turkman patronage, not only because the styles they contain correspond in date to the period of Black Sheep rule, but also in view of the fact that, unlike them, all manuscripts known to be of Herāt or Shiraz origin are stylistically consistent in their illustrations. . . . Court painting under the Turkmans began by using the current styles of Shīrāz and Herāt indiscriminately,

and often combined in the same manuscript, sometimes together with a *simpler* style that probably originated in the northwest. . . . The court style remained virtually unchanged when the White Sheep Turkmans took over in 1467.[12]

The hallmark of this school was a high degree of style hybridization, which explains the manifestation of Herātī / Tabrīzī / Shīrāzi and Ilkhanid illumination styles in TSMK A2830. Even the chain-like design reminiscent of Mamluk manuscript illumination is found in Turkman manuscript illustration.[13] There is only one place where such a mixture of illumination techniques could have come together and that is Tabriz between 1460 and 1470 under the Aqqoyunlu chief Uzun Ḥasan.[14] A peculiarity of Turkman scribes was that, like the author of TSMK A2830, they omitted colophons. This habit, along with their eclectic and confusing mixing of illumination styles, has contributed to the immense difficulty of identifying their manuscripts. "Previously most Turkman court painting was attributed to Herāt, and its 'commercial' branches written off noncommittally as 'provincial' or 'western Iranian.' But as the unrivalled collections in Istanbul became gradually more widely known and published, a clearer picture emerged, though a fair amount of detail remains to be filled in, owing largely to the uninformative nature of the colophons in the manuscripts concerned."[15]

The conclusion I have arrived at is that TSMK A2830 should be considered as a wholly Turkman production from the period of the Aqqoyunlu.[16] Even its cover, which unlike those of the Ottoman cluster manuscripts, has survived, matches those attributed to Tabriz from the period (figs. 12.9 and 12.10).[17] In short, the manuscript is an excellent example of the multitude of styles that can coexist on the illuminated folio of an Aqqoyunlu manuscript. The remaining puzzle is how a manuscript made in Turkman territory might have reached the Ottoman ruler in Istanbul.

How TSMK A2830 Reached Meḥmed II

In 1472 Meḥmed's son, Prince Mustafa, kidnapped Yūsuf Mirza, the nephew of the Aqqoyunlu chief Uzun Ḥasan, who was Meḥmed's main eastern Anatolian and western Iranian rival. Meḥmed used the opportunity to pressure the Aqqoyunlu leader for a heavy ransom.[18] Specifically included in the ransom demand was a request for "rare books and murakka' [calligraphy] albums."[19] Although the chronicles do not indicate whether Uzun Ḥasan complied with the demand, the extant manuscript evidence suggests that it was carried out. In addition to TSMK A2830, at least thirty similar manuscripts in the Topkapı Saray library contain identical opening medallions with dedications to Meḥmed and employ similar illumination techniques. These, I suggest, can and should be seen as the Aqqoyunlu chief Uzun Ḥasan's fulfillment of at least the manuscript portion of Fâtih Meḥmed II's ransom demand.[20]

On the basis of the earliest colophon date of the Ottoman cluster copies (1474), an

A

Fig. 12.7. *A*, Tīmûr Lang hunting, from Sharaf al-Dīn ʿAlī Yazdī's *Ẓafarnāma*. 1436. Shīrāz. Gouache with gold and ink on paper. 26 × 17 cm. Courtesy: Arthur M. Sackler Gallery, Smithsonian Institution, Washington, DC. Lent by the Art and History Trust Collection: LTS1995.2.17. *B*, Comparison of three-dot Timurid symbol. Close-up of three-dot pattern on lapis lazuli background behind the foliated design framing the opening folios of TSMK Ahmet 2830 (upper image) and close-up of Tīmûr Lang's parasol (lower image) with the three-dot pattern symbolic of the Timurid flag. Courtesy: Topkapı Saray Museum, Istanbul. Ahmet 2830, fols. 2a–1b. Arthur M. Sackler Gallery, Smithsonian Institution, Washington, DC. Lent by the Art and History Trust Collection: LTS1995.2.17.

early 1470s date for the arrival of TSMK A2830 at Topkapı Saray may be postulated. If TSMK A2830 was part of a ransom payment from Uzun Ḥasan in 1472, the idea of an Aqqoyunlu provenance is reinforced. Furthermore, the fact that the innermost band of the medallions on the frontispiece of the manuscript contains the same type of delicate gold sprays as those that adorn the maps inside would indicate that the manuscript was specially prepared for Meḥmed from cover to cover. This further reinforces the hypothesis that TSMK A2830 was intended as part of a prince's ransom.[21]

Fig. 12.7. (continued)

B

Fig. 12.8. *Shamsa* bearing dedication to Shams al-Dīn ibn Ziyā al-Dīn al-Zūshkī from *Manafi-i Hayawan* (Benefits of Animals) by Ibn Bukhtīshū'. Note parallels in design with the outer scallop pattern of the *shamsa* medallions of TSMK Ahmet 2830 (fig. 12.3). Circa 1297–1300. Ilkhanid. Maragheh, Iran. Gouache and ink on paper. Courtesy: The Morgan Library and Museum, New York. M.500, fol. 2a.

We can thus surmise that the Europeans were not the only ones aware of Meḥmed's passion for maps. Uzun Ḥasan must also have reckoned that an illustrated geographical manuscript would please Meḥmed, the well-known map aficionado. Judging by the number of copies of the Ottoman cluster that Uzun Ḥasan's gift generated, it would seem that the Aqqoyunlu chief was right.

Catching Meḥmed's Fancy

Someone took a great fancy to TSMK A2830 when it arrived at Topkapı Saray. This is clear not only on the basis of the number of surviving copies of the Ottoman cluster but because patterns from the maps of TSMK A2830 were used in later Ottoman illuminated artwork.

The Freer Gallery of Art in Washington, DC, owns a single sheet painting (13 × 19.1 cm) titled *Portrait of a Turkish Painter* (hereafter "Freer Portrait") (fig. 12.11), which has been determined by Esin Atıl to be a traced copy of the *Portrait of an Ottoman Youth*, presently housed at the Isabella Gardener Museum, Boston (here-

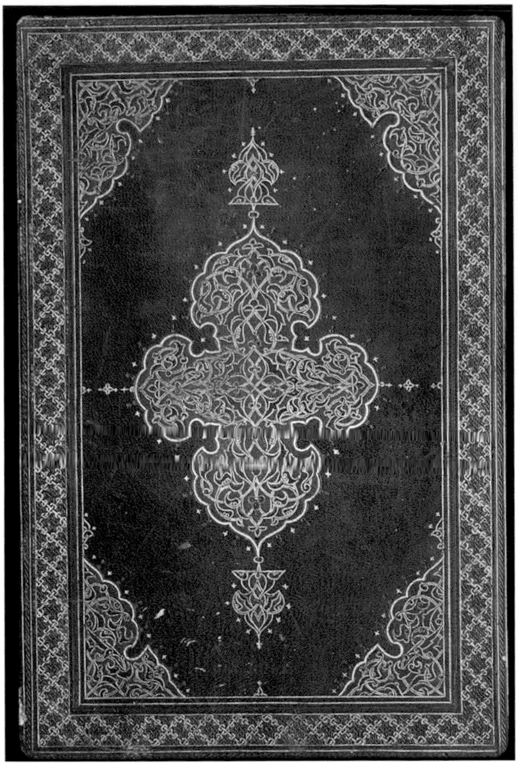

Fig. 12.9. Front cover of TSMK Ahmet 2830. Burgundy leather with lobed centerpiece and quarter corner pieces decorated in gold arabesque with touches of blue. Circa 1470s. Aqqoyunlu. 31.5 × 21.4 cm. Courtesy: Topkapı Saray Museum, Istanbul. Ahmet 2830, cover. Photo: Karen Pinto.

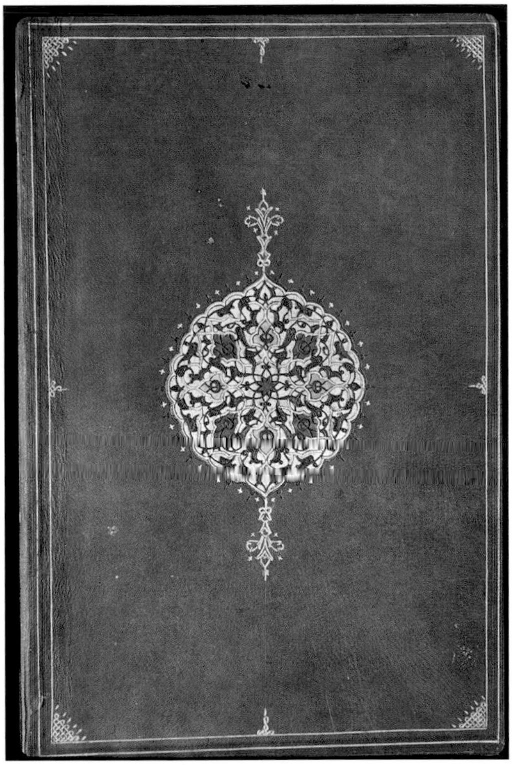

Fig. 12.10. Inside cover of TSMK Ahmet 2830. Polychrome arabesque roundel in gold against a dark-chocolate leather background. Circa 1470s. Aqqoyunlu. 31.5 × 21.4 cm. Courtesy: Topkapı Saray Museum, Istanbul. Ahmet 2830, inside cover. Photo: Karen Pinto.

after "Gardner Seated Scribe") (fig. 12.12). "When they are superimposed, the outlines fit perfectly, indicating that the Freer miniature is a traced copy of the Gardner one, the only alterations being the details as decorative motifs and colors applied to areas within the general outline."[22]

The "Gardner Seated Scribe" is the work of an anonymous European painter, whereas the "Freer Portrait" with its variations is the product of an Eastern hand. The identity of the painter and copyist of these two portraits is the subject of debate.[23] What is striking about the "Freer Portrait" is that the garment of the figure is decorated with a diaphanous gold fish pattern on a blue background, bearing a distinct re-

A

Fig. 12.11. *Portrait of a Turkish Painter*. Note the garment was originally painted with red pigment (and shows through), but this was repainted in a lapis lazuli blue with a gold fish pattern (*B*, detail enlarged), closely resembling the gold fish pattern used to decorate the seas on the world map in the manuscript TSMK Ahmet 2830. Circa 1470–1480s. Gouache and gold on paper. 19 × 13 cm. Courtesy: Freer Gallery of Art, Smithsonian Institution, Washington, DC. Purchase, F1932.28.

semblance to the gold fish pattern employed in the decoration of the maps of TSMK A2830. Close examination reveals that the garment of the scribe in the Freer copy was originally painted red. The red pigment shows through in places where the blue paint has flaked off, indicating that someone requested that the garment be repainted.[24]

This coincidence suggests that TSMK A2830 was a hit when it arrived in Istanbul, and that somebody took such a great liking to it, particularly to the diaphanous gold fish pattern, that he / she wanted the motif repeated in the commissioned copy of the "Gardner Seated Scribe."[25] I surmise that this "somebody" was none other than Fâtih himself.[26] From the introductory medallions, we know that TSMK A2830 was intended for Meḥmed; from his interest in portrait painting, it is presumed that the two portraits in question emanate from his private collection; and from his biographers, we learn of both the aesthetic refinement of his taste in art and the active interest that Meḥmed took in his manuscript and painting collection, seeking always to expand it. Therefore it seems logical to presume that of all the manuscripts that Meḥmed received as gifts from Uzun Ḥasan, TSMK A2830 was one of his favorites, and that he sought to replicate not only the entire manuscript with its full complement of maps but also the details of their decoration.

The fondness for this design of delicate see-through fish on a plain, dark background continues briefly past Fâtih and is seen again in the 1498 copy of Hātifī's *Khusraw wa Shīrīn*, produced in Bāyezīd II's Istanbul and presently housed at the Metropolitan Museum in New York (Ms. 69.27, fols. 2a and 67b). After the fifteenth century the pattern disappears from the Ottoman illumination vocabulary, vanishing into the depths of Topkapı Saray's library as did TSMK A2830 and the KMMS Ottoman cluster.[27] From the evidence cited above, it can be surmised that when TSMK A2830 was received at Meḥmed's palace in Istanbul sometime in the late 1460s / early 1470s, it was not just glanced at and cast aside. Patterns from it influenced late fifteenth-century artistic products at Topkapı Saray.

B

Fig. 12.11. (continued)

Fig. 12.12. *Portrait of an Ottoman Youth,* attributed to Gentile Bellini. When this painting is superimposed on the *Portrait of a Turkish Painter* (fig. 12.11a), the outlines fit perfectly, indicating that the Freer miniature is a traced copy of the Gardner one. Gouache, gold, and ink on paper. 18.2 × 14 cm. Courtesy: Isabella Stewart Gardner Museum, Boston/Bridgeman. P15e8.

Meḥmed and the Ottoman Cluster

Why would Meḥmed, keen collector of European paintings and mimetic maps, commission copies of a manuscript like TSMK A2830 with its obviously nonmimetic maps? Could he have found the stylized maps in TSMK A2830 quaint and curious in comparison to the mimetic ones that he was accustomed to collecting? Perhaps he recognized that despite the geometrical shapes of the continents and seas, there was a basic delineation that matched the Ptolemaic ones, and that they were not in fact inaccurate, only schematic. This does not explain the multiplicity of manuscripts that make up the Ottoman cluster. We know that when Meḥmed liked a particular pattern, he had it copied. In this case, he appears to have had the full manuscript copied. It seems unlikely, given his eclectic tastes and enjoyment of both Western art and Eastern poetry that he suddenly felt he had to espouse the traditional Islamic way of viewing the world, especially since he had by this time closely perused Ptolemy's Greek-language *Geography* now at TSMK G.I. 27.[28] The most persuasive explanation lies, I suggest, in the difference between Meḥmed's private intellectual disposition and his public face as ruler. What we gain from his patronage of the KMMS Ottoman cluster is an understanding of the image of the world that Meḥmed wished to promote among his subjects.[29]

Logically one would ascribe the copying and promotion of "traditional Islamic" material to Meḥmed's son Bāyezīd II. He was the one credited with being a pious Muslim, the one who is supposed to have revoked his father's Westernizing practices and tastes. Yet the colophons of the earliest Ottoman KMMS copies place them in the latter half of Meḥmed's reign.

Perhaps a clue to the solution of this conundrum can be found through a contemporary analogy. When one looks back at the alcohol and gambling prohibition in present-day Muslim Pakistan, the automatic reaction is to think that it must have been promulgated by Zia ul-Haq, known for his intense piety. The irony is that it was in fact the socialist leader Zulfiqar Ali Bhutto who, during his final troubled years in power, promulgated a series of prohibitions banning alcohol and gambling in an attempt to mollify the Islamicists. In the end this didn't work and Bhutto was still beheaded. The point is that—contrary to intuition—the earliest attempts to institute sharīʿa law in Pakistan were begun under an earlier, more secular leader. Similarly, it is reasonable to consider that while historians tend to see Bāyezīd II as the Islamicizer, as the one who turned his face away from the West and toward things Eastern, this process might easily have begun earlier, during Meḥmed's later years.

ʿĀlī Qūshjī and Meḥmed's Postconquest Mosque Complex

It is also conceivable that someone else encouraged Meḥmed to commission copies of TSMK A2830, possibly a scholar who believed in classical Islamic learning and specifically in the KMMS geographical series. Such a person could have been ʿĀlī Qūshjī, a high-ranked senior scholar of science and astronomy in the court of the Timurid ruler Ulugh Beg at Samarqand.[30] Seeking to escape the chaos that followed the collapse of the Timurids, Qūshjī, the chronicles tell us, visited Meḥmed sometime in the early 1470s, at which point Meḥmed appointed him resident professor of science in charge of the new madrasa (school) at Aya Sofya with a handsome wage of 200 akçe a day. It is also recorded that it was Uzun Ḥasan who sent Qūshjī to Meḥmed, which leads us to wonder if ʿĀlī Qūshjī may have been the bearer of Uzun Ḥasan's ransom payment. Since Qūshjī was trained in the classical sciences and the KMMS was part of this popular geographical tradition, especially among the Timurids of Iran and Central Asia as is evinced by the extant KMMS manuscripts, it is likely that it could have been Qūshjī who first brought the KMMS genre to Uzun Ḥasan's attention and subsequently to Meḥmed's.[31]

Qūshjī died on December 19, 1474. Although he could not have held the teaching post at Aya Sofya for long, he was in Istanbul at a critical moment during the founding of Meḥmed's first library complex and the acquisition of books to fill it. Qūshjī would have had time to make recommendations and supervise the commissioning of the first set of Ottoman KMMS copies for Meḥmed's public library.[32]

Ibrahim Erünsal establishes that Meḥmed commissioned copies of large numbers of books to be placed in the new libraries that he established in Istanbul following the conquest.[33] It is likely that the earliest KMMS copies were commissioned for one of the early postconquest libraries either by Meḥmed or at the advice of a scholar whom he respected. It would be very useful if one could determine, on the basis of the colophons, which other manuscripts were commissioned for Meḥmed's public libraries. This would give us a more complete picture of the kind of information Meḥmed presented to his people.[34]

We know from the accession numbers of the two Ottoman cluster copies at the Sülemaniye library in Istanbul (Aya Sofya 2971a and Aya Sofya 2613) that they were part of the Aya Sofya collection. From this, one may presume that they were either absorbed into this collection at some later date or that they were commissioned from the outset for the library at Aya Sofya, which Meḥmed had converted into a mosque and a madrasa soon after taking over Constantinople.[35] If the copies were intended for Aya Sofya, it would bolster the theory of ʿĀlī Qūshjī's connection with these KMMS copies, since it was around this time that Meḥmed appointed him professor of science at the Aya Sofya madrasa. Perhaps Meḥmed thought that his new scholar would appreciate it if copies of the geographical manuscript he had brought with him as a gift were made available for use in his classes. Or perhaps Qūshjī himself requested that copies of certain manuscripts be made. We will not know the precise story until more information comes to light. If it can be proven that Ali Qūshjī was responsible both for introducing Meḥmed to this particular kind of medieval Islamic carto-geographical book and for encouraging him to commission copies for use in his public libraries (Aya Sofya and Fâtih), then it raises questions about the patronage of the KMMS vision of the world by the crème de la crème of Islamic science: the medieval astronomers.

Comparison of World Maps in TSMK A2830 and the Ottoman Cluster

The earliest world map of the Ottoman cluster, Aya Sofya 2971a (see fig. 4.5), is not a tracing from TSMK A2830 (see fig. 12.1) but a freehand copy that was subsequently faithfully traced in the other manuscripts of the Ottoman cluster (refer to chapter 11 and compare figs. 4.5 and 12.1). Everything is slightly different. Hence the distinctive angle of the Mediterranean in the Ottoman cluster maps in contrast to those of TSMK A2830, the mother map. The Indus River (marked Mihrān on the world map) is sometimes squiggly, as in the mother map, and sometimes straight. This can be explained on the basis of later KMMS Ottoman cluster manuscripts being traced from an earlier exemplar in the cluster rather than directly copied from the mother manuscript; lines and shapes tend to be straightened out or further exaggerated in the process of transmission. The two inland seas, the Caspian and the Aral, retain their keyhole appearance with minor variations. The Tigris and Euphrates Rivers are both smaller, although they retain their TSMK A2830 structure. The Euphrates noticeably does not meet up with either major sea but hangs almost as a frontier between them, reduced to a pronounced crescent shape. In both the mother world map and the world maps of the cluster, the Euphrates acts as the boundary separating the Arabian Peninsula from the rest of the world.

Taken as a whole, the world maps in TSMK A2830 and in the Ottoman cluster manuscripts are almost identical. Territorial boundaries on the cluster maps are marked with the same shapes as those employed by TSMK A2830, with the occasional shrinking and expanding in places. The space on the Ottoman cluster world map assigned to the Maghrib, to Egypt, to the Saharo-Sahelian sectors lying to their south, and to the Iranian territories in the east are visibly reduced, whereas the area accorded to Abyssinia has been slightly increased. The bulbous head of the Arabian Peninsula has become smaller, significantly reducing the area assigned to the Arab tribes as well as the area assigned to Iraq. The areas along the edges of the map have expanded, so that the wastes around the northern and southern extremities (appropriately termed *barārī*, meaning "open-country / steppe / desert") are larger.[36]

Part of the explanation for this expansion is found in the simple fact that the maps of the Ottoman cluster are on a larger size of paper (19 cm on a side or more, compared with 13.2 cm in TSMK A2830). The enlargement is uneven, however. For instance the territory marked "Bilād al-Rūm"—that is, Byzantium—has been allocated disproportionately more space in the world maps of the Ottoman cluster than it had in the mother map. Stretching across all Anatolia and Syria, and incorporating as it does almost the whole of the Levant, "Bilād al-Rūm" can be read as synonymous with a desire to expand the boundaries of the Ottoman Empire.

The tribal belt on the western flank of the Black Sea, composed of, in order of occurrence, the Sarīr, the Khazar, the Burṭās, and the Rūs, has also expanded and now presses into territory assigned to "Bilād al-Rūm."[37] The area accorded to the Bulgars, designated "Bulghār al-Dakhīl," the Inner Bulghār, has been reduced in keeping with the fact that by this point Meḥmed had incorporated most of Bulgaria into the Ottoman Empire. The swathe of land assigned to the Slavs ("al-Ṣaqāliba") along the northern end of the Bosphorus crossing over from Asia to Europe has been significantly elongated in Aya Sofya 2971a. This too is a telling change, because by 1474 Meḥmed had overrun most of the lower Danube region and had designs on all of it, up to and beyond Buda.[38]

Of particular interest is the way in which the space accorded to "al-Arḍ al-Kabīra min al-Rūm" (the Land of Greater Byzantium) on the European flank has grown to take over almost the entire European triangle. The semicircular areas assigned to the Ifranj (Franks) and Andalus (Muslim Spain) have shrunk considerably. This change, more than any of the others, bears the mark of Meḥmed's territorial ambitions in Europe. Being heir to the Byzantines meant controlling most of Europe and Asia Minor. This is what the expanded boundaries of Rūm can be read as representing on the world maps of the Ottoman cluster.

Noticeable too are the sizes of the landmasses in the Ottoman cluster versions of the world map. Africa is bigger, longer, more pointed, and unmistakably swordlike in shape. Asia too is bigger, with significant expansion along its extremities. As a result India, Tibet, China, and especially the northern wastelands ("Barārī al-Shamāl") have been given more space. Europe is visibly larger. However, the hint of European contact with the westernmost tip of North Africa present in the TSMK A2830 world map has been removed in the Ottoman rendition.

The world map of the Ottoman cluster was adroitly reproportioned to impress on the viewer the greatness and expanse of "Bilād al-Rūm" and "al-Arḍ al-Kabīra min al-Rūm"—and the Ottoman Empire as successor to Byzantium—in comparison with all other territories of the world. The message is reinforced by the space accorded to the lands designated "al-Ṣaqāliba" and "Bulghār al-Dakhīl," creating the impression of Ottoman control almost to the territorially voluminous northern steppes ("Barārī al-Shamāl") and the lands of the Turks' eponymous ancestors who, in the eleventh and twelfth centuries, as we know from the history of Turkic migrations from Inner Asia, made their way westward to Anatolia.[39] What matters, according to the world map of the Ottoman cluster, is that the Ottoman Empire dominates the image of the world as the new Byzantium with its nominative implication of a neo-Roman Empire.[40]

Meḥmed's Public Face

In *Late Antique, Early Christian and Medieval Art,* Meyer Schapiro argues that

> *if a work is a copy, it will betray the period of copying in intimate features of the copyist's own style.* This is a principle that has been confirmed countless times in studies of literature and art. . . . Weitzmann himself says of medieval art in general: "The style of the model has quite often a more powerful influence on a work of art than the individuality of an artist, but are in a broader sense those of a special workshop or a distinct locality at a given time. He believes that the minute differences by which one distinguishes medieval works are "rarely the characteristics of the individuality of an artist, but are in a broader sense those of a special workshop or a distinct locality at a given time." Taken together here, these principles are discouraging, if not puzzling; but in practice, *the "broader sense" may lead us to the truth, if we have enough evidence and a sufficiently searching analysis.*[41]

Place of copy is one of the most telling features of any manuscript—where it was made, the ambiance of the atelier in which it was produced, the guiding principles employed, all these factors influenced the final product. What can the illustrations of the Ottoman cluster tell us about the ambiance in which they were copied? One would presume that the ambiance in the atelier must have been based in part on the mood among the people and in part on the mentalité of their chief. Is there an answer buried in the flourishes and exaggerations that can lead us to an understanding of the mind-set of Meḥmed's Istanbul palace atelier and beyond it to Meḥmed himself? Have the artisans of the maps deliberately communicated certain messages through the forms?

One difference between the world map of the mother manuscript (TSMK A2830) and that of the Ottoman cluster manuscripts is the size of the Indian Ocean in comparison to the Mediterranean. In the Ottoman cluster maps, the Mediterranean is reduced in size compared to TSMK A2830, while the Arabian Peninsula has been elongated (compare fig. 11.4 with fig. 12.2a and fig. 11.6 with fig. 12.2b). We could put this down as an updating on the basis of firsthand observation were it not for the fact that the Ottomans were not actively plying the Indian Ocean until the sixteenth century.[42] Nothing in the rendition of the world in the Ptolemy manuscript housed at Topkapı Saray would have enabled Ottoman manuscript illuminators to understand this in the last quarter of the fifteenth century. Nor could they acquire this information from European maps, since the enlargement of the Indian Ocean on European maps also does not occur until the sixteenth century.

Or we can choose to read more into the image and discuss the symbolic meaning of the changes in the depiction of the Indian Ocean. Note the exaggeration of the distinctly menacing, hook-like end of the Indian Ocean—that is, the Red Sea—in the Ottoman cluster maps. In these, the Red Sea seems almost to have broken through the meager land barrier that separates it from the Mediterranean. Into this we can read careless copying or deliberate design. The latter would mean that Meḥmed's artists saw a convenient symbol that they could exploit in order to stamp the classic Muslim world map with the Fâtih's burning desire to overrun Europe and expand the boundaries of the Ottoman Empire. Such an adjustment to the Indian Ocean's outline would have been the perfect visual metaphor for Meḥmed's territorial ambitions: the "Muslim" Indian Ocean first menaces then overwhelms the "Christian" Mediterranean.[43]

This visual symbolism would also have matched the metaphor Meḥmed used to adorn the gate of his new palace (Topkapı Saray) in his new capital (Qusṭanṭīniyya—-i.e., Istanbul): "The ruler of the two seas and the two continents"—an inscriptional metaphor that persisted for centuries.[44] Cemal Kafadar reads into these words a stripping of the "ghazi ideal" from the Ottoman mantle: "Being a ghazi was not the primary component of the Ottoman ruler's multiple identity anymore; he was first and foremost a sultan, a khan, and a caesar."[45]

One cannot forget that Meḥmed was also a self-styled *Muslim* Alexander, the one who would go the other way—from east to west.[46] He must have considered himself the greatest warrior in the name of Islam, because it was he who finally captured the long sought after prize of Constantinople, designated the Ottoman *Kızıl Elma* (Red Apple), a city perceived as the nerve center of the Christian world and coveted by Muslim conquerors for eight centuries.[47]

The world maps of the Ottoman cluster reflect this sentiment, even at the cost of demoting the Arabian Peninsula in size and thus stature and making its importance to Muslims secondary to the Ottoman agenda of empire. Later, Selīm I made good on the legacy of "ruler of the *two* seas" initiated by his grandfather, when he overran the Mamluks in Syria and Egypt (1516–1517) in his attempt to reach and control the *other* major sea—the Indian Ocean. Thus, my reading of the Ottoman KMMS maps concurs with İnalcık when he says: "The conquest of Constantinople turned Meḥmed II overnight into the most celebrated sultan in the Muslim world. He began to see himself in the light of a world-wide empire. He believed in the absolute character of his power, and wished Istanbul to become the center of the world in all respects. . . . He saw in all three titles, the titles of Khan, Ghazi, and Caesar, gates leading to dominion over the whole world."[48]

This reading of the rationale for the production of the KMMS Ottoman cluster fits especially well with Heath Lowry's seminal reevaluation of the *ghazi* thesis as a fabrication of state-sponsored chroniclers from the late fifteenth century onward.[49] Nowhere is Meḥmed's Alexander ethos better expressed than in the representation of the Ottoman empire on the KMMS Ottoman cluster world maps as "Bilād al-Rūm" combined with "al-Arḍ al-Kabīra min al-Rūm" in Europe. This may be read as an affirmation of the idea that the Ottomans saw themselves as heirs to Byzantium. In fact, by co-opting the name and location of the former Byzantine Empire on the map, Meḥmed accorded to himself and his empire the highest position in the Muslim world akin to the Roman Empire.

If one can see in these Ottoman cluster world maps the expression of Meḥmed's imperial ideals, then it is not difficult to see why he would have commissioned these KMMS copies for the public libraries of his *new* capital. They presented an ideal format with which to stimulate support among the Ottoman populace for his conquests—another spoke in the wheel of a smart ruler's propaganda machine. Meḥmed's artists could manipulate these iconic KMMS images in ways that would not have been possible with the more mimetic contemporary European productions. Uzun Ḥasan's gift of TSMK A2830 thus presented Meḥmed with an excellent opportunity to project his territorial ambitions, while simultaneously enabling him to project a public image of being a good Muslim ruler who patronized well-known scholars and promoted the copying of classical Islamic works. "We know that in his private associations Mehmed had considerable sympathy for heretical ideas. But it is equally certain that as chief of state he strictly and undeviatingly enforced the Sunnite brand of Islam, which he himself publicly observed."[50]

When we focus exclusively on Meḥmed's admiration, desire, and use for European maps and art, we miss this crucial other dimension. The picture that the former provides us is of Meḥmed's inner, private life and personal aesthetique. Focus on it occludes the public dimension.[51] We are forgetting that if he wanted to project himself as a Muslim caesar, he had to create the aura of territorial conquest as well as that of being a patronizing ruler and upholder of Islamic tradition. He may have enjoyed the European art of portraiture, but he could not fill his new architectural creations in Istanbul with murals of himself, the way his predecessors, the Byzantine emperors, had done because—according to Islamic ideals based on emulation of the Prophet Muhammad—portraits, sculptures, and images, especially of oneself, were considered self-aggrandizement and shunned.[52] He would have had to keep his interest in European portraiture and European cartography away from the public eye.[53]

Like all of us, Fâtih Meḥmed II had multiple faces. He was torn between his attraction to Western objects and his devotion to his Eastern, Muslim heritage. Was the latter imposed deliberately by the demands of the role he took upon himself for the Muslim world or was it genuine eclecticism? When one reads the accounts of the scintillating debates between major Muslim scholars that went on for days, sponsored by Meḥmed, or of the personal interest that he took in fledgling Muslim scholars, or for that matter his love of Persian poetry to the point of dabbling in it himself,[54] one cannot help but feel that Meḥmed's Muslim side was as genuine as the one of his Renaissance European tastes. Much remains to be done to uncover this other side along with the public dimension of Meḥmed's persona, and one can only hope that the Ottoman cluster will help to clarify this picture and other related puzzles.

13

Conclusion
Mundus est immundus

This world of images and signs, this tombstone of the "world" ("Mundus est immundus") is situated at the edges of what exists, between the shadows and the light, between the conceived (abstraction) and the perceived (the readable/visible). Between the real and the unreal. Always in the interstices, in the cracks. Between directly lived experience and thought. And (a familiar paradox) between life and death. It presents itself as transparent (and hence pure) world, and as reassuring, on the grounds that it ensures concordance between mental and social, space and time, outside and inside, and needs and desire. . . . The more carefully one examines space, considering it not only with the eyes, not only with the intellect, but also with all the senses, with the total body, the more clearly one becomes aware of the conflicts at work within it, conflicts which foster the explosion of abstract space and the production of a space that is other.

HENRI LEFEBVRE, *The Production of Space* (p. 389)

Maps are not territory; they are spaces, spaces to be crossed and recrossed and experienced from every angle. The only way to understand a map is to get down into it, to play at the edges, to jump into the center and back out again. We need to trace and retrace its lines by eye and by hand and question it's every dot until the liminal palimpsest below the surface reveals itself to yield clues of the elusive social mentalité within which the map was born. We must lay bare the ideograph in order to grasp the keys that it holds. Only then can we use maps as alternate doorways into history.

This book is about the many ways of seeing maps, Islamic maps of the world in particular, that take us to different places, spaces, and gazes. These Islamic maps are unique in the way they cross time and space, from medieval to early modern, from one end of the Islamic world to another, retaining from India to Spain across the span of eight centuries an iconographic singularity of form that enigmatically resists cartographic encroachments from modernity. Nuances of the cartographers lurk in the crevices of map lines and patterns of illumination that when studied reveal surprising new insights into medieval and early modern Islamic history.

This book illuminates three ways of map analysis. The first segment analyzes ingrained cosmographies that lie behind the innocuous blue band that encircles each Islamic world map and for that matter every imago mundi. By tracing the roots of the Encircling Ocean deeply, one segment of this book aims to establish the importance of iconographic and iconological exercises for decoding forms on maps. Chasing the roots of the encircling ocean proves that it is a motif with no clear beginning or end. We discover that its ring form is the basic stamp of all imago mundi—even those that we do not usually think of as maps. It is the quintessential metaform used to brand the power of dominion on paper, skin, and stone.

Another segment of this book penetrates the history behind one place on the map, Bilād al-Būja (Beja), to understand how and why this particular place, so obscure in our history books, was given a privileged double berth on Islamic world maps. Given that the surface of every map is constrained by the dimensions of the material on which it is made, every chosen place leaves behind a historical trace. These traces reveal that the Beja are one of the final frontiers of Islam where pagan rituals mix with Islamic practices. Out of this inquiry the temporal imagination emerges as one of the dominant architects of space. It presents itself as an imagining that is triggered as much by extreme *otherness* as it is by a subtle reflection of *self*. Immortalized by Herodotus as Troglodytes, Pliny as Blemmyes, and Rudyard Kipling as "Fuzzy-Wuzzy," the Beja are unique in the annals for having captured our imagination from the earliest days of recorded history.[1]

The third segment of this book addresses a cluster of six of Iṣṭakhrī's *Kitāb al-masālik wa-al-mamālik* map manuscripts to reveal a surprising source of patronage: Ottomans of the late fifteenth century. When probed, this set reveals ways in which these maps were exploited to promote the conqueror of Constantinople, Sultan

Meḥmed II's, ambitions of a world empire to his subjects. This is in contradistinction to the well-honed research on Meḥmed's active interest in and collection of more mimetic European cartography. The study of the Ottoman cluster points to the distinction between public and private dimensions we must make when studying cartographic collections. No one expected to find Ottoman patronage of Islamic maps, yet this Iṣṭakhrī set proves that interest in classical forms of Islamic mapping continued well into the Ottoman and early modern periods. It also proves that the counterintuitive thrives in maps and history. We just have to uncover it.

It is my hope that through these three different approaches this book will prove useful in revealing new ways of reading maps. In the same way that a digital file can be compressed and yet mysteriously contain a larger set of data encoded in its symbols, so it is with maps. We can apply the methods in this book as decryption keys to unfold from maps a larger picture—that of the history surrounding them. This book is deliberately *not* a comprehensive study of all aspects of Islamic maps. It asserts that Islamic cartography, like all cartography, is best understood through in-depth analysis of its many dimensions. This book explores only a few of these dimensions as a contribution to assembling an understanding of the tantalizing world of Islamic mapping.

Acknowledgments

One of the great pleasures of writing this book was having the good fortune to work with so many terrific people. I have been the grateful recipient of wisdom, patience, editorial assistance, goodwill, encouragement, and friendship throughout every aspect of this work, and for that help I would like to extend a heartfelt thank-you to everyone who participated.

Foremost, my gratitude goes out to my amazing advisor (Hoja of hojas) Richard W. Bulliet. Without his unwavering support, this book would never have been realized. Thank you Professor Bulliet for always being there for your students. You are one of Columbia University's greatest gems.

I am also grateful to the late Olivia Remie Constable, who died tragically young in 2014. Her class on the Mediterranean led me to the discovery of medieval Islamic maps. Many other scholars, peers, and students at Columbia were equally inspirational and encouraging. I would like to acknowledge Zümrüt Alp, Mohsen Ashtiany, Alexander Barreto, Stuart Borsch, the late Lucy Bulliet, Mokhtar Ghambou, David Weiss Halivni, Hossain Kamaly, Alidad Mafinezam, Jean-Marc Oppenheim,

Larry Potter, Dagmar Riedel, Ramzi Rouighi, Nerina Rustomji, Robert Scott, Reeva Simon, Anders Stephanson, Yanki Tshering, Marc van der Mieroop, Parvaneh Pourshariati, and Neguin Yavari for their support and insights.

I thank the former provost of the American University of Beirut (AUB), the late Peter Heath, for his financial support and encouragement of my work. Sadly, he too was taken from us too early. A William and Flora Hewlett Junior Fellowship Faculty Research Grant gave me time off—albeit in the middle of the 2006 Israeli-Lebanon war and subsequent evacuation—to work on this book, and a University Research Board Grant made it possible for me to initiate a digital database of the maps. At AUB and in Beirut, I also received encouragement from fellow academics and friends: Hamelkart and Inana Hamati Ataya, Sana Ghaddar, Syrine Hout, Mostafa Itani, Tarif Khalidi, David Koistinen, Lucy and Jeremy Koons, Baker Maktabi, Karla Mallette, Clare and John Meloy, and Rosangela Silva. Of special mention are my wonderful AUB students, in particular Souad Hammoud, Lama Hoteit, Kassem Jouni, and Hussein Naim, who assisted me in my research. I cannot thank these students enough for all their hard work and good humor over the years.

At Gettysburg College, I am eternally grateful for the inspiration, editorial advice, and unwavering encouragement of Lisa Portmess of the Philosophy Department. Without Lisa's faith this book may never have seen the light of day. I am also grateful to Dan DeNicola who organized a semester-long private seminar on place and space theory that provided collegial inspiration among like-minded thinkers. Carolyn Snively of the Classics Department collected images from Topkapı Saray in Istanbul and encouraged me in every way to complete this book. Without the administrative assistance of Rebecca Barth, I may never have been able to see my way through the thicket of payments for the 150 images in this book. The provost, Christopher Zappe, generously covered the costs associated with acquiring the images for this book. I was blessed in the final stages by the wonderful friendship and enthusiastic encouragement of the late and great historian Bill Pencak (1951–2013) who was visiting from Penn State. Sadly he too did not live to see this book in print. In the History Department, I thank my colleague Abou Bamba for his camaraderie and encouraging words. I am also grateful to Natalie Hinton of Musselman Library who patiently processed my many interlibrary loan requests and to the Middle Eastern librarian, Jeremy Garskof, who hunted down and ordered many a useful reference. Eric Remy of IT graciously shared his Java programming expertise. I thank Felicia Else of the Art History Department; Marc Beard, Megan Sijapati, and Deborah Sommers of the Religious Studies Department; Susan Russell of Theater Arts; and Suzanne Flynn of the English Department for their friendship and support. A number of Gettysburg students contributed to map-related research. I express my gratitude to Robert Shaw Bridges III, Kelsey Chapman, Katlin Davis, Adaeze Duru, Elizabeth Elliott, Rachel Fry, Dallas Grubbs, Nathan Hill, Thomas Knapp, Jon Mateer, Josh

Poorman, Olivia Yongwan Price, Matthew Redman, Jesse Siegel, Julian Weiss, Gregory Williams, and Gwendolyn Williams for their diligent contributions and good humor over the years.

My new colleagues at Boise State have been extremely supportive, going so far as to give me my first year off so that I can complete my book projects. I am especially grateful to fellow historian and dean Shelton Woods and my chair, Joanne Klein, for their generous support of my work, and to Barton Barbour, John Bieter, Lisa Brady, Jill Gill, Katherine Huntley, Lynn Lubamersky, Lisa McClain, Nick Miller, Todd Shallat, Emily Wakild, David Walker, and Michael Zirinsky for their words of encouragement.

Over the years I have benefited from the advice, learning, and support of many wonderful thinkers and colleagues in the field, including Fred Astren, Jere Bacharach, Eleazar Birnbaum, Elizabeth Bishop, Jonathan Bloom, Sonja Brentjes, Emily Burnham, Giancarlo Casale, Brian Catlos, Roberto Dainotto, Andrew Devereux, Jean-Charles Ducène, Pınar Emiralioğlu, Felipe Fernández-Armesto, Cornell Fleischer, Matthew Gordon, Andreas Görke, Tony Greenwood, Mattia Guidetti, Claus-Peter Haase, Douglas Haynes, Paul Heck, Yuka Kadoi, Cemal Kafadar, Cigdem Kafescioğlu, Andreas Kaplony, Ahmet Karamustafa, Nancy Khalek, Tarif Khalidi, David King, Sharon Kinoshita, Gil Klein, Manfred Kropp, Selim Kuru, Martin Lewis, Yuen-Gen Liang, Alan Mikhail, Zacharie Mochtari de Pierrepont, Parviz Morewedge, Roy Mottahedeh, Farouk Mustafa, Gülru Necipoğlu, Victor Ostapchuk, Ralph Pedersen, Helen Pfeifer, George Quinn, Sumathi Ramaswamy, Brett Rogers, Eric Ross, Justin Rudelson, Kerry Segal, Engin Sezer, Fuat Sezgin, Larry Simon, Zeren Tanındı, Daniel Terkla, Baki Tezcan, Kären Wigen, Myriam Wissa, and Travis Zadeh.

Soon after embarking on my work in Islamic cartography, I discovered the writings of the late J. B. Harley (1932–1991). Though he is no longer with us, I thank him in absentia for all the great *carto*-inspiration. In his stead the late David Woodward (1942–2004) was a bastion of support. Many of the members of Maphist (now ISHM) befriended me and offered advice, and I gained many valuable insights from the International Conference on the History of Cartography (ICHC) gatherings. In particular, Barbara Belyea, Tony Campbell, Angelo Cataneo, Andrew Cook, Catherine Delano-Smith, Veronica della Dora, Kathryn Ebel, Matthew Edney, Tom Goodrich, Alice Hudson, Christian Jacob, Joel Kovarsky, Marcia Kupfer, Imants Ļaviņš, Keith Lilley, Neil Safier, Alessandro Scafi, Kirsten Seaver, George Tolias, Zsolt Török, and Kees Zandvliet generously shared ideas, work, good humor, and plenty of encouragement. During an SSHRC postdoc at the University of Alberta, I benefited from the insights of map aficionados Andrew Gow and the late Ronald Whistance-Smith.

None of my work on Islamic cartography would have been possible without access to the precious manuscript collections in the libraries where I conducted research. In Istanbul, where I spent the bulk of my research time, I am indebted to Filiz

Cağman and Nevzat Kaya, the former directors of Topkapı Sarayı Müzesi Kütüphanesi and Sülemaniye library, and the present directors, A. Haluk Dursun and Ümer Kuzgun, for granting me unimpeded access. Most valuable of all, they permitted me to photograph their map collection. This was one of the greatest gifts as it has permitted me to build up a large database of Islamic cartographic images that has provided me with countless years of work. If only all libraries in the world would follow suit and generously place the researcher above financial considerations. I am also grateful to the staff in both libraries who cheerily fetched manuscripts irrespective of number. Esra Müyesseroğlu of Topkapı Sarayı Müzesi has graciously handled all my high-resolution image and permissions requests. I would also like to acknowledge the assistance of the custodians of the Istanbul University Library, Bayezit Library, and M. Serhan Taysi of Millet Kütüphanesi, as well as the other forgotten little libraries in Istanbul and Anatolia that I surprised with a visit. Finally, I am indebted to Havva Hanim, administrator of the Archaeology Museum Library. I owe her thanks not only for her support in seeking out rare and unusual manuscripts but especially for her good humor and eagerness to go above and beyond the call of duty to help out researchers.

I was welcomed into the research wing of the Islamic Conference in Istanbul (IRCICA) by the then director, Dr. Ekmeleddin Ihsanoğlu. I am indebted to Salih Ishakbeyoğlu who arranged for renewals of my Turkish research permit and visa. Tahsin Tahaoğlu and I worked together on numerous occasions at Topkapı Saray, and I am grateful for his assistance with ornamental Kufic and other paleographic conundrums.

I am also indebted to the legendary photographer, the late Josephine Powell (1919–2007), from whom I received a crash course in the art of macro photography, an essential tool for all scholars of material history. İlker Alp, another talented photographer, filled in the gaps and assisted me with some of the photography.

Outside Istanbul, I was generously hosted by librarians everywhere. Dr. J. J. Witkam, curator of the Islamic manuscripts collection at the Leiden University Libraries, arranged for the loan of microfilms and generously provided me with room and board at his own house so that I could examine the manuscripts in Leiden. Dr. Arnoud Vrolijk continues the Leiden tradition of generosity to researchers and has been most helpful with the arranging of permissions and high-resolution images. We share a common interest in maps.

The former custodians of the Oriental Manuscripts Collection at the Bodleian Library in Oxford—Doris Nichols and Colin Wakefield—went out of their way to make my research stints productive. Doris Nichols provided me with room and board at her own house on my second visit and took a keen interest in the prepublished copies of my work that I left in her care. Their successors, Gillian Grant and Alasdair Watson, have been equally helpful.

My research stint at the British Library was facilitated by Tony Campbell, the former director of the Map Room, and Dr. Andrew Cook of the India Office Library. In addition, Colin Baker and Isa Waley, curators of the Arabic, Persian, and Turkish collections at the India Office Library, assisted in manuscript access and brought new acquisitions to my attention.

In St. Petersburg, Russia, I am grateful to Dr. Irina Popova for granting me permission to consult the manuscripts at the Institute of Oriental Manuscripts of the Russian Academy of Sciences; Dr. Serguei Frantsouzoff for arranging the permission and facilitating my visit; and Dr. Oleg Bol'shakov, Dr. Aly Ivanovich Kolesnikov, Alla Sizova, Amalia Stanislavovna Zhukovskaya, and Adilia Yulgusheva for helping to make it a productive and memorable research trip.

At the Chester Beatty Library in Dublin, I am grateful to the curator of Islamic manuscripts, Elaine Wright, for her generosity and willingness to allow me to consult a broad range of relevant manuscripts. I thank Frances Narkiewicz for the many years of correspondence and the expeditious arranging of high-resolution images and publication permission. I thank Hyder Abbas for his excellent camaraderie during the hours that I spent in the library reading room and for going out of his way to check manuscript acquisition details in the archives.

In the United States, I was welcomed into the Middle East section of the Library of Congress in Washington, DC, by Christopher Murphy, the head of the Turkish Section. Jeff Spurr, former curator of the Aga Khan collection of Islamic slides at the Fogg Library at Harvard University, has assisted my research in every way possible. The curators of the Islamic collection at the Morgan Library and Museum were also very helpful. I would also like to thank the custodians of the Islamic collections at the Freer Museum of Art in Washington, DC, in particular Cory Grace, who has gone out of his way to arrange for the speedy delivery of high-resolution images and publication permission, and Tim Kirk, who arranged for me to view key manuscripts.

I am grateful to the custodians of other manuscript libraries in Berlin, Bologna, Paris, Strasbourg, Madrid, and Cairo for arranging for slides, microfilms, and reproduction permission. The staff at Columbia University's library have been extremely helpful over the years. In particular, I would like to thank Kitty Chibnik, Rudolph Ellenbogen, Renata Johnson, Martin Messik, Lars Meyer, Richard Walters, and, especially, Robert Scott, who introduced me to the wonders of Nikon slide scanning.

Financial support was crucial. For this, I would like to thank Columbia University for its generous scholarships, including a Traveling Fellowship and a Whiting Foundation write-up grant, and the Social Science Research Council of New York for early recognition of my work with an Ibn Khaldun Prize followed by a fellowship for doctoral research in Turkey, which, together with a grant from the American Research Institute in Turkey, enabled me to conduct extensive research with the rich manuscript collection in Istanbul, Turkey. The Alfred Howell fund helped pay for the slide and microfilm tariffs. Subsequently, a Friends of J. B. Harley Research Fellowship in the History of Cartography at the British Library made it possible for me to complete research in London and Oxford, and an SSHRC postdoctoral fellowship at the University of Alberta gave me the opportunity to start work on this book. AUB funded my research trip to examine manuscripts at the Bibliothèque nationale de France in Paris, Gettysburg College funded my trip to the Institute of Oriental Manuscripts of the Russian Academy of Sciences in St. Petersburg, and Boise State covered the costs for my visit to the Chester Beatty Library in Dublin. The Pinto, Alp, and Bulliet families also chipped in generously. Recognition by the National Endowment of Humanities for my work gave me the shot in the arm needed to finish this book so that I could start another titled *The Mediterranean in the Islamic Cartographic Imagination*. The publication of this book with its numerous color images would not have been possible without the generous subvention provided by Nina Ansary.

Publishing this book with the University of Chicago Press is a dream come true. For this, I owe an immense debt of gratitude to the editorial director Christie Henry for taking an interest in my book proposal, to the senior project editor Mary Laur for her patience and gentle guidance, to Logan Ryan Smith for handling the logistics, and especially to Kelly Finefrock-Creed for her careful and meticulous copyediting. Abby Collier ushered this manuscript through the reviewing and acceptance process, and I am grateful to her and to the two anonymous reviewers whose feedback greatly improved the final version of this book.

The debt I owe to family and friends for their support throughout this process is immeasurable. In Istanbul, Zümrüt Alp and her wonderful family provided me with a home away from home. In the United States, my friends George Bourozikas, Tenagne Haile-Mariam, Steven Mines, Jose Clemente Orozco, and Photini Sinnis have been rocks in my life for more than two decades. My many friends in Pakistan, especially Naheed and the Moini family and Nyla Qayoom, have been constant in their support over the years.

My brother and sister-in-law, Alex and Sandra Pinto, saw the initial version of this book to fruition. I am eternally grateful to them for their loving support of my educational endeavors over the years. I am also indebted to my nephew, Simon Pinto, for his assistance with the scanning of images. Thanks also go out to Stephanie Pinto and Mike Mathyk for assistance with book loans from the local Calgary library. In-laws in Didsbury, Alberta, in particular Marilyn Richards and the late Esther Snyder, and Barb Snyder in Texas, were extremely supportive.

Special love and thanks go out to my brilliant husband, Devon Richards, whose deft editorial touch gilds every page of this book, and to our beautiful daughter, Safiye, for all the loving joy and companionship she has brought into my life.

My amazing mother, the late Adele B. Pinto (1928–2008), assisted with everything from monetary support to language assistance in Russian and French. I regret that I was not able to share the final published version with her. Even though my father, Lt. Col. Felice G. Pinto (1918–1984), passed away well before I began the PhD journey, he would have been proud to hold this book in his hand. He instilled in me at a young age a deep love of learning that made all of this possible.

I alone am responsible for any omissions and errors. *Wa allahu aʿlam.*

Notes

Chapter 1

1. Based on the 40-plus extant copies of the main body of atlas-like carto-geographical manuscripts (comprising the work of Iṣṭakhrī, Ibn Ḥawqal, and Muqaddasī), which on average contain 20 (+/−) maps apiece, we can enumerate approximately 800–840 maps. If we add to these maps the world and Qibla maps found in the many Ibn al-Wardī and Qazwīnī manuscripts, then the number of maplike images exceeds 1,000. This figure does not include the maps that occur in the Idrīsī manuscripts, which can range from 20 to 70-plus per manuscript. If we include Ottoman, Safavid, and Mughal cartographic productions along with the Idrīsī maps, I estimate we have extant around 2,000-plus maps produced in the Islamic world. If we expand the definition of what constitutes a map and include the maplike pictures to be found in miniature art along with Kaʿba tile maps and wall paintings, then we can add another thousand Islamic *carto*graphic possibilities to that number. (See chapter 3 for visual examples of some of these other possibilities.)

2. A Muslim geographer, Muqaddasī (about whom readers will hear more during the course of this book), tells us that he was once

sitting with a famous merchant on the seashore near Aden staring out at the Persian Gulf and Indian Ocean. When the merchant found out that Muqaddasī was preoccupied with trying to understand the many conflicting accounts of the shape of this sea, he "smoothed the sand with the palm of his hand and drew a figure of the sea on it. It was neither a *taylasān* nor a bird. But he showed it having gulfs, tongues, and numerous bays. Then he said: 'This is a representation of this sea and, in a word, there is no doubt that it has two arms, one of which extends to Wayla, the other to al-Qulzum, and a gulf extending from the other side to 'Abbādān. It has no other form but this." In spite of knowing otherwise, Muqaddasī chose to present what he refers to as a "rather general map," to avoid including any information that may be disputed. Muqaddasī, *Best Divisions for Knowledge*, 11.

3. There was a sixty-year hiatus between Konrad Miller's extensive reprint of medieval Islamic maps in the six-volume *Mappae Arabicae* and *HC 2.1* of the ongoing *History of Cartography* series. At the same time, I completed a master's essay titled "Surat Bahr al-Rum: The Mediterranean in the Muslim Cartographical Imagination." Exceptions to this sixty-year hiatus include the work of J. H. Kramers, in particular his Arabic edition of Ibn Ḥawqal's *Ṣūrat al-Arḍ* (Picture of the Earth) and subsequent French translation, *Configuration de la terre*, published posthumously in 1964 with translations of the maps; Beckingham's brief article "Ibn Hauqal's Map of Italy"; and the revised entry on *kharīṭa* by Maqbul Ahmad in *EI2*. I discuss these and other works at length in the secondary source discussion in the next chapter. Recent examples of this "belittling" can be seen in the work of Silverstein, "Arabo-Islamic Geography," 59; Sezgin, *Mathematical Geography*, 127; and King, *World-Maps*, 37–38.

4. Ahmet Karamustafa seconds this evaluation in *HC 2.1* when he states that "Orientalists could find little scientific basis for them and so failed to take them seriously" (9). Even though scholarship on Islamic cartography has increased since 2000, the emphasis continues to be primarily on veracity, authenticity, and place-name identification, with little effort to understand the mentalité and historical stories concealed within the names and shapes on these maps. Exceptions are Kaplony's "Comparing al-Kāshgarī's Map," Herrera's "Granada en los atlas de al-Sharafi," and Rapoport's "View from the South."

5. Harley, "Deconstructing the Map," 238.

6. Among the numerous prehistoric examples of maplike paintings, some of those in the prehistoric Lascaux caves in France, dating to more than 17,000 years ago, have been interpreted as star charts. See http://issuu.com/lightmediation/docs/the_lascaux_cave___a_prehistoric_sky-map_3390?mode=embed&documentId=080926150256-7f1d39d0e87c4e28a7515a2712d1e473&layout=grey (accessed March 5, 2015); http://www.lascaux.culture.fr/#/en/00.xml (accessed April 26, 2015); and Curtis, *Cave Painters*. See also Fernández-Armesto, *Pathfinders*, 16–37.

Depending on our definition of "map," language itself could be considered a form of mapping. Like mapping, language involves the use of symbols. The pictographs of the Chinese language can, for instance, be read as a series of maplike symbols. Studies on the mapping of language and text abound since the linguistic turn of the 1990s. See, for instance, Conley's *Self-Made Map* and Padron's *Spacious Word*.

7. Zadeh and Antrim were the first scholars to use this translation for *Kitāb al-masālik wa-al-mamālik*. I am grateful to them and their mentor, Roy Mottahedeh (Gurney Professor

of History at Harvard University), for coining this translation for the title of the KMMS manuscript series, which until their use of "routes" for *masālik* and "realms" for *mamālik* was usually translated as "Book of Roads and Kingdoms." See Zadeh, *Mapping Frontiers*, 15; and Antrim, *Routes and Realms*.

8. See chapter 3 for a detailed rationale for the usage of KMMS.

9. Should we define these maps as Middle Eastern or Islamic? There is no easy solution to this naming dilemma. "Middle Eastern" is a term for the region that came into widespread use with American ascendancy in world politics following World War II. It replaces "Near East," which was used by Europeanists from the eighteenth and nineteenth centuries. Technically, the word makes no sense. "Middle" of what "East"? It only makes sense if one thinks of Europe in relation to America as the "Near East." In which case the use of "Middle East" is sensible as an expression of American hegemony in the region. There is an effort to bring "West Asia" into use as a replacement for "Middle East." It makes more geographic sense and provides an alternative to the resonances of American hegemony in the term "Middle East." But "West Asia" is not a widely recognized term. The alternate term used more frequently for describing the medieval period in the region is "Islamic." The issue with using "Islamic" is that it implies that everything that uses the term is related in some way to the religion of Islam whereas there are many aspects of material and cultural history of the region, including maps, which are not obviously related to the religion of Islam. Marshall Hodgson sought to implement a solution by introducing the term "Islamicate" to refer to aspects of Islamic culture that "would refer not directly to the religion, Islam, itself, but to the social and cultural complex historically associated with Islam and the Muslims, both among Muslims themselves and even when found among non-Muslims." Hodgson, *Venture of Islam*, 1:59. Hodgson's approach has regained popularity and is now considered the more politically correct approach. I did consider using "Islamicate maps" but held back because of the cumbersome nature of the word and the resonance that it carries of connections to the Islamic caliphate. The fallback term to describe aspects of Muslim culture (medieval and modern) is "Islamic" or "Middle Eastern," and these are what I use in this book, fully cognizant that these terms are fraught with problems. I use these terms in the Barthian sense as empty, parasitical forms in which "meaning leaves its contingency behind . . . and only the letter remains." Barthes, *Mythologies*, 117–18. For more details on the use of these terms, see Pinto, "Middle East," 1522–23; and Pinto, "What Is Islamic about Islamic Maps?"

Chapter 2

1. I first began working on this tradition in the early 1990s. Subsequently, the History of Cartography Project at the University of Wisconsin–Madison published, through the University of Chicago Press as part of its *The History of Cartography* (*HC*) series, the volume *Cartography in the Traditional Islamic and South Asian Societies* (*HC 2.1*). I completed and deposited my doctoral dissertation on the subject of the medieval Islamic world maps, "Ways of Seeing.3," at Columbia University in 2002. Since then study of the Islamic carto-geographical tradition has grown rapidly. Toward the end of this chapter I discuss recent scholarship on the subject.

2. Recent contributions on the subject of Islamic sciences have significantly advanced the

picture. See, for instance, Roshdi, *Arabic Science*; Gutas, *Greek Thought*; Saliba, *Islamic Science*; and Dallal, *Islam, Science*, which are important examples of this trend. Outside of David King, *World-Maps*, and Fuat Sezgin, *Mathematical Geography*, though, few historians of Islamic science address the question of Islamic mapping as a scientific endeavor.

3. The nineteenth- and twentieth-century list of scholarship on the Islamic geographers is voluminous. For an in-depth discussion that names the key architects and lays out the Orientalist passion for the medieval Arabic geographical tradition, see Tolmacheva, "Medieval Arabic Geographers." In addition, see Zadeh, "Beyond the Walls of the Orient," in *Mapping Frontiers*, 148–77, for an excellent critique of Orientalist scholars who went to great lengths to falsely prove Sallām the Interpreter's mid-ninth century encounter with the Great Wall of China during his journey to find Gog and Magog.

4. It is not my purpose to completely dismiss the body of Orientalist contributions on the subject of Islamic geography. From the earliest work of the sixteenth-century French Arabist Guillaume to the work of the prodigious twentieth-century Russian scholar Ignatiy Iulianovich Krachkovsky, with giants such as Reinaud, de Meynard, de Goeje, Kramers, Mžik, and Minorsky, they have made significant contributions to the subject. These contributions have, however, been written up in great detail in a multitude of sources including *EI1*, *EI2*, *HC 2.1*, and Sezgin, *Mathematical Geography*, 1:3–21. Given that literature surveys are the bane of most readers, I decided not to repeat surveys of earlier literature that are comprehensively addressed elsewhere. My purpose in this review of scholarship on Islamic geographical material is to focus on new avant-garde approaches that make a deliberate break with past assessments and fall into what can be described as a postmodern genre.

5. Up until Miquel's work, the most extensive study was Krachkovsky's *Arabskaya geograficheskaya literatura*.

6. For an excellent collection of essays studying the role of *adab* in Arabic belles lettres, see Kennedy, *On Fiction and "Adab."*

7. Silverstein, *Postal Systems*, 5, 64.

8. Silverstein, "Medieval Islamic Worldview," 285.

9. Touati, *Islam and Travel*, 3, 258–64.

10. Ibid., 128–55.

11. Emily Burnham's PhD dissertation, "Edges of the Earth," extends Zadeh's discussion of the unknown through an in-depth examination of seven geographical texts. What she shows is that the unknown is a significant part of the geographical texts. In doing so, she calls into question the conclusions of Touati, Silverstein, and others who see the bulk of the geographical texts as focusing on the Islamic world to the exclusion of other areas especially the unknown edges of the world. One hopes that she publishes this very useful study soon.

12. Antrim, *Routes and Realms*, 70–72.

13. Silverstein, "Arabo-Islamic Geography," 59.

14. Zadeh, *Mapping Frontiers*, 10 (quotation), 24, 41, 123, 131, 134–35, 144, 147, 165, 174; plates 4, 6a–b, 7–8, 11–12.

15. Touati, *Islam and Travel*, 132–55.

16. Antrim, *Routes and Realms*, 114–25.

17. Ibid., 109. Antrim's justification for this comes in a footnote on page 175 in which she

asserts, "The copying of maps was not necessarily any less stable than the copying of written texts (if anything, it may have been more so.)" This runs counter to received wisdom on maps in the history of cartography and, for that matter, modern art theory. See, for instance, Bryson, *Vision and Painting*, 13–35, in which he explains how the nineteenth-century Rankean approach to images creates the myth of the "Essential Copy" when "no single Essential Copy can ever be made, for each strange trace left behind by the habitus was, relative to its social formation, itself an Essential Copy already" (15).

18. Miquel, *La géographie humaine*, 20–30, included hand-drawn black-and-white sketches of maps from one manuscript but does not list the manuscript number or location.

19. Miller, *Mappae Arabicae*, 2:184, in which Miller also ridicules the coarseness of the cartographic portrayals.

20. Because Sezgin focuses on European connections to Islamic mapping, his compendium, *Mathematical Geography*, is not a replacement for Miller's *Mappae Arabicae*.

21. Posthumously completed by G. Wiet and published as Ibn Ḥawqal, *Configuration de la Terre*.

22. For a selection of some of Kramers's key articles, see Kramers, *Analecta Orientalia*. For an example of Kramers's worst article on the subject, see "Geography and Commerce."

23. I discuss the maps in the next chapter. Von Mžik also studied the connections between Ptolemy and Arab geography in "Ptolemaeus und die Karten."

24. Mžik, *Al-Iṣṭaḫrī und seine Landkarten*.

25. Baghdad, for instance, is misplaced in Africa, but Mžik does not comment on this.

26. Ahmad, "Kharīṭa," 1079.

27. Beckingham, "Map of Italy," 78.

28. See chapter 3 for further discussion on the Ma'mūnid map and Sezgin's questionable 'Umarī map thesis.

29. I have examined this manuscript at length. This fourteenth-century rendition of the world cannot be a direct copy of the missing ninth-century Abbasid globe. The expert in mathematical geography, David King, concurs. His conclusion is summed up in his caption to 'Umarī's world map: "Remarkable but entirely superfluous longitude and latitude grid and an unhappy definition of the climates." King, *World-Maps*, 34. King explains that "there is no need to assume a highly ingenious cartographic projection: the copyist of the manuscript just drew the lines and circles in the simplest possible manner that would have the meridians meet at the poles. The proof of this—and of the fact that he was nothing more than an amateur in mathematical geography—is that the divisions on his scale for the seven climates—which correspond to latitudes from 0° (!!) to about 50° rather than from about 16° to about 50°—are also uniform. In the accompanying text Ibn Faḍlallāh explains that whereas some people assign an irregular progression of latitudes to the climates he is going to use a uniform one!" Ibid., 36–37.

30. Sezgin aggressively counters anyone who tries to point out this crucial discrepancy. See, for instance, his attack in prose on Tibbetts's attempt to set the record straight. He accuses Tibbetts and the editors of the *History of Cartography* series, particularly *HC 2.1*, of being dishonest and of drawing a "primitive and erroneous picture" of the achievements of Islamic cartography. Sezgin, *Mathematical Geography*, 1:18–21 (quotes, pp. 19, 21), 101–7, 133–36. The significant contributions of *HC 2.1* are discussed later in this chapter. Instead of utilizing the

findings of *HC 2.1* in his update, Sezgin relies on the outdated works of nineteenth- and early twentieth-century Orientalists whose findings he can harness to confirm his theories. He leaves out crucial contributions to the subject of mathematical cartography by David King. (See the discussion on David King's contribution to the study of Islamic cartography later in this chapter.)

31. The only major addition to this new voluminous set is Sezgin's recognition of the debate on whether Ptolemy accompanied his original treatise with maps, and of the fact that there are no Ptolemaic manuscripts with maps until the thirteenth century, an issue that he had missed in his 1987 book *Contribution of the Arabic-Islamic Geographers*. He expands his discussion of Bīrūnī maps but is unable to admit that Bīrūnī's world map bears less resemblance to the world as we know it than KMMS models even though it influenced a broad range of manuscripts from the *'ajā'ib* (fantastic/wondrous) and encyclopedic traditions. Readers should refer to chapter 3 of this book for a discussion on Bīrūnī's map and the variety of manuscript traditions that use variations of it. Upon viewing this map, they will be able to tell that it cannot be classified as an improvement on the KMMS world map. It is drawn from a different perspective, with the eye of the cartographer located in Central Asia. It is in fact one of the hardest medieval Islamic world maps to decode.

32. For instance, Sezgin cites and reproduces Brunetto Latini's world map of circa 1265 (*Mathematical Geography*, 3:144, map 55), but his only argument for claiming that this is "the oldest indirect European imitation of the [Islamic Ma'mūnid] world map" is that this world map contains place markers without labels, a sign according to Sezgin that "its copyist painted the map depiction, but obviously failed to concern himself with the place names written in an alien language." The use of "obviously" requires explanation because it is *not* obvious. There is no effort to discuss how the map matches up with the Islamic mapping tradition or Sezgin's contemporary reconstruction of the Ma'mūnid map or 'Umarī's. Sezgin, *Mathematical Geography*, 1:xviii. This is one of hundreds of maps that Sezgin throws at readers without adequate evidence of their connection to the Ma'mūnid map. Strangely, volume 2 of Sezgin's new study is still not in print. Only volumes 1 and 3 are available.

33. Sezgin, *Mathematical Geography*, 1:127.

34. See Witkam's reservations on Sezgin's Frankfurt facsimile series, which he has called "a *Fehlleistung* in the field of scholarly publishing on an as yet of unheard of scale." Witkam, "Arabic Manuscripts in Distress," 180.

35. For examples, see Kimble, *Geography in the Middle Ages*; Crone, *Maps and Their Makers*; and Tooley, *Maps and Map-Makers*.

36. Qibla charts can best be described as way-finding diagrams and instruments for locating the direction to Mecca. The applied mathematics of geodesy is used to locate points on the surface of the earth in order to determine the magnitude of either the whole earth or a sizeable portion of its surface.

37. King and Lorch, "Qibla Charts." David King has a plethora of works on the subject. See, for instance, King's "Sacred Direction in Islam," "Science in the Service of Religion," "Architecture and Astronomy," *Some Medieval Qibla Maps*, "Two Iranian World Maps," *World-Maps*, and *Islamic Astronomy*, among many others. For a complete listing of David King's rich body of works on the subject of Qibla maps, zīj tables, and Islamic sciences in general, see http://www.davidaking.org/Publications.htm (accessed March 11, 2015).

38. For more detail on this thesis, see King, "Two Iranian World Maps."

39. Heck, *Construction of Knowledge*, 97–99.

40. Tibbetts, "Balkhī School,"120–21.

41. I caution scholars who use Tibbetts's *HC 2.1* articles not to rely exclusively on the manuscript information that he lists.

42. The downside is that Brauer's work has some Arabic translation and transcription errors, as well as some errors with the dating of maps. The only other person who has worked on the question of boundaries in Islamic carto-geography is Antrim (*Routes and Realms*, 114–25). Both works should be approached with care.

43. Bloom, *Paper before Print*, 151.

44. Savage-Smith has also published two articles with her former assistant on the *Book of Curiosities* project, Yossef Rapoport, whose specialty is Islamic law and everyday life in the medieval Islamic world, in particular women, slaves, and peasants: "Medieval Islamic View of the Cosmos" and "The *Book of Curiosities*." In addition, Savage-Smith and Rapoport worked jointly on a cutting-edge website for the manuscript, which permits viewers to see the entire manuscript and its images in exceptionally high resolution, along with relevant translations: cosmos.bodley.ox.ac.uk (accessed April 26, 2015). A version of this has been published by Brill: *Eleventh-Century Egyptian Guide*, ed. and trans. Rapoport and Savage-Smith. Savage-Smith's earliest article on the manuscript, also published jointly with Jeremy Johns, is "*The Book of Curiosities*." There are controversial dating issues related to the maps of the *Book of Curiosities* manuscript. Savage-Smith claims that the maps are coterminus with the eleventh-century date of the manuscript, but Jeremy Johns intimates a later twelfth-/early thirteenth-century date through the image captions of their joint article. See chapter 3 for further discussion about the maps in the *Book of Curiosities* and their lingering dating issues. It would be extremely useful if the Bodleian Library conducts a carbon-dating test on their *Book of Curiosities* manuscript in order to resolve the question of its dating.

45. Rapoport continues to support Savage-Smith's dating of the *Book of Curiosities* and has pushed it even earlier to c. 1050.

46. Alai, *Maps of Persia*, 3–9.

47. Elsevier originally asked me to do this piece, but because of the 2006 Israeli-Lebanese war and a hasty evacuation, I had to decline; and I am grateful to Brentjes for taking over this entry and producing such a solid overview that is of service to everyone with an interest in the history of Islamic cartography.

48. Bouhdiba's brief article "La tradition iconographique" displays such a superficial and faulty grasp on Islamic cartography that I excluded it from the main discussion. There are also some dissertations on the subject that have not been published to date. Including Tarek Kahalaoui's dissertation, "The Depiction of the Mediterranean in Islamic Cartography (11th–16th Centuries)," which follows on my master's essay "Surat Bahr al-Rum," and uses the like-named *Eastern Mediterranean Cartographies* chapter as its launching point. In addition, there are the recent dissertations of Imants Ļaviņš, "Depiction of Scandanavia," and Emily Burnham, "Edges of the Earth," which make significant contributions to the subject.

49. See *L'océan atlantique musulman* and "La Mèditerranèe Musselmane" as examples of Picard's work. Since I do not address the Mediterranean at length in this book, I have not cited all the examples of Picard's extensive work on the subject.

50. See the listings in the bibliography for Jean-Charles Ducène. For a full list of Ducène's work, see https://ephe-sorbonne.academia.edu/DucèneJeanCharles (accessed March 26, 2015).

51. Additional articles on Islamic mapping in *JMISR* include the brief but valuable one by the Islamic astronomy expert, Paul Kuntizsch, "Celestial Maps and Illustrations in Arabic-Islamic Astronomy." The other article by Yossef Rapoport is "The *Book of Curiosities*: A Medieval Islamic View of the East," which does not incorporate the depth of analysis of his more recent work.

52. Brotton, *Twelve Maps*, 54–81. For a discussion on how Idrīsī's maps have skewed the discussion on Islamic cartography, see chapter 3 of the present volume.

53. This is the case even if the book does not focus on the map pictured on its cover. See, for example, the cover of *GE*, which reproduces an attractive section from a KMMS Mediterranean map; the cover of Antrim's *Routes and Realms*, which shows the earliest extant KMMS world map; Seed, *Oxford Map Companion*, with the Mediterranean map from the *Book of Curiosities*; and Akerman's *Finding Our Place*, depicting an Ottoman map from Pīrī Re'īs's *Kitāb-ı Baḥrīye* (Book on Maritime Matters).

54. "Concepts in the History of Cartography" was J. B. Harley's first published attempt to expand the boundaries of the history of cartography. Published jointly with M. J. Blackmore, it took up an entire issue of *Cartographica*.

55. Selected examples of Harley's work include "Ancient Maps Waiting to Be Read," "Meaning and Ambiguity," " Iconology of Early Maps," "Monarchs, Ministers, and Maps," "Maps, Knowledge, and Power," "Secrecy and Silences," "Historical Geography," and "Why Cartography Needs Its History" (with David Woodward). Refer to the bibliography for key works of Roland Barthes, Robert Darnton, Jacques Derrida, Terry Eagleton, Michel Foucault, E. H. Gombrich, Frederic Jameson, Thomas Kuhn, Erwin Panofsky, Edward Soja, Yi-Fu Tuan, and Hayden White.

56. Harley, "Deconstructing the Map," 231.

57. Ibid., 3.

58. Harley and Zandvliet, "Art, Science, and Power," 17. For more of Harley's work published posthumously, see Laxton, *New Nature of Maps*.

59. *HC 1*, xvi.

60. See, for instance, Jacob's *La description de la terre habitée*, "Géométrie, graphisme, figuration," "Mapping in the Mind," "Il faut qu'une carte soit ouverte ou fermée," "Quand les cartes réfléchissent," "De la terre à la lune," *Sovereign Map*, "La carte du monde," "Toward a Cultural History of Cartography," and "Will There Be Histories of Cartography after the *History of Cartography* Project?"

61. This is a favorite topic of Monmonier's, and since then he has published a number of other books in a similar vein with stimulating titles, such as *Air Apparent*, *From Squaw Tit to Whorehouse Meadow*, and *No Dig, No Fly, No Go*.

62. See Cosgrove's *Iconography of Landscape*, *Mappings*, *Apollo's Eye*, and *Geography and Vision*. Lewis and Wigen's *Myth of Continents* belongs in this illustrious group.

63. This is but a sampling. Theoretical discussions on maps are an active genre. There are many other thought-provoking works, such as Lilley, *City and Cosmos*; Safier, *Measuring the New World*; Scafi, *Mapping Paradise*; della Dora, *Imagining Mount Athos*; and Cattaneo, *Fra Mauro's Mappa Mundi*.

64. Nuti, "Mapping Places"; Whittington, *Body-Worlds.*

65. Conley, *Self-Made Map,* 3.

66. Birkholz, *King's Two Maps.*

67. Barnes and Duncan, *Writing Worlds*; Duncan and Ley, *Place/Culture*; Gregory, *Geographical Imaginations*; and Lewis and Wigen, *Myth of Continents.* With *Representing Place,* Casey joined the ranks of premier map theorists. His work rivals that of Christian Jacob. In keeping with his phenomenological interest in place, Casey has authored a number of other books on place, including *Getting Back into Place* and *Fate of Place.*

68. Schama has been exploiting images for history since his "Domestication of Majesty."

69. Bulliet's new book *The Wheel* is forthcoming 2016.

70. Because of the premodern nature of this book, I have not addressed the use of material culture in modern Middle East history. Due to the paucity of written sources, pre-Islamic history relies on material culture for evidence. Thus, Robert Hoyland employs a wide range of sources from tombstones to rock drawings, seals, stelae, and even votives in his *Arabia and the Arabs.* I have excluded the work of Islamic art historians because they too rely on images and material culture for their work, which makes an exhaustive list impossible within the confines of this brief literature survey. Four books with a strong historical bent warrant mention: Jonathan Bloom's exemplary *Paper before Print* (discussed earlier in this chapter); the pathbreaking work of Finbarr Barry Flood, in particular *Objects of Translation* and his contribution to our understanding of the medieval interaction of Hindu-Muslim culture; Nasser Rabbat whose work on architecture and material culture, *Mamluk History through Architecture,* has deservedly been recognized for broadening the picture of medieval Egyptian history; and, of late, the efforts of Emine Fetvacı to read history through Ottoman miniatures in *Picturing History at the Ottoman Court.*

71. See also Ramaswamy's earlier article on the subject, "Catastrophic Cartographies."

72. Felipe Fernández-Armesto's *1492, Near a Thousand Tables, Civilizations,* and *Millennium.* As a rule, most of Fernández-Armesto's books involve the use of maps for history. He does not, however, write history exclusively from the point of view of maps the way Ramaswamy and I do.

Chapter 3

1. Even in 1890, Guy le Strange reported that "the geographer Idrisi, is perhaps better known in the west than any other Arab cartographer. As long ago as 1592 the text of his book was printed in Rome." Le Strange, *Palestine,* 7.

2. See *HC 2.1,* 173–74, for a listing of extant Idrīsī maps and manuscripts. The Bodleian Ms. Arab c. 90, *Book of Curiosities,* also contains an Idrīsī map on fols. 27b–28a. The presence of this Idrīsī map suggests a later twelfth- or thirteenth-century date for the maps in the *Book of Curiosities* manuscript because to the best of our knowledge Idrīsī completed his *Kitāb nuzhat al-mushtāq,* also called the *Kitāb Rujār* (Book of Roger), in 548/1154. For the dating dispute, see discussion of the *Book of Curiosities* manuscript later in this chapter along with note 7.

3. For an example of the Pīrī Re'īs map of South America, see *PDH,* 58–59. Pīrī Re'īs is also known for an extensive geographical treatise, the *Kitāb-ı Baḥrīye* (Book on Maritime Matters).

The copy at TSMK Hazine 642 contains 215 detailed maps of the Mediterranean, illustrating islands, coasts, and harbors.

4. The most controversial study is Hapgood, *Maps of the Ancient Sea Kings*, which was popularized by von Däniken in *Chariots*. For more detail on the Pīrī Reʾīs map, see S. Soucek, *Turkish Mapmaking*; and McIntosh, *Piri Reis Map*.

5. In an effort to rectify this imbalance, I have written an article examining the connections between Pīrī Reʾīs's world map and Islamic cartography: "*Searchin' his eyes*."

6. An excellent example of this phenomenon is the four-part BBC radio series on medieval mapping (*The Medieval Ball*, aired December 2000), in which Islamic cartographic contributions were included in a separate session devoted exclusively to the subject.

7. In the earliest article on the Bodleian's Ms. Arab c. 90 (*Book of Curiosities*), "A Newly Discovered Series," Johns and Savage-Smith date the manuscript, based on the textual subject matter, as sometime between 1068 and 1071. The captions that accompany the maps, however, list the date as "Undated 13th century?" The dating confusion continues. Savage-Smith uses "An Eleventh-Century Egyptian View" in the subtitle of a recent article, and Rapoport pushes the date back even earlier to c. 1050 in his January 2015 Warburg Institute talk entitled "The World Map in the Fatimid Book of Curiosities (c. 1050): Mathematical Geography between Late Antiquity and Islam." This in spite of the fact that the manuscript contains an Idrīsī-type world map (see note 2) and a section of Ibn Bassām al-Tinnīsī's thirteenth-century account on Tinnīs. Rapoport and Savage-Smith dispute the date of Tinnīsī's work in their subsequent Brill facsimile *Eleventh-Century Egyptian Guide* (470) based on an article by Lutfallah Gari which uses the *Book of Curiosities* as proof for the earlier date. This, as Houari Touati points out in his book review of *Eleventh-Century Egyptian Guide*, amounts to circular reasoning and undermines the dating of the manuscript. A radiocarbon test on the *Book of Curiosities* is urgently needed in order to resolve the serious question of the accurate dating of this manuscript.

8. Scholars who have worked on this manuscript confirm this assessment when they say: "Despite the originality of its maps of the Mediterranean, the second book of *The Book of Curiosities* seems not to have had any influence on later cartographical developments." Johns and Savage-Smith, "A Newly Discovered Series," 20.

9. For more information on Qazwīnī's *ʿAjāʾib al-Makhlūqāt* manuscript, see von Bothmer, "Die Illustrationen des Münchener Qazwini"; and Berlekamp, *Wonder*.

10. For a succinct general introduction to Bīrūnī and his works, see *EI3*, s.v. "Bīrūnī," by Michio Yano, accessed January 20, 2014, http://referenceworks.brillonline.com/entries/encyclopaedia-of-islam-3/al-biruni-COM_25350. For a brief but terrific biography of Bīrūnī, see Lawrence, "Al-Biruni."

11. For a fascinating discussion on how to digitally reconstruct a world map from Bīrūnī's latitudinal and longitudinal (zīj) tables, see Ļaviņš, "World Map Reconstruction."

12. Tibbetts, "Later Cartographic Developments," 141–42. Note that figure 6.4. of *HC 2.1* lists the date of Ms. Or. 8349 incorrectly as 420/1029 whereas the correct date is 635/1238 as Tibbetts mentions in the text accompanying this map. Ramsey Wright published a copy of this manuscript along with a translation of the Persian version of Bīrūnī's manuscript. Bīrūnī, *Art of Astrology*, trans. Wright. See my discussion of figure 6.20 in chapter 6.

13. Who the "Wiznīk" are is a mystery. The name can also be read as "Waznīk" or "Wuznīk."

The name sounds Polish and could be the area of their origin. Note that in the Qazwīnī rendition shown in figure 3.4 "Waznīk" is not listed although the Baltic Sea (Baḥr Warank) is.

14. Yāqūt, *Introductory Chapters*, 44n1; Kowalska, "Sources," 41–88; and *EI2*, s.v. "al-Ḳazwīnī," by T. Lewicki.

15. Ibn al-Wardī–type world maps are even found in some fifteenth-century KMMS manuscripts. See, for instance, TSMK A3012, fols. 2b–3a.

16. There is considerable controversy over the authorship of *Kharīdat al-ʿajāʾib wa-farīdat al-gharāʾib*. According to Ben Cheneb in *EI1*, it is no more than a plagiarized copy of Najm al-Din Aḥmad ibn Ḥamdān ibn Shabīb al-Ḥarrānī al-Ḥanbalī's *Jāmiʿ al-funūn wa-salwat al-maḥzūn*. We are told that Ḥarrānī lived in Egypt about 732/1332, but beyond that, information on him and his book is limited. The editors of *EI2* add some information to the *EI1* entry. They tell us that according to Ziriklī, Ibn al-Wardī is correctly referred to as Ibn al-Wurūdī. In addition, the editors suggest that the real author could have been Zayn al-Dīn Ibn al-Wardī (d. 749/1349), a philologist, historian, and poet from late thirteenth century/early fourteenth-century Aleppo, or even ʿUmar ibn Manṣūr ibn Muḥ. ibn ʿUmar Ibn al-Wardī al-Subkī, who is named as an author in a Vatican copy of the *Kharīdat* manuscript. In other words, there is no consensus on who first penned *Kharīdat* nor whether it was written in the fourteenth or fifteenth century. All we can say is that it appears to have been written somewhere in the Levant or Egypt. As with the KMMS manuscripts where original authorship, especially of the maps, is also in question, we need to set aside impossible-to-resolve authorial questions for more telling questions of use and milieu. Like that of the KMMS, the life of the *Kharīdat* manuscript ends up being much larger than the original authors could ever have imagined. From the point of social history, readership and popularity is of much greater value than singular origins. For this reason, the editorial comment in the *EI2* entry that the *Kharīdat* is "without any scientific value" is a disservice. We cannot let our value judgments from the point of view of the twentieth and twenty-first centuries cloud our historical understanding of the later Islamic Middle Ages because that will prevent us from understanding with an open mind the reasons why *Kharīdat* was so popular for so many centuries as indicated by the numerous copies that exist of this book today—numbering in the many hundreds. It is probably because of this plethora of extant copies that the *Kharīdat* had, as the *EI2* editors put it, "a certain vogue among [eighteenth- and nineteenth-century] orientalists." See *EI2*, s.v. "Ibn al-Wardī"; and *EI1*, s.v. "Ibn al-Wardī," by Ben Cheneb. Fragments were translated and a variety of editions published by Hylander, Tornberg, M. Fraehn, and De Guignes. See also Nebes et al., *Orientalische Buchkunst*, 165. Just like all the pseudo-Ptolemy map manuscripts falsely ascribed to Ptolemy that abound, this issue should not prevent us from studying the maps in Ibn al-Wardī manuscripts.

17. I am working on a book-length study of the subject.

18. A Ṭabarī manuscript from Venice opens with a clime-type map of the world. This particular manuscript is Balʿamī's Persian translation of Ṭabarī's *Taʾrīkh*. Biblioteca Nazionale Marciana, Venice, Fondo G. Nani 78, Inv. Cod. Marc. Or. 128.

19. For a translation, see Yāqūt, *Jacut's geographisches Wörterbuch*, 1:28–29, fig. 4; or *HC 2.1*, 146, fig. 6.7.

20. Tibbetts, "Later Cartographic Developments," 146–48; Olsson is trying to broaden and nuance the study of "climes in medieval Islamic scholarship," by arguing that there is "no

uniform understanding of the climes among Muslims as has been noted by modern scholars." Olsson, "World in Arab Eyes," 491.

21. For an example of this map, see *HC 2.1*, 147, fig. 6.8.

22. For an example of this Idrīsī-type map, see Ibn Khaldūn, *Muqaddimah*, 1: frontispiece and 109–11, which is discussed by Kahalaoui, "Towards Reconstructing."

23. TSMK B. 411, fols. 141b–142a. For brief discussions on this map, see Lentz and Lowry, *Timur and the Princely Vision*, 149–50, which refers to this as "a map of the Timurid World"; Tibbetts, "Balkhī School," 126 and 128; and Roxburgh, *Persian Album*, 115. For a brief but excellent discussion on the development and use of graticules in Islamic maps, see Bloom, *Paper before Print*, 152–54. Bloom does not mention this Timurid map, but his suggestion that the idea for the use of grids comes from Chinese influence due to increased contacts between Iran and China during the thirteenth and fourteenth centuries is highly plausible.

24. For a witty discussion of the poor skill of the *şehnameciler*, in particular Lokmān, from the perspective of the sixteenth-century historian Muṣṭafā ʿĀlī, see Fleischer, *Bureaucrat and Intellectual*, 105, 239, and 249. As Fleischer aptly puts it, "Elegant dynastic propaganda rather than justification of action or presentation of an officially sanctioned version of events was the purpose of the imperial *şehname*-writer" (239). For a recent corrective to Fleischer's view, see Fetvacı, "Office of the Ottoman Court Historian," 7, 16, where she argues that "Ottoman *şehname*s commented on contemporary events, promoted the political agendas of courtiers as well as the sultran, and helped characterize their patrons and creators in highly nuanced ways. Neither passive ornaments nor formulaic eulogies of the sultan, these texts exerted significant agency in shaping the perspectives of their elite audiences in the Ottoman palace. They were of central importace for shaping history itself." Fetavacı rescues Loḳmān's reputation with her assertion that "when we examine Lokman's output of both image and text, it becomes obvious that the manuscripts are as much about members of the Ottoman court as they are about the sultan. Thus, the former understanding of the *şehnameci* as the eulogizer of the sultan is not quite accurate: Lokman clearly operated on behalf of a wider circle of courtly power-wielders."

25. The *Zübdetü't-tevārīḫ* manuscripts include a visual panoply of forty-plus lavishly painted images illustrating world history according to Islam from the earliest prophets to the sitting Ottoman sultan. The copy at Türk ve İslam Eserleri Müzesi (Museum of Turkish and Islamic Arts), Istanbul, TİEM 1973, presented to Sultan Murad III (r. 1574–1595) contains forty miniatures as does the copy that Loḳmān prepared for the grand vizier, Siyavuş Pasha (d. 1602). But the copy prepared in 991/1583 for the chief black eunuch, Meḥmed Agha (d. 1590), contains forty-five illustrations. Renda studied three of the most resplendent extant versions of *Zübdetü't-tevārīḫ*, including the one at Chester Beatty (Ms. 414), extensively in her doctoral dissertation, "Üç Zübdet-üt Tevarih." The elaborate visual program represents some of the most famous Ottoman miniatures from the late sixteenth and early seventeenth centuries and has been extensively studied by Ottoman art historians, most recently in *Picturing History* by Fetvacı, who trawls through Loḳmān's visual program in an attempt to make the history of the period come to life.

26. Bağcı et al., *Osmanlı Resim Sanatı*, 133, discuss this scroll, but the matter of its naming is clarified in Eryılmaz Arenas-Vives's dissertation, "*Shahnamecis* of Sultan Süleyman," 229.

27. For examples of this map, see *HC 2.1*, 221, fig. 11.15; and Minorsky, *Chester Beatty Li-*

brary, plate 13. For a discussion on the *Tārīḫ-i Hind-i Gharbī*, including its maps, see Goodrich, *Ottoman Turks and the New World.*

28. I stress the "imperial" context because KMMS-type world maps continue to be popular among the general public in other Ottoman-copied Arabic manuscript traditions, such as the so-called pseudo–Ibn al-Wardī manuscripts, which are a hybrid cosmography cum geography series, that were extremely popular during the sixteenth, seventeenth, and eighteenth centuries judging from the large number of extant copies.

29. Extensive explication of this map is not possible here without taking away from the overall focus of this chapter. This map is so rich that a discussion could be devoted entirely to it, as I did in "Ottomans Mediating Islamic Cartographic Space," prepared for "Imagining Imperial Topologies: Spatial Experience in the Ottoman World and Beyond," at Stanford University, May 16–17, 2014, and to be published as "Imagined Spaces, Nostalgic Topologies, and Territorial Anxieties: The Curious Case of the KMMS-Type World Map in the Tomar-ı Hümayun," in *Ottoman Topologies: Spatial Experience in an Early Modern Empire*, ed. Ali Yaycioğlu and Cemal Kafadar (Stanford University Press, forthcoming).

30. For a mid-sixteenth-century example and for a discussion of the development and use of this tradition, see Harawī, *Lonely Wayfarer's Guide*. For a listing of the three different types of images depicting the Kaʿba, see Ettinghausen, "Die bildliche Darstellung der Kaʿba," an albeit dated but still very useful source. J. M. Rogers has also worked on this material. See Rogers, "Two Masterpieces," 12–13, for images and discussion on the 951/1544–1545 Topkapı Saray scroll that was prepared as a proxy pilgrimage for Süleyman's favorite son, Şehzade Meḥmed. See also Rogers, "Itinieraries," 243–45. Aksoy and Milstein have written an immensely original and intriguing article about pilgrimage certificates and their paper, uses, and patterns: "Illustrated Hajj Certificates." Milstein has also written a detailed article on twenty-plus *Futūḥ al-ḥaramayn* manuscripts. Roxburgh, "Pilgrimage City," presents a synthesis of work done on the subject up to 2008. For a discussion of the *Dalā'il al-Khayrāt* series, see Bağcı et al., *Osmanlı Resim Sanatı*, 278–79, which also contains a color double-plate example of the depictions of Mecca and Medina. So numerous are these copies that chances are if you are looking for an illuminated manuscript to buy from the *sahaflar* (book market) in Istanbul, you will likely be offered a nineteenth-century copy of the *Dalā'il al-Khayrāt.*

31. There is no extensive scholarly study on Kaʿba tile maps (also referred to as Qibla tile maps). The only scholarly article to discuss them in the context of other illustrations of the Kaʿba is the dated but still very useful "Die bildliche Darstellung der Kaʿba" by Ettinghausen. The only other investigation of these images is a brief discussion by Shiela Canby in "Pilgrimage and Prayer." Numerous examples can be accessed through a simple web search under the heading "Kaaba tile." Here are a few examples: Victoria and Albert Museum, London, http://collections.vam.ac.uk/item/O79334/tile-unknown/ (accessed March 11, 2015); Aga Khan Museum Online Gallery, http://www.akdn.org/museum/detail.asp?artifactid=1624 (accessed March 11, 2015); Louvre, http://cartelfr.louvre.fr/cartelfr/visite?srv=car_not_frame&idNotice=22704&langue=fr (accessed March 15, 2015).

32. The 846 CE and 885 CE versions of Ibn Khurradādhbih's *Kitāb al-masālik wa-al-mamālik* were edited by Michael Jan de Goeje as part of a single textual version. The reference to the sacred geographical Qibla scheme occurs on page 5. See also Silverstein, *Postal Systems*,

64, 89, 91, and 93–94, for a discussion of Ibn Khurradādhbih's *Kitāb al-masālik wa-al-mamālik* and de Goeje's edition.

33. David King is the reigning expert on this subject. His publications on the subject are too numerous to list fully. See, for instance, "Qibla Charts" (with Lorch), "Sacred Direction," *Some Medieval Qibla Maps*, and *World-Maps*. For a complete listing of David King's works on the subject of Qibla maps, zīj tables, and Islamic sciences in general, see http://www.davidaking.org/Publications.htm (accessed March 11, 2015).

34. As Çağman and Tanındı astutely point out, between 1226 and 1540 Jerusalem did not have any walls. Çağman and Tanındı, *Topkapı Saray Museum*, 69; Gruber and Colby, *Prophet's Ascension*, 52.

35. For ways in which this image could have been inspired by an eighth-century Umayyad one, see my forthcoming article "Fit for a Prince."

36. See, for instance, Tibbetts, "Beginnings," 95; Bloom, *Paper before Print*, 145–46.

37. Five pages after this quote is another one that talks about the *Ṣūrat al-Ma'mūniyya* and its other maps, suggesting instead a book along the lines of a KMMS manuscript with its accompanying regional maps. Mas'ūdī, *Kitāb al-tanbīh*, 33, 38.

38. See chapter 7 for Islamic examples of images of the world with multiple Encircling Ocean bands signifying cosmographic layers of the universe.

39. See, in particular, Sezgin's *Contribution* and *Mathematical Geography*. Sezgin, a historian of Islamic science, is urgently trying to prove the existence of an extremely mimetic Islamic mapping tradition from the ninth-century onward. See discussion in chapter 2 on Sezgin's work. Note that Sezgin refers to 'Umarī as Ibn Faḍlallāh according to his first name, not his last. For additional discussion on Sezgin's theory, see chapter 2 and note 36 of this chapter. More work needs to be done on the maps in the 'Umarī manuscript in order to better understand the sources of influence. For a broader perspective on the encyclopedic nature of 'Umarī's work, refer to the work of Elias Muhanna, in particular "Arabic Encyclopaedism," 343–47.

40. Gutas, *Greek Thought*, 53–60, where he focuses on the question of the Bayt al-ḥikma. See also Micheau, "Scientific Institutions," who emphasizes—erroneously according to Gutas—the Bayt al-ḥikma as a clearinghouse for translations. Both these views are significant revisions of earlier scholarship that saw the Bayt al-ḥikma as a center of learning and scientific discoveries.

41. Mas'ūdī, *Kitāb al-tanbīh*, 183–85; Sezgin, *Mathematical Geography*, 1:80–81. I have used Sezgin's translation here with a few adjustments to indicate the difference between sites and cities in order to make sense of Mas'ūdī's listing of 4,530 places, which works out if one is including all known sites and not just the cities. Sezgin takes this to mean cities only and therefore presumes that the copyist of the Mas'ūdī manuscript made an orthographical error of 4,000. In fact, in *Kitāb al-masālik*, 5, Ibn Khurradādhbih mentions the existence of 4,200 cities, and Mas'ūdī, who lived and wrote about half a century later, logically increased this number. My point is that even when using Sezgin's translation the interpretation is the same. It makes sense to read this as Mas'ūdī talking about the KMMS series from the point of view of the bulk of the extant Islamic mapping tradition. It doesn't make sense to presume that Mas'ūdī is describing the maps in 'Umarī's manuscript, of which the earliest copy dates to almost five centuries after Mas'ūdī's lifetime, when closer in time and far more numerous is the KMMS

mapping tradition. Scholars applying a positivist, Rankean approach to the history of cartography are unable to see these KMMS maps as "correct" or "good" but rather view them as quaint miniatures with quirky shapes. From this perspective, which is still pervasive in the field of the history of cartography, maps are worthwhile only if they are built on the principle of mimesis. People who view maps in this light cannot see the development of the history of mapping—Islamic and otherwise—logically as part of a larger continuum. See, in addition, chapter 2 for my discussion of and citations from Sezgin's books on this subject.

42. For more details on the KMMS maps and a range of visual examples, see chapter 4.

43. Mas'ūdī, *Kitāb al-tanbīh*, 184.

44. Sezgin, *Mathematical Geography*, 1:95–96.

45. Zuhrī, *Ja'rāfīya* (sometimes referred to as *Kitāb al-jughrāfiyā*), 306. For a discussion on the naming discrepancy of Zuhrī, *Kitāb al-ja'rāfīya*, see Pinto, "*Searchin' his eyes*," 74–77. The reader should note the difference in my translation from that of Sezgin's in *Mathematical Geography*, 1:79. In some places Sezgin's translation is not faithful to the original text and the difference in meaning is significant, such as in the line: "Because the earth is spherical but the *ja'rāfiyā* [presumably a reference to a foldable map] is simple [uncomplicated] and its form expands the way astrolabes and collections of eclipses unfold," which Sezgin translates as "since the Earth was round, but the geographical depiction was flat. They moved it into the plane, as they had done with the astrolabe and with the eclipses contained in their records." The latter half of Sezgin's translation makes no sense and is not in line with what the Arabic text says.

46. It is puzzling that Sezgin cannot see that Zuhrī is referring to the KMMS mapping tradition. Instead, Sezgin uses these references in Zuhrī and Mas'ūdī to argue that they are referring to a spherical globe of the kind that he has reconstructed based on the zīj (astronomical) tables of the mathematical geographer Khwārazmī. See Sezgin, *Mathematical Geography*, 3:4, map 1b, for Sezgin's reconstruction, which is influenced by his own contemporary knowledge of the shape of the world just as nineteenth- and early twentieth-century reconstructions of ancient Greek maps based on textual descriptions were influenced by the reconstructors' modern mapping knowledge. (See, for instance, the reconstructions of the maps of Hecataeus, Dionysius Periegetes, and Strabo in Bunbury, *History of Ancient Geography*, 1:148, 2:238, 490; and the map of Dicaearchus in Cortesão, *Portuguese Cartography*, 1: fig. 16.) It is for this reason that these reconstructions are no longer accepted as representative of ancient Greek mapping by historians of cartography. For a lengthier discussion on the problems of ancient Greek mapping and the lack of extant examples, see chapter 6. The basis of Sezgin's enterprise rests on two elements: the coincidence of the name of Khwārazmī's manuscript, *Kitāb ṣūrat al-Arḍ*, ignoring that it is also a title used by some KMMS manuscripts; and Zuhrī's mention of an inexplicable dark patch in the area of the Indian Ocean where it connects with the Encircling Ocean on the map of the *ja'rāfiyā* (presumably a reference to the name of the book or perhaps a map, although this too is a highly debated matter). Taking these two elements as his foundation, Sezgin creates his own reconstruction of the Ma'mūnid map, asserting that it looked like the world map in the late fourteenth-century 'Umarī manuscript located in the Topkapı Saray library (TSMK A. 2797). In doing so, Sezgin ignores the four maps in the Khwārazmī manuscript—discussed in the latter half of this chapter. In particular, he provides an inadequate explanation of the map of the Encircling Ocean in the Khwārazmī manuscript. See

Sezgin, *Mathematical Geography*, 1:122. On the matter of the possible meanings of the word *ja'rāfiyā*, see Pinto, "*Searchin' his eyes*," 74–77.

47. Hamdhānī, *Kitāb al-buldān*, 283; also, Ṭabarī, *Ta'rikh*, 2:1199. See also Tibbetts, "Beginnings," 90; Bloom, *Paper before Print*, 144–45; Sezgin, *Mathematical Geography*, 1:64.

48. For more on Meḥmed II and his commissioning of KMMS copies, please refer to chapters 10–12.

49. The earliest studies are Saxl, "Zodiac"; Beer, "Astronomical Dating"; and the response by Hartner, "Qusayr 'Amra." More recent studies include those by Savage-Smith, *Islamicate Celestial Globes*, 16–17; and Savage-Smith, "Celestial Mapping," 12–16. Fuat Sezgin lists the cupola star chart of Quṣayr 'Amra as the earliest extant celestial atlas in his latest book. Sezgin, *Islam Uygarliginda*, 14. Savage-Smith and Sezgin both misdate this star chart to the period of Walīd I (705–715). Following a thorough cleaning and decipherment of a key inscription on the site, it has been confirmed that the bathhouse was commissioned by the Umayyad prince al-Walīd ibn Yazīd prior to becoming the eleventh Umayyad caliph, al-Walīd II (r. 743–744). http://www.artdaily.org/index.asp?int_sec=2&int_new=55730#.UBnZoUS5Irg (accessed August 16, 2015).

50. Lohuizen-Mulder, "Frescoes in the Muslim Residence," 133.

51. Ettinghausen, *Arab Painting*, 32.

52. Lohuizen-Mulder, "Frescoes in the Muslim Residence," 126; Fowden, *Quṣayr 'Amra*, 71–72.

53. For more on the discovery and interpretation of this image, see Pinto, "Fit for a Prince" (under review).

54. Balādhurī, *Futūḥ*, 371.

55. Ya'qūbī, *Kitāb al-buldān*, 239.

56. See, for instance, Necipoğlu, "Plans and Models."

57. For additional details on this, see chapter 7. For arguments on how Baghdad may have been laid out according to the plan of an imago mundi, see Wendell, "Imago Mundi." For earlier theses on the cosmological origins of the round city of Baghdad along with counterarguments, see Lassner, *Shaping*, 169–83.

58. This reference comes from Sezgin, *Mathematical Geography*, 1:77.

59. Maqrīzī, *Itti'āẓ*, 2:292. See also Halm, *Empire*, 374; Bloom, *Paper before Print*, 146–47. One wonders if the recently discovered *Book of Curiosities* is related in some way to this reported map of the Fatimid caliph Mu'izz.

60. Unfortuantely, Reinaud does not provide a reference for this silk map. It sounds like an early version of a KMMS map but there is no way of confirming this. Abū'l-Fidā, *Géographie d'Aboulféda*, 1:CCLXIII.

61. Ibn al-Nadīm, *Fihrist*, trans. Dodge, 2:672, but he does not mention whom it was made for.

62. Bibliothèque nationale et universitaire, Strasbourg, Cod. 4247. See Mžik's facsimile edition of Khwārazmī's *Ṣūrat al-arḍ*. The manuscript's colophon dates it to 428/1037, almost two centuries after the death of Khwārazmī (d. ca. 332/847), so we cannot be sure if Khwārazmī included maps with his original manuscript or if the maps are a later addition. The manuscript contains four blank pages, suggesting that four more maps were intended. For the latest discus-

sion on the Khwārazmī manuscript, see the doctoral dissertation of Imants Ļaviņš, "Depiction of Scandanavia," 14–17, and Ļaviņš, "Islamic Geographical Tables."

63. Leitner has attempted to extract the "world" map out of Khwārazmī's table, but his results have to be viewed with a considerable amount of skepticism because while one can plot coordinates on a grid, it is another matter to draw forms around those coordinates. Wilhelm Leitner draws the forms around the coordinates so that the map resembles early Renaissance "Ptolemaic" versions of the world, and on the basis of this, Leitner argues that the map of Khwārazmī influenced Christopher Columbus! See Leitner, "Abu Ga'far."

64. This is Sezgin's reading. Sezgin, *Mathematical Geography*, 1:122.

65. It could also be read as *fawwāra* ("spring," "fountain," or "jet of water"). Sezgin identifies it as *qawāra*, which also does not have an exact dictionary equivalent. Ibid.

66. Tibbetts, "Beginnings," 105–6, attempts to identify the two unknown maps. He suggests that the one with the mysterious title of *Jazīrat al-Jawāhir* (Island of the Jewel) could be the island of Sri Lanka or Taprobane. Confirming my view, Tibbetts identifies the anonymous sea map in "Beginnings," 106, with the caption "Baḥr al-Muzlim" (another name for the Baḥr al-Muḥīṭ, or Encircling Ocean), which he identified in an earlier publication as a map of the Indian Ocean. Tibbetts, *South-East Asia*, 68. It is hard to prove or disprove these identifications since this map contains labels that are difficult to decipher. See Ļaviņš, "Islamic Geographical Tables," for a different interpretation of this mysterious map of the sea.

67. This is the prize manuscript of a small, but delightful, library in Istanbul: Millet Genel Kütüphanesi, Ms. Ali Emiri, Arabi 4189.

68. For a rich, detailed, and extraordinarily original investigation of this map that discusses possible variations between this copy and the original as indicated through the text, see the work of the reigning specialist on this map, Andreas Kaplony, "Comparing al-Kāshgarī's Map," 137–53 and 209–25.

69. One of the earliest extant examples is Ibn Khurradādhbih's *Kitāb al-masālik wa-al-mamālik* (written around 231/846), discussed earlier in the chapter. Other early examples with the same title occur in works by Marwazī (d. 274/887), Sarakhsī (d. 286/899), and Jayhānī (ca. tenth century). See Silverstein, *Postal Systems*, 64–65 and 92–94, for a discussion on the way in which the "science of geographical writing in Arabic is indebted to the postal itineraries collected in the *Dīwān al-Barīd*" (93); and Heck, *Construction of Knowledge*, chap. 3, on how these geographies grew out of the administrative apparatus in tandem with the hegemonic goals of the state. Ducène discusses the question of the naming of Iṣṭakhrī manuscripts in "Quel est le titre veritable."

70. Touati, *Islam and Travel*, 119–55, argues otherwise. One can agree with the case that Touati makes for the "primacy of visual observation" for Ibn Ḥawqal and Muqaddasī, but Masʿūdī is a harder sell.

71. Occasionally there are variations among the regional maps: some illustrators include two maps of the Nile; others include up to three maps of the Mediterranean. For the purposes of this book, it suffices to know that all the KMMS geographical manuscripts—with the notable exception of the Muqaddasī series—begin with a map of the world, unless the manuscript is incomplete and the initial pages, including the world map, are missing.

72. The extent to which the maps are an integral part of the text has been the subject of

some debate recently. See Karamustafa, "Introduction," 4–5, who states that "they served a didactic or illustrative function subservient to the main textual narrative," so long as we keep in mind that the "graphic representation holds its own and cannot be explained away through textual comprehension."

73. The term "Atlas of Islam" is employed by Konrad Miller in his opus magnum on medieval Islamic maps, *Mappae Arabicae,* and taken up later by J. H. Kramers; see, for instance, his "La question." See also de Goeje, "Die Istakhrī-Balkhī Frage." I prefer to use KMMS for two reasons: (a) because it is shorter, and (b) because it represents a break from the nomenclature of the early twentieth-century and marks a reinvestigation of this tradition that makes a break with past analyses.

74. The earliest extant geographical manuscript containing maps dates from 479/1086, approximately a hundred years after the originals were first put down. The earliest extant manuscript is a copy of Abū 'l-Qāsim ʿAlī al-Naṣībī Ibn Ḥawqal's *Kitāb ṣūrat al-Arḍ*, housed at the Topkapı Saray library (Ahmet 3346) in Istanbul.

75. I refer to these four main authors of the cartographically illustrated KMMS geographical series as Balkhī, Iṣṭakhrī, Ibn Ḥawqal, and Muqaddasī in keeping with the tradition of Islamic historiography and for the benefit of the reader. Some scholars prefer to refer to Muqaddasī as ʿMaqdasī instead. I prefer to use the more common, recognizable usage of his name, Muqaddasī. This has the added advantage of distinguishing his work from that of another encyclopedic geographical historian, al-Muṭahhar al-Maqdisī, author of *Kitāb al-Badʾ wa-al-taʾrīkh.*

76. With the exception of *Kitāb tafsīr ṣuwar kitāb al-samāʾ wa-l-ʿālam lī Abū Jaʿfar al-Khāzin* (Book of Interpretation of the Images of Abū Jaʿfar al-Khāzin's "Book of the Heavens and the Universe"), which Barthold, preface, 18, interprets as proof that Balkhī was *not* responsible for making the maps. Ibn al-Nadīm, *Fihrist*, trans. Dodge, 1:302–3, credits thirteen more *tafāsīr* (commentaries on the Qurʾān) to Balkhī's name; and in his *Irshād*, 6:141, Yāqūt lists works compiled by Balkhī's grandson but does not list any geographical texts. The new *EI3* entry on Balkhī does not bring any new information to light. *EI2*, s.v. "al-Ḳazwīnī," by T. Lewicki. *EI3*, s.v. "al-Balkhī," by Hans Hinrich Biesterfeldt, accessed May 27, 2015, http://referenceworks.brillonline.com/entries/encyclopaedia-of-islam-3/al-balkhi-abu-zayd-COM_23114.

77. Barthold points to a likely source for this misnomer when he notes in contradistinction to de Goeje that Ṣafadī borrowed his information from Yāqūt. Barthold, preface, 17.

78. This is a matter that has been tackled fruitlessly from the nineteenth century and de Goeje onward. See note 74 and note 54 in chapter 2. Ducène addresses the problem of the titles of Iṣṭakhrī manuscripts in "Quel est le titre veritable."

79. This is an issue that has haunted the study of these maps since the nineteenth century (see chapter 1) and continues in some quarters today. Picard, "Mèditerranèe Musselmane." It cannot be the case that maps are inherently more stable than texts as Antrim suggests in *Routes and Realms*, 175. If anything, the reverse applies. For while handwriting can vary and errors creep in with copies, the variations in shape, form, and allocation of space are distinct from one stemma of map to the other. It is through this distinction and difference, in fact, that these maps can be dated and classified. Were it not for the differences in the copies of miniature paintings over the centuries the entire edifice of medieval and early modern art history would collapse.

80. The regional maps of the Mediterranean in this manuscript are especially striking for their mimesis. I will analyze and discuss the maps of the Mediterranean in a different book, *The Mediterranean in the Islamic Cartographic Imagination*, written under the auspices of an NEH fellowship. In this book, I restrict my focus to the world maps.

Chapter 4

1. This Timurid Persian map is the only medieval Islamic KMMS world map oriented with west on top. For purposes of continuity of understanding—so that the reader does not have to absorb a new direction for understanding the KMMS world map—I have oriented it like the other KMMS world maps, with south on top.

2. Panofsky, *Studies in Iconology*, 3.

3. This problem of identifying the KMMS images as maps exposes one of the major weaknesses of Panofsky's model: all his examples are drawn from the contemporary Western world. He does not factor into his analysis objects that are not easily identifiable in Western culture.

4. There are a few rare exceptions, such as those in Sallust's *The Jughurthine War* manuscripts at the Bodleian Library, Oxford (Ms. Rawl. G. 44, fol. 17v). Also some early Renaissance maps, such as the famous mid-fifteenth-century Fra Mauro map, are oriented like Islamic maps with south on top.

5. According to the Peters Projection, the most accurate projection of the world, Africa is much larger than Europe.

6. The overextension of Africa through a merging with Australia is a concept that exists in ancient Greek geographical literature as well.

7. This penchant for mimesis is referred to in theoretical circles as the "Natural Attitude," and it has come under strident criticism. For an explanation of why this mode of analysis is considered problematic, see Bryson, *Vision*, 13–36.

8. On Arab and Muslim navigation of the Indian Ocean and the farthest south that Muslim sailors traveled along the East African littoral, see Hourani, *Arab Seafaring*, 80–81; Toussaint, *Indian Ocean*, 52; A. Lewis, "Maritime Skills," 246–47; and Tibbetts, *Arab Navigation*. For information on Arab and Muslim navigational skills of the period, see Arunachalam, "Technology"; Tolmacheva, "Arab System"; and Tibbetts, "Red Sea." I am grateful to Dr. Ralph Pedersen for generously sharing his insights and expertise on premodern Indian Ocean maritime matters along with many useful references.

9. There is an alternate, infrequently seen form, which employs an oblong shape instead of a circle. There are only three extant oblong KMMS world maps and two are of questionable authenticity. Therefore this book focuses on the most common model: the classical "round" model.

10. Although none of the rivers are marked, it is generally assumed that the major river draining into the Caspian on this map is the Volga. It is mentioned in a number of Arabic and Persian sources, including *Ḥudūd al-ʿālam*, 75. The other two rivers draining into the Volga are difficult to identify. According to the anonymous *Ḥudūd*, the one shown on the eastern flank of the Volga is the Irtish—a river in Siberia that ends up in present-day Kazakhstan. In

his second edition of the *Encyclopedia of Islam* entry on the river, C. E. Bosworth argues that it cannot be the Irtish, but must instead be a reference to another river further west (possibly the Yayik, which rises in the Urals and flows to the Caspian), since it emptied into the lower Volga. *EI2*, 12:458, col. 2. The eastern tributary of the Volga, referred to in the *Ḥudūd* as the "Rūs" is even harder to identify. Minorsky follows on Toumansky and suggests that it could be the upper course of the Volga or the river Don. I think its more likely to be the Aras River (Araxes of yore), which flows from Erzurum in eastern Turkey to the Caspian, and is called al-Rass in Arabic. This would explain the mistaken connection shown by the cartographer between the Caspian and the Black Seas via the Volga River.

11. Note that Pinna's information on the world map in this manuscript is faulty. Pinna, *Il Mediterraneo*, 47. Jean-Charles Ducène discovered another late nineteenth-century Iṣṭakhrī copy in the Maktabat ʿAbd al-ʿAzīz, Medina. It does not, however, contain a world map. I am grateful to Ducène for his generosity in sharing images from this manuscript with me. For Ducène's discussion of this manuscript, see his article "Un nouveau manuscrit." Note that the correct number for the manuscript in Maktabat ʿAbd al-ʿAzīz, Medina, is ʿArif Ḥakamt Jughrāfiya 910/8.

12. The question of the injunction against images is a highly debated one. Other than the injunction of *shirk* (joining of partners with God), the Qur'an does not prohibit the use of images, but there are a number of ḥadīth that rail against the makers of images and condemn them to hell. This does not seem to have hindered the vibrant Islamic miniature tradition, which contains detailed figural and animal depictions that fly in the face of possible injunctions, at least from the twelfth century onward. For a fuller discussion of this issue, see Grabar, *Formation*, 72–98; and Elias, *Aisha's Cushion*, 1–138.

13. "Aqqoyunlu" is correctly transcribed as "Aqquyunlu," but "Aqqoyunlu" is currently the more prevalent spelling. "Akkoyunlu" appears in older works.

14. For more on the matter of the "Essential Copy," see Bryson, *Vision*, 1–12.

15. The Muslim caliphs understood that the islands of Cyprus, Crete, and Sicily were crucial stepping-stones for ships traversing the Mediterranean, and therefore the cartographers pegged them as strategically important for controlling the Mediterranean. Crete was the target of Muslim incursions as early as the period of the first Umayyad caliph, Mu'āwiya (660–680 CE). Control of Cyprus was key for controlling the Levantine coast. This is why Cyprus has rarely if ever been independent in its history. The first Muslim raiding parties landed on the island in 632, soon after the death of the Prophet Muhammad, during the first wave of conquests under the first caliph, Abū Bakr. Sicily was so admired in Muslim circles that by the twelfth century it was being referred to affectionately as the "daugther of Andalusia." Muslim raids started in the mid-seventh century, although the island was not fully occupied and incorporated into the Aghlabid province of Ifrīqiya (equivalent of Roman North Africa) until the early tenth century.

In the Persian Gulf, the least well known of the three islands is the one labeled "Lāft." I have been able to identify this as a reference to the largest Iranian island of Qishm because the main town is called "Lāft." Qishm is not only the largest island in the Persian Gulf; it is also located right at its mouth with the Indian Ocean and guards the Straits of Hormuz. It must have been a crucial island for shipping, yet because "Lāft" was not noticed before on the medieval Islamic maps, its significance was not realized until now. The island of Baḥrain, familiar to us today as a major business center off the coast of Saudi Arabia—and lately for the brutally suppressed protest movement—was one of the most important islands in the Persian Gulf in the medieval period because of its natural deepwater port and its pearls. Now oil has superseded pearling. Khārak, better known in petroleum circles as Kharg, lies opposite the islands of Baḥrain and Tārūt, along the Iranian flank. It is positioned close to the Hormuz Straits (entrance to the Persian Gulf from the Indian Ocean) and as a result has always been a key and heavily contested site. Like Baḥrain, it too was known in the medieval period for its high-quality pearls. Today Kharg is one of the key ports for Iranian oil.

16. Bagrow, *History of Cartography*, plate A. It was reprinted in resplendent color in *HC 2.1*, plate 7.

17. The Leiden KMMS manuscript has a folio size of 41.5 × 59.3 cm whereas most KMMS manuscripts have an average page size of around 22 × 14 cm. In other words, the Leiden KMMS manuscript is more than double the size of the average manuscript. This along with its elaborate illumination and ornate Kufic calligraphy suggests that this manuscript was most likely made for presentation to a king or a high-level dignitary.

18. No one has yet examined the possible connection between these two manuscripts.

19. I am grateful to Dr. Emilie Savage-Smith of the University of Oxford for sending me a single-page copy of her notes on this manuscript. It was useful to receive confirmation of the date, which I had also determined as AH 696. Savage-Smith's detailed analysis on the paper is thorough, but she does not address the parallels with the Leiden manuscript.

20. The copy at Sülemaniye library of TSMK's A3347 needs a detailed technical examination of the paper and inks as does the plainer version of this map found in Paris at the Bibliothèque nationale de France (Arabe 2214, fols. 52b–53a). The world map is oriented unusually with north on top instead of south. The illustrator does not differentiate the seas, which are all painted in an olive green.

21. By "classical" I mean the so-called Balkhī school of mapping, which I refer to as the KMMS mapping tradition. It is the primary subject of this book and comprises the work of the carto-geographical treatises attributed to Iṣṭakhrī, Ibn Ḥawqal, and Muqaddasī. For more information on the KMMS acronym and the problems with the attribution of the KMMS mapping tradition to the Balkhī school, please refer to the discussion in the introduction of this book.

22. See, for instance, the recent the intriguing work of Silverstein on premedieval Islamic geographic connections in "Islamic Worldview," 273–85.

Chapter 5

1. Note that I have adjusted for Rushdie's sexism and inserted my own gaze by adding "[S]" at the start. I thank Adrianne Wortzel for bringing this quote to my attention through her unusual performative article on mappamundi. Wortzel, "Sayonara Diorama," 419.

2. It should be noted that in medieval Europe *imago mundi*, meaning "image of the World" in Latin, was occasionally used to name geographical treatises, not just world maps. On this

issue, see Woodward, "Medieval *Mappaemundi*," 301–3; and Scafi, "Defining Mappaemundi," 345–54.

3. I have conducted two other iconographic exercises on the meanings underlying the form of the Mediterranean Sea and on the meaning behind the forms of the maps of the Maghrib (North Africa and southern Spain): "Surat Bahr al-Rum" (in *Eastern Mediterranean Cartographies*) and "Passion and Conflict."

4. During a 2013–2014 National Endowment of the Humanities fellowship, I will complete a book-length study on the Islamic maps of the Mediterranean, including an extensive iconographic analysis of the maps of the Mediterranean.

5. Acknowledging all the problems inherent in "prehistoric" dating, I adhere on this front to convention: that is, that the "prehistoric" period ends in the Middle East (specifically Mesopotamia and Egypt) at 3,000 BCE, whereas in the western Mediterranean and northern Europe it continues right up until the arrival of the Romans. Scandinavia is particularly problematic since there the Iron Age is said to continue right up until the eighth or ninth century CE.

6. Keep in mind Catherine Delano-Smith's caution that "the besetting difficulty (common to any interpretation of the meaning of prehistoric art) is that while a design may incorporate certain symbols, this does not necessarily mean that the artist who painted it intended a particular symbolic interpretation" ("Cartography in the Prehistoric Period", 88). Delano-Smith's article contains a wealth of additional prehistoric images with encircling ocean–like features.

7. The parallels between these forms and the later medieval Islamic maps is striking. See, in particular, the Ibn al-Wardī map examples, in chapter 7. The encircling triangles on the Babylonian map resemble the representation of the Muslim cosmic mountains of Jabal Qāf on the Ibn al-Wardī maps. For an alternate reading, see Polcaro, "Tuleilat al-Ghassul Star Painting," 276–83.

8. Schmidt, "Göbekli Tepe, Southeastern Turkey"; and Schmidt, "Stone Age Sanctuaries."

9. Allen, "Egyptian Concept," 24.

10. Maspero, *Histoire ancienne*, 16–19 (quote); and Ball, *Egypt*, 1–3. Plumey seconds this view in his article "Cosmology." He refers to the "Great Celestial River" as the "Great Circular Ocean" (20). A version of this story is also cited in some medieval Arabic geographical texts, often in the form of a ḥadīth-type narration. See chapter 7.

11. Plumey, "Cosmology," 25–26. See also Moers, "Pharaonic Egypt"; and Allen, "Egyptian Concept," 24–25.

12. Al-Kisā'ī, *Tales*, 7.

13. Tirmidhī, *Sunan*, 44:3298. Cited again below in the section on Indic Worlds.

14. Plumey, "Cosmology," 39.

15. The sarcophagus is on permanent exhibition at the Metropolitan Museum of Art alongside the temple of Dendur. For a detailed discussion on this image and other cosmographic examples, see Allen, "Egyptian Concept," 26–29. Allen cites this as "one of the earliest maps of the world as round" (29).

16. See Delano-Smith, "*Imago Mundi*'s Logo," 209; Unger, "Cosmos Picture," 1; Horowitz, "Babylonian Map," 154.

17. Horowitz, "Babylonian Map," 156. It is precisely this kind of hook-like end that distinguishes the Persian Gulf in the medieval Islamic maps.

18. For a detailed description of these places, defined ambiguously as *nagu*, which can have a variety of meanings, see Horowitz, "Babylonian Map," 156–64.

19. Heidel, *Babylonian Genesis*, 97, relates it to the primeval water cosmologies of Egyptian, Phoenician, and Vedic literature. Horowitz, "Babylonian Map," 156, raises the possibility that *Marratu* could be a reference to a "bounded" Encircling Ocean as opposed to the cosmic sea. It is unclear whether this form indicates that the Babylonians thought of the world as a flat disk surrounded by the "Bitter/Salty" Ocean or floating in it. Delano-Smith, "*Imago Mundi*'s Logo," 209.

20. For a discussion on these so-called Heavenly Islands, see chapter 7.

21. See Horowitz, "Babylonian Map," 156–64.

22. Unger, "Cosmos Picture," 3–5; Horowitz, "Babylonian Map," 159–60. Particularly telling is the *Labbu* myth, which speaks of a fantastic sixty-league-long *basmu* (viper) created in this sea that rings the Babylonian world map. This description matches the description in the medieval Islamic geographical texts, which also speak of terrifying fish sixty-days long that inhabit the Encircling Ocean (See next section for details.)

23. "Enûma Elish," 80–81; translator's parentheses. Heidel also raises the example of the Tiamat-Apsû cosmic struggle over the primeval waters that resulted in the creation of the world. Heidel, *Babylonian Genesis*, 98–101. This story is preserved in the *Babyloniaca* of Berossus. Horowitz, "Babylonian Map," 159. A similar story of the creation of the world from the primeval ocean is found in the Muslim tradition as well and is addressed in chapter 7.

24. Unger, "Cosmos Picture," 1–7; Delano-Smith, "*Imago Mundi*'s Logo," 209–11. For a recent discussion on early Mesopotamian mapping, including the Babylonian clay tablet map, see Michalowski, "Masters."

25. Delano-Smith, "*Imago Mundi*'s Logo," 211.

26. For imaginative interpretations, see Delano-Smith's article "Cartography in the Prehistoric Period," 54–101, as well as her article "Prehistoric Cartography in Asia."

27. See, for instance, the material in *Persian Art*—in particular Ackerman, "Symbol and Myth."

28. There appear to be parallels between the artistic motifs used in Susian and prehistoric Egyptian art that remain unexamined. See, for example, the painted clay urn with flamingos and ibexes from Nagada II (ca. end of fourth millennium BCE) reproduced in Woldering, *Art of Egypt*, 18.

29. Ackerman argues through these motifs that Mazdean cosmology predates the Achaemenids. Ackerman, "Symbol and Myth."

30. Tibbetts, "Beginnings," 93–94; Ahmad, "Djughrāfiyā," 576. For information on Sassanid geography and the importance of the Eran-Shahr and *kishwar* system, see Daryaee, *Ērānšahr*, 2–7; and *EIr*, s.v. "Haft Kešvar," by A. Shahpur Shahbazi, accessed April 30, 2015, http://www.iranicaonline.org/articles/haft-kesvar. Miri soundly establishes the connection between the Sassanian and Islamic geography in her recent book on Sassanian historical geography and administrative organization. What Miri notes, in particular, throughout her book is that Islamic geographic material is essential for filling in the gap of extant Sassanian material. To fully understand the latter, we need the former. See Miri, *Sassanian Pārs*.

31. It is thought that the symbol derives from Assyrian art in which the winged disk was

the symbol of the supreme deity. It acquired further stylistic input from renewed contacts with the Egyptians. See Porada, *Ancient Iran*, 158 and 167. Hittite seals, however, are richer in examples of the Ahura Mazdā image. Given that the form of the world in the KMMS world maps resembles that of a bird, it further bolsters the possibility of a connection with the Ahura Mazdā motif.

32. Dating the start of the Parthian empire is problematic because although the Arsacids were in control of the eastern Iranian province of Parthava from at least the late third century BCE, the empire itself is dated only from the reign of Mithradates I (ca. 171–138/137 BCE) because he was the one who succeeded in establishing control over other smaller principalities in Iran and Mesopotamia and eventually wrested control from the Seleucids in Iraq. Curatola and Scarcia, *Persia*, 47–55.

33. Ghirshman, *Persian Art*, 87.

34. Extant small Parthian ware is limited. Porada, *Ancient Iran*, 199; Curatola and Scarcia, *Persia*, 56; *EIr*, s.v. "Iranian Pre-Islamic Elements," by Maria Vittoria Fontana, accessed May 1, 2015, http://www.iranicaonline.org/articles/art-in-iran-xii-iranian-pre-islamic-elements-in-islamic-art.

35. Lassner, *Topography*, 132–36. See chapter 7 for more information on the round city of Baghdad. For a discussion of the Abbasid caliph Manṣūr's design for the city of Baghdad that he founded in 762 CE and the way in which he went to great length to model it on Ctesiphon, the former Sassanian capital, located 37.3 km southwest of Baghdad, and Sassanian imperial tradition for round cities, see Gutas, *Greek Thought*, 50–53.

36. See, for instance, my discussion on the Bibliothèque nationale de France, Paris, Supplément Persan 332a in chapter 3. For a discussion on this, see Pancaroğlu, "Medieval Persian Cosmography."

37. "From the same very early Islamic times, and possibly still earlier, some primitive geographical notions must have come to the Arabs, notions that were rooted in a very ancient past and transmitted by Jewish and Christian circles and were mostly of oriental origin. They were often more of a cosmographical than of a geographical nature and related for instance to the extension of the earth . . . to the encircling ocean . . . to the paradisiacal origin of some rivers. . . . Hereto belong also traditions on an original division of the earth, and speculations on its forms, which is said to be that of a bird in a traditional saying." Kramers, "Djughrāfiyā," 63–64. See also Ahmad, "Djughrāfiyā," 576–77; and Tibbetts, "Beginnings," 90–91.

38. TSMK A 3346, fol. 62b. The question of the bird in the map is another part of the iconographic story of the KMMS world map that I plan to fully address in another publication: Pinto, "What Is Islamic about Islamic Maps?"

39. Most scholars of the art of the Sassanid period make this point. See, for instance, Porada, *Ancient Iran*, 226.

40. Fortunately, I am not alone in positing Sassanian connections for Islamic cartography. Although Kramers did not raise the parallels in connection with Sassanian silverware and round cities, he did suggest that Sassanian Iran was one of the main influences for the KMMS mapping tradition. See Kramers, "L'influence." For examples of Sassanian silver platters, see Harper, *Royal Hunter*, 31. In addition, late medieval Persian literature following the renaissance of the vernacular celebrates Sassanian kings, such as Khusraw Parviz. See Pancaroğlu, "Medieval Persian Cosmography," 32–33.

41. Berlekamp, *Wonder*, 93–95.

42. This is another concept that found its way into Islamic cosmography in the form of the Jabal Qāf (mountains of Qāf). In chapter 7, I discuss the implications of these mythical mountains on Islamic geographic and cartographic writing, particularly in the context of Sufism. Streck and Miquel discuss the connections at length in their *EI1*and *EI2* "Qāf" entries, and Kramers and Ahmad discuss Iranian connections with Islamic geography, generally, in their *EI1* and *EI2* "Djughrāfiyā" entries.

43. Boyce, *Zorastrianism*, 16–17.

44. Ibid., 33.

45. Gutas, *Greek Thought*, 40–60 (quotations at pp. 34 and 29, respectively). Gutas's entire book is geared toward addressing this phenomenon. For more discussion on the *Bayt al-ḥikma* and how it affects our understanding of the KMMS tradition, see my earlier discussion in chapter 3.

46. For a set of examples of Sassanian salvers with Encircling Ocean forms, see Ringbom, "Three Sasanian Bronze Salvers," 3032.

47. Jacobs, "Jewish Cosmology," 66.

48. Ginzberg, *Legends*, 5:15.

49. See, for instance, W. F. Warren's 1909 reconstruction in Jacobs, "Jewish Cosmology," 27.

50. Sarna, *Understanding Genesis*; sketch reproduced in Blacker and Loewe, *Ancient Cosmologies*, 69.

51. Ginzberg, *Legends*, 5:26–27.

52. Jacobs, "Jewish Cosmology," 71.

53. Cosmas Indicopleustes quoted in M. Campbell, *Other World*, 54.

54. Yāqūt, *Mu'jam*, 1:31; Yāqūt, *Introductory Chapters*, 19–20.

55. Bautch, *Geography of 1 Enoch*, 1–2, 162–85.

56. Scott, "Perceptions of the World," 192.

57. Leviathan-like fish recur in the Islamic context in *'ajā'ib* literature, as well as on the maps that occur in some of these manuscripts.

58. For more on this, see Wensinck, "Ocean," 2.

59. For more information on this, see Nazmi, *Muslim Geographical Image*, 50–52, 60–61.

60. Pingree posits the *Siddhānta* as the origin of the title of one of the earliest set of zīj tables, namely Fazārī's *Zīdj al-Sindhind al-kabīr*. *EI2*, s.v. "Sindhind," by D. Pingree. Other scholars have also noted the linguistic parallel between *Siddhānta* and *Sindhind*. See Krachkovsky, *Arabskaya*, 66; Nazmi, *Muslim Geographical Image*, 69–70.

61. Gutas, *Greek Thought*, 24.

62. There is significant debate about the actual location of Arīn (*EI2*, s.v. "al-Ḳubba," by V. Minorsky). Arīn is sometimes identified as a town on the border of Mūltān in Sind (Nazmi, *Muslim Geographical Image*, 71), although Mercier claims that he has definitively identified Arīn in Madhya Pradesh. (Mercier, "Meridians of Reference," 97).

63. TSMK A3346, f. 1b. Kramers, in Ibn Ḥawqal, *Configuration*, 2, translates this as "al-Quwadhiyan." Aside from missing the transcription for the double *alif*, Kramers has misread *rā'* as a *dhāl*.

64. For a flavor of the depth and breadth of Bīrūnī's discussion on Hindu cosmography, see Bīrūnī, *Alberuni's India*, 1:152–318.

65. Tibbetts alludes to this possibility of Indic influence when discussing the Persian *kishwar* system. He says, "There is a possibility that the idea was ultimately derived from Babylonian sources, although there are resemblances to the Indian cosmographic concept of Mount Meru and the lotus petals." He is vague and noncommittal and seems to have been influenced by the second half of *HC 2.1* in which his own work appears, but he stops short of making a definitive statement. Tibbetts, "Beginnings," 93–94.

66. Zimmer, *Indian Art and Civilization*, 17; Minkowski, "Dharma," 12–14 and 20–22.

67. The connection of the celestial river to the Encircling Ocean is another common motif of most cosmographic myths. For instance, Rāmāyaṇa, 1:38–44.

68. Bīrūnī, *Alberuni's India*, 1:273, 296–97. The latter shows Bīrūnī's diagrammatic rendition of the tortoise-circle.

69. Wilford, "Sacred Isles," 376. For caution regarding Wilford's text, see Minkowski, "Dharma," 9–12.

70. Schwartzberg, "Cosmographical Mapping," 336.

71. The Vedic scholar Dr. Lucianne Bulliet passed away tragically in 2011. The inclusion of this picture is a way of remembering her contribution to this book.

72. Bīrūnī, *Alberuni's India*, 1:228–38, 251–56.

73. See *HC 2.1*, 296, for this image.

74. Ibn Ḥawqal, *Configuration*, 1:2, 5. Note that Kramers transliterates the name of this place as "Quwādhīyān" whereas careful checking of the original manuscript at Topkapı Saray Museum A3346 reveals that the name is actually "Qawārīyān." Antrim suggests that "Al-Quwādhiyān" (*sic* continuing the orthographic error from Kramer's French translation) "refers to a town located just northeast of Balkh on the eastern side of the Oxus River in what is today Tajikistan," although the precise location and the basis for this deduction are not given. Antrim, *Routes and Realms*, 176n22.

75. See Bīrūnī's discussion on Hindu cosmography, in *Alberuni's India*, 1:152–318.

76. Minkowski, "Dharma," 22–24. For an enlightening discussion on the straddling of the Purāṇic and Siddhāntic models of the physical universe, see Plofker, "Cosmologies of India."

77. The Jain cosmographical tradition, accompanied by a wealth of images, is discussed in Schwartzberg, "Cosmographical Mapping," 367–75.

78. One may postulate connections with the Upanishads, which has numerous homologies between the human body and cosmos. I am grateful to the late Dr. Lucy Bulliet for pointing this out, as well as for her expert advice on this section.

79. "Male" version reproduced in Berthon and Robinson, *Shape of the World*, 46; for the "female" version, see *HC 2.1*, plate 28.

80. Bīrūnī, *Alberuni's India*, 1:229–30.

81. Mandala (*dkyil 'khor* in Tibetan) means "sacred circle that protects the mind. It is the purified circle of an enlightened being, an environment wherein the endless compassion of the enlightened one is expressed. By extension, a mandala is also a model of the ordinary universe, which is brought together in the its visualization as a gift to be given away to all enlightened beings. This universe is not the entirety of all universes, there being infinite numbers and kinds scattered throughout the infinite void. . . . The universe consists of a sphere of energy that has

emerged from the infinite void. Its bottom is filled with oceans and a golden earth, and an immense mountain rises up in the center." Thurman, *Inside Tibetan Buddhism*, 54.

82. Bīrūnī, *Alberuni's India*, 1:243–50.

83. Qur'ān 65:12; Bukhārī, *Ṣaḥīḥ* 43:632–34, 54:417–18, 420; Muslim, *Ṣaḥīḥ* 10:3920–25; Tirmidhī, *Sunan*, 15:13; among many others. Bīrūnī discusses the seven earths and seven ocean covers in Hindu cosmography at length in multiple sections of his book. *Alberuni's India*, 1:228–56.

84. Hamdhānī, *Kitāb al-buldān*, 3.

85. Tirmidhī, *Sunan*, 44:3298.

86. Bukhārī, Muslim, and Tirmidhī all carry multiple references to this admonition. Bukhārī, *Ṣaḥīḥ*, 46:13–15, 59:6–9; Muslim, *Ṣaḥīḥ*, 22:171, 174, 176; and Tirmidhī, *Sunan*, 16:34.

87. Al-Kisā'ī, *Qiṣaṣ al-anbiyā'*, 5–6. Translation, Kisā'ī, *Tales*, 8–9. For further details, see chapter 7, note 107.

88. Ṭabarī, *From the Creation to the Flood*, 207–8; translator's parentheses. The complete ḥadīth is cited in chapter 7.

89. Exactly when and how is indeterminable, but the extant material record suggests that rectilinear forms were prominent from the Neolithic period onward in China. In the earliest Chinese civilization, Shang art is characterized by "the imposition of abstract, balanced, geometric patterns over entire surfaces. Even Shang domiciles, palaces, temples, and tombs were invariably square or oblong, governed in orientation by the four cardinal directions and dominated in design by a persistent attempt at symmetry." Chang, *Ancient China*, 291. See also Henderson, "Nonary Cosmography."

90. Schuyler Cammann speculates that the form was "conceived in an effort to apply to the greater world the plan of the simple magic square of three." Quoted in Henderson, "Chinese Cosmographical Thought," 204. Henderson recounts the Chinese tradition of attributing the idea to one of the legendary sage-kings of high antiquity, Fu Xi, who is said to have come up with the idea after observing the pattern on the shell of a turtle emerging from a river. Henderson expands on his ideas in "Nonary Cosmography."

91. Needham suggests that there may be a relationship between the *kai thien* (*gaitian*) and the Babylonian double-vault theory. Needham, *Shorter Science*, 2:84.

92. *Zhou Bi Suan Jing*, quoted from Yee, "Taking the World's Measure," 118.

93. Foucault explores the nonconforming nature of Chinese thought in *The Order of Things*, which was spawned—as Foucault tells us—by his Borgian encounter with "a certain Chinese encyclopedia" (xv).

94. I agree with Cordell Yee's skepticism on this debate. He points to research that suggests that the Han mathematicians and astronomers were aware of the spherical shape of the earth. For instance, the *huntian* theory attributed to Zhang Heng (78–139 CE) states that "the enveloping heavens are like a chicken egg, and the celestial form is round like a crossbow pellet; the earth is like the egg yolk, alone occupying the center." Yee, "Taking the World's Measure," 118. Aside from the specific parallels that this view presents with some of the statements of the medieval Islamic cartographers, Zhang presents a completely different view that raises the intriguing question of why Chinese cartographers persisted in the "square" earth concept in spite of evidence to the contrary.

95. Needham, *Shorter Science*, 3:83. Berthon and Robinson, *Shape of the World*, 38, provide a variation on this: "Rain falling off the earth flowed down to its four edges from which it fell off to form the *surrounding ocean*."

96. *Chu Chih-fan* [*Zhu Zhi-fan*] (Records of Foreign Peoples) by Chao Ju Kua [Zhao Ru Gua] (Sung geographer, 1225 CE), quoted from Needham, *Shorter Science*, 3:31–32.

97. "Mingtang Yueling lun," in Cai Yon, *Cai Zhonglang ji* (second century CE), quoted in Henderson, "Chinese Cosmographical Thought," 212. See also M. E. Lewis, *Construction of Space*, 260–273.

98. *Ch'onhado* is the Korean variation of the Chinese term *tianxia*.

99. For an excellent discussion of this debate, see Ledyard, "Cartography in Korea," 256–67.

100. See ibid., 264. The problem is that none of Zhou Yan's writings survive intact.

101. Park, *Mapping the Chinese and Islamic Worlds*, attempts to fill in this lacuna but leaves much uncovered.

102. For more on water-based cosmogonies, see Eliade, "The Waters and Water Symbolism," in *Patterns*, 188–213; and Eliade, "Symbols and History," in *Images and Symbols*, 151–78.

Chapter 6

1. There are only two extant examples of ancient Greek drawings from the classical period: the so-called Romance Papyrus (Bibliothèque nationale de France, Paris, Co. Suppl. Gr. 1294) and the Heracles Papyrus Oxford, Sackler Library, Oxyrhynchus Pap. 2331). In the 1990s, researchers discovered sketches on a papyrus fragment known as the Artemidorus Papyrus, now housed at the Museo Egizio in Turin. Views on this papyrus range from genuine to nineteenth-/twentieth-century fraud. The maplike sketches are so vague and fragmentary that they are hard to identify. Guesses range from all Spain to part of it, to possible details of the river Ebro or a view of Huelva in southwest Spain. See Brodersen and Elsner, *"Artemidorus Papyrus,"* in particular Talbert, "The Map," 63–64.

2. Reconstructions of ancient Greek maps on the basis of textual sources are problematic because we cannot avoid the influence of present-day cartographic knowledge. Textual readings of the ancient Greek sources suggest that there was significant dispute about the size of the continents. Some scholars presumed Europe, for instance, to be much larger than Africa and Asia, but this is not reflected in the reconstructions of Bunbury and others. Bunbury, *Ancient Geography*, 1:148. Richard Talbert concurs with my view on the matter of reconstructions. See Talbert, "The Map," 57.

3. Aujac, " Theoretical Cartography," 130.

4. Urbinas Graecus 82, Bibliotheca Apostolica, Rome. Dated to the late thirteenth century, it is thus far the earliest known copy of Ptolemy's *Geography*.

5. There is a voluminous literature on both sides of this debate. See, for instance, Polaschek, "New Light"; and Bagrow, "Origin." The suggestion that Ptolemy had to draw a map first in order to compile his eight-thousand-strong latitudinal and longitudinal listing of places is one of the many strange arguments used to justify the thesis that Ptolemy accompanied his

treatise with long since lost maps. A. Jones, "Ptolemy's *Geography*," 122–23. Aside from the fact that it would have been well nigh impossible for any scholar prior to satellite mapping and Google Earth to construct a map—or even a series of maps—with eight thousand places, Muslim scholars in the medieval period filled many volumes (called zīj tables) with the latitudes and longitudes of thousands of places. They did not use maps to compile these tables, so why Ptolemy would have needed to do so is puzzling. For more on zīj tables, see the work of David King, in particular *EI2*, s.v. "Zîdj," by D. A. King and J. Samsó. Sezgin includes a comprehensive overview on the debate surrounding the thorny question of whether Ptolemy's *Geography* was accompanied by maps in which he suggests that it would not have been possible to draw the rivers on Ptolemaic maps because of the lack of sufficient data points and that the source of the extant models only date back to the fourteenth-century cartographic interpretations of the Byzantine monk Maximos Planudes. Sezgin, *Mathematical Geography*, 1:38–56.

6. See Aujac, *Claude Ptolémée*. Albeit a terrific article in other respects, Alexander Jones falls prey to the pitfall of using Renaissance maps to explain what Ptolemy's second century CE maps would have looked like. Jones, "Ptolemy's Geography," 123–27.

7. For illustration, see Willcock, *Companion*, 210. Georgia Irby discusses this shield too although she avoids reconstructing it. Irby, "Mapping the World," 87–88.

8. Homer, *Iliad*, 18.478–83, 486–89, 607–8.

9. Cosgrove, *Apollo's Eye*, 29–36. This shield captured the Western imagination from the Enlightenment onward, and to date, elaborate reconstructions of it are still being produced. *Gentleman's Magazine*, 19 (1749), 392, engraving used by Cosgrove, *Apollo's Eye*, 33. See also the thread "Achilles' Shield" on the *Pirates and Revolutionaries* blog, which provides a rich array of the many interpretive renditions to date (http://piratesandrevolutionaries.blogspot.com/2009/03/achilles-shield-shield-of-achilles-as.html, accessed March 16, 2015).

10. Yāqūt also tells us that his fellow geographers were aware of other ancient notions of the shape of the world, such as a table, a drum, a dome, and a stone cylinder or pillar, although he does not tell us the exact ancient sources of these ideas. Yāqūt, *Muʿjam*, 1:31; Yāqūt, *Introductory Chapters*, 19–20.

11. Aristophanes, *Clouds*, 31–32.

12. On the basis of this 423 BCE comic portrayal of the political impact of maps, Irby argues that "large scale maps were known in Athens from the fifth century onward, and they were symbolically powerful." Irby, "Mapping the World," 81–82.

13. Ibid., 82.

14. Herodotus, *Histories*, 1:4.36.

15. Aristotle, *Meteorologica*, 2.5.362b.13; and the Stoic philosopher Geminus (first century BCE) followed suit. Geminus *Introduction to the Phenomena*, 16.4.5.

16. Plato, *Timaeus*, sec. 1:33C–35A.

17. Lilley, *City and Cosmos*, 28–30.

18. Strabo, *Geography*, Book 1, 1.8.

19. I thank my colleague Dr. Brett Rogers, former member of the Classics Department at Gettysburg College, now at the University of Puget Sound, for this crucial grammatical clarification.

20. Separating what was just a textual account or a reference to an actual journey from

what was actually a map is yet another example of how hard it is to analyze Greek maps, if one cannot know if these "circuits around the world" actually existed in visual form. For an excellent discussion on the linguistic problems involving *periodos ges*, see Romm, *Edges of the Earth*, 26–31. Georgia Irby suggests that these geographic treatises may have been accompanied by maps. But she concurs that in the absence of extant examples, "we cannot know how many Greek maps were produced, or what exactly their content and purpose may have been." Irby, "Mapping the World," 83–84.

21. See, in particular, the voluminous and imaginative work of Christian Jacob. Two examples among many are "Inscrire la terre habitée"; and "La carte du monde." For an overview of the Greek mapping tradition, see the two stimulating articles by Irby and A. Jones in *AP* ("Mapping the World" and "Ptolemy's *Geography*," respectively).

22. See note 5 above.

23. *EI2*, s.v. "Ḵharīṭa (or Ḵhāriṭa)," by Maqbul Ahmad.; Tibbetts, "Beginnings," 94–95.

24. For a detailed discussion on this matter, see chapter 3.

25. *EI2*, s.v. "Ḵharīṭa (or Ḵhāriṭa)," by Maqbul Ahmad. There is no clear agreement on this matter. Kramers maintains that the primary influence was Iranian; Barthold, preface, 9, argues for Greek. Nazmi proposes that Iranian influence held sway up to the twelfth century and that Greek took over thereafter. Nazmi, *Muslim Geographical Image*, 75–77. Nazmi is, however, basing his post-twelfth-century thesis on the basis of the work of Idrīsī, Ibn Saʿīd, and Qazwīnī. When, in fact, KMMS maps after the twelfth century display more Iranian influence rather than less.

26. Tibbetts, "Later Cartographic Developments," 137.

27. See Dilke, "Service of the State," 201–11. For a specific discussion on the Severan Marble Plan of the imperial city of Rome, carved in 203–211 CE, see Trimble, "Severan Marble Plan."

28. See, for instance, the stone Dilke identified as a Roman map of Gaul. Dilke, "Service of the State," 207. For further examples of questionable cadastral surveys, see Dilke, "Roman Large-Scale Mapping," 213, 219, 221, 223–31. Talbert objects strongly to Dilke's identification and labels it spurious. For a flavor for the debate between scholars on this matter, see Talbert, "Twenty-First Century Perspectives." See also Bekker-Nielsen, "Terra Incognita," 155–57, for further support of Dilke's argument.

29. Salway, "Mapping in Roman Texts," 204–10.

30. Whittaker, "Seeing Like a Roman," 99–102.

31. Talbert, *Rome's World*, 133–36. Disagreeing with Talbert, Emily Albu asserts that the Peutinger map is actually a Carolingian product. See Albu, "Imperial Geography"; and Albu, "Rethinking the Peutinger Map." In "Urbs Roma," 182–87, Talbert imagines a possible location for the Roman prototype of the Peutinger map.

32. Dilke, "Service of the State," 207–8; Berthon and Robinson, *Shape of the World*, 27; Talbert, "Twenty-First Century Perspectives," 13–14, where he discusses the debate between Dilke and Janni and Kai Brodersen's book *Terra Cognita*, in which Brodersen argues that the Agrippa map was never a world map but only an extensive text, *res gestae*. With the publication of *Rome's World*, two years later in 2010, Talbert goes further and states definitively—as if the Janni and Brodersen dispute never existed—that "large 'world' maps were unquestionably to be found on public display in Rome and elsewhere, and as such were of potential value for the making of the

Peutinger map's original. Most famous of its type was the map of the *orbus terrarum* commissioned by M. Vipsanius Agrippa." Talbert does not explain what caused him to switch to such a definitive assertion for the existence of the Agrippa map. Talbert, *Rome's World*, 136–37. For a different approach, see Nicolet, *Space, Geography, and Politics*, 95–114 and 167–221.

33. Eumenius translated by Richard Talbert, *Rome's World*, 138; Talbert's brackets; emphasis my own.

34. For an interesting discussion and analysis of the dating and provenance issues, see Carder, *Codex Arcerianus A, Wolfenbüttel*, 1–35. See also Campbell, *Roman Land Surveyors*.

35. See Woodward, "Medieval *Mappaemundi*," 299–301, for an overview of these three traditions.

36. Whittaker, "Seeing Like a Roman," 101.

37. Edson and Savage-Smith's book, *Medieval Views of the Cosmos*, is to date the only attempt to juxtapose Christian and Islamic mapping traditions. Even in this case there is less of an attempt to study one tradition in light of the other and more of a straight juxtaposition of chapters on each tradition separately. Still the authors are to be credited for making an effort to work together and for breaking ground on this important subject. I have conducted extensive research on this subject and intend to publish my findings in a book tentatively titled *Islamo-Christian Cartographic Connections*.

38. The literature on the medieval European mapping tradition is voluminous. As such, this section cannot be a comprehensive recapping of the medieval European tradition. I have aimed at bringing to the attention of the reader relevant highlights.

39. There is only one known medieval European world map example that does not feature an Encircling Ocean: the diagrammatic Y-O Isidorian mappamundi sketch on the inner fly of the St. Gall manuscript. Even with this map, it is possible to argue that the thick encircling black line was intended to be representative of the Encircling Ocean. The image resembles the work of a student scribbling what his teacher was explaining in class. Discussions on this St. Gall map and manuscript are too numerous to list completely. For starters, readers may consult the reference to it in Woodward, "Medieval *Mappaemundi*"; and for a recent discussion on the manuscript, see Lozovsky, "Medieval St. Gall," 65–82. Viewers can examine the image online courtesy of the Stiftsbibliothek: St. Gall, Codex 237, fol. 1r, http://www.e-codices.unifr.ch/en/csg/0237/1/small (accessed March 16, 2015).

40. Some scholars refer to this as the Tanais or Don River, but given the enormous importance of the Bosphorus in comparison to the rivers, along with the fact that the Byzantine empire straddled it, it seems logical to presume that this was customarily meant to represent the Bosphorus.

41. The best discussion on the meaning of the "T" symbol in T-O maps is by J. T. Lanman, "T in T-O Maps."

42. Evelyn Edson traces the T-O format back to the Roman geographers Sallust and Lucan. But since the manuscripts containing maps that she uses for her argument date to the eleventh and fourteenth centuries, the argument is hard to prove. Edson, *Mapping Time*, 18–35.

43. I speak here of the T-O map in the Escorial manuscript, R. II. 18, and of the palimpsest map in the St. Gall manuscript. Regarding the Escorial manuscript, Williams, "Isidore, Orosius," 13, argues that the ink matches the text and that it should be dated to a few decades

after the death of Isidore in 632. Dalché, on the other hand, argues in "De la glose à la contemplation," 749–52, that all three T-O diagrams in this manuscript belong to a later addition. The T-O map in the Gall manuscript is subject to a similar dating controversy. The form of this map is a Y-O map, which is considered a later innovation. Evelyn Edson returns to the problem in "Maps in Context." As with the dating of medieval Islamic manuscripts, the dating of the early medieval European maps is complicated by the lack of clearly dated colophons, causing the dates assigned to vary significantly from one publication to another.

44. Isidore, *Etymologies*, 277; translator's parentheses and brackets.

45. I say most because the palimpsest Y-O map in the Stiftsbibliothek St. Gall manuscript (Codex 237, fol. 1r) does not include a double-lined circle for the Encircling Ocean. See earlier discussion in note 39 above.

46. For more examples and a further discussion on List T-O maps, see Edson and Savage-Smith, *Medieval Views of the Cosmos*, 41.

47. Derolez suggests influence from the Late Antique cosmography of Ps.-Aesthicus. As noted by Derolez himself, the copy that Lambert probably consulted is an eleventh-century copy that is still in the library. Derolez asserts the influence of the Late Antique work of Orosius's *Historiae adversus paganos* but does not comment on the existence of a Late Antique copy that Lambert could have consulted. Derolez, *"Liber floridus,"* 48–49.

48. For an excellent example from a Bede manuscript housed at the Bodleian library, see http://bodley30.bodley.ox.ac.uk:8180/luna/servlet/detail/ODLodl~1~1~38065~118519:De-natura-rerum-?sort=Shelfmark%2CFolio_Page%2CRoll_%23%2CFrame_%23&qvq=q:bede;sort:Shelfmark%2CFolio_Page%2CRoll_%23%2CFrame_%23;lc:ODLodl~1~1&mi=28&trs=101# (accessed March 16, 2015).

49. See, for instance, Dalché, "Maps in Words," 227.

50. Both these maps have been the subject of studies too numerous to list here. For a recent summary on the central place that this Psalter map occupies in the canon of medieval European visual studies and the many studies on it, see Whittington, "Psalter Map." For the Hereford mappamundi, see *HWM*.

51. Contrary to naming practices, medieval European cartography is heavily laced with ecumenical symbols whereas Islamic cartography, in spite of the name, has no overt religious symbols. I have discussed this in "Traces of the Diabolic and the Divine in Islamic Maps," and "What Is Islamic about Islamic Maps?"

52. For an in-depth discussion of the Hereford map, consult Scott Westrem's place-by-place analysis of the layout in *Hereford Map*, 34–37.

53. These examples are too numerous to enumerate. For more information, see von den Brincken, "Jerusalem on Medieval Mappaemundi." For an intriguing study on the representation of Jerusalem from the Old Testament to the Beatus manuscripts and Crusader representations, see Kühnel, *Earthly to the Heavenly Jerusalem*. Especially useful also is Kühnel's *Real and Ideal Jerusalem*. Keith Lilley discusses the connections between the heavenly depiction of Jerusalem and Jerusalem on earth at length. He too sees the connection between the double-ringed circular marker for Jerusalem, the world, and the cosmos. See, in particular, Lilley's chapter "Urban Mappings: From the Heavenly to the Earthly City," in *City and Cosmos*, 15–40.

54. Another example from a Lambert manuscript: "Ruler Encased by Encircling Ocean," Lambert, *Liber floridus*, ca. thirteenth century, Paris, Bibliotheque Nacionale, Ms. Latin 8865,

fol. 45b. Albert Derloez, the reigning expert on this Ghent autograph copy of the *Liber floridus*, tells us that the inscriptions allude to the world historical significance of Augustus and his role in the creation of world peace through his edict ordering a census and the putting down of the civil wars. He states that "the map in the ruler's hand, replacing the customary sphere of the world, constitutes another link between history and geography." Derolez, *Liber floridus*, 126.

55. Gurevich, *Categories*, 163. To show the influence of the double-ringed circle, Lilley brings to our attention medieval English municipal seals situated within imago mundi–type stamps. Lilley, *City and Cosmos*, 22. Kline raises similar examples with the seals of Edward II. Kline, *Maps of Medieval Thought*, 59.

56. Charles Burnett explores the "strangeness of Hildegard of Bingen's cosmology" in "Science of the Stars."

57. For a discussion of this image and other images in Hildegard manuscripts, see Caviness, "Hildegard as Designer." See also Denis Hüe's interesting discussion on imago mundi images that incorporate anthropomorphic forms in "Tracé, écart," 134–35.

58. Edson, "Maps in Context."

59. Hiatt, "Map of Macrobius," 149.

60. Ibid., 166–68. Hiatt discusses the problems of dating and developing a stemma for the tradition.

61. For details on this debate, see the most up-to-date article on it by the reigning expert on the subject of the Beatus tradition: Williams, "Isidore, Orosius."

62. Many books on the history of cartography include reprints of Beatus maps. By far the best and most comprehensive reprint of all the Beatus manuscripts worldwide is the five-volume set by John Williams, *The Illustrated Beatus*.

63. For an excellent discussion on this map and other geographical Mozarabic manuscripts, see Aillet, *Les Mozarabes*, 167–69.

64. To see this image, please consult http://www.themorgan.org/collection/Las-Huelgas-Apocalypse/13 (accessed May 2, 2015).

65. For starters, see Netton, *Muslim Neoplatonists*; Morewedge, *Neoplatonism*; and Adamson and Taylor, *Arabic Philosophy*. There is no question that medieval Islamic maps are just as affected by Neo-Platonist ideas as medieval European ones. The Encircling Ocean is the most obvious manifestation, but there are others. This book is not the place to delve further into aspects of Neoplatonic influences on Islamic cartography. I do, however, have plans to explore the impact of Neoplatonic thought on medieval Islamic mapping at a future date. I touched on some of these Neoplatonic overtones in an honorary talk that I gave in the Virginia Garrett lecture series at the University of Texas, Arlington, in October 2006, entitled "Traces of the Diabolic and the Divine in Islamic Maps."

66. Lilley, *City and Cosmos*, 35.

67. Ibid., 37–40 and 126. On the one hand, Lilley argues that this image of "God as Architect" reinforces the Neo-Platonic Timean one. On the other hand, Lilley argues that "the compass as an instrument used by God in Creation to trace out cosmological form would seem to owe more to medieval Hebrew sources" (126). This dichotomy mirrors some of my findings in this and the previous chapter.

68. See http://warburg.sas.ac.uk/vpc/VPC_search/record.php?record=17890 (accessed March 16, 2015).

69. *Holy Bible: New Revised Standard Edition with Apocrypha* [NRSV] (Oxford: Oxford University Press, 1977). All subsequent biblical scripture references are to this verson.

70. Isidore, *Etymologies*, 273.

71. Wright, *Geographical Lore*, 59–60.

72. Ibid., 200 (Bernard of Clairvaux), 201 (Hildegard of Bingen).

73. Wright does allude, albeit briefly, to possible connections between ancient Egypt and Persian cosmology, but no one has followed up on this. Ibid., 58.

74. Ibid., 59.

75. Isidore, *Etymologies*, 276.

76. Kline, *Maps of Medieval Thought*, 7–48.

77. Ibid., 10–13.

78. Ibid., 26–29.

79. Readers interested in further exploring the double-ringed circle in the medieval Christian imagination should consult Kline, *Maps of Medieval Thought.*

80. I use "imago mundi" as the way to describe the symbol even though, as I acknowledge in chapter 5, note 2, there are some occasions in which "imago mundi" was used to title a book on geography of the world. I do not believe that the exceptions should preclude the use of the term to label a symbol that most readers familiar with the history of cartography would recognize.

81. Lilley, *City and Cosmos*, 7–94.

82. In "Géographie Arabe et Géographie Latine," Dalché argues for the latter. He seems not to be aware, however, of the Isidorian world map with Arabic notations discussed later in this section. Nor does Dalché take into consideration cartographic connections outside of the twelfth century.

83. For an example of a Macrobian world map showing multiple southern continents, see Hiatt, "Map of Macrobius," 152, fig. 3, and 157, fig. 5.

84. Radiocarbon dating by the DFG-ANR Project Coranica, 2015, at the request of Karen Pinto and Arnoud Vrolijk, curator of the Oriental Manuscripts and Rare Books Special Collections, at Leiden University Libraries. Study of this map and the manuscript within which it is housed is part of an ongoing project.

85. This manuscript is incomplete and therefore doesn't contain a colophon. The notation on the opening folio, however, indicates two dates: one of AH 846/1442 CE and the other, more significant one that lists the name and author of this manuscript (*Kitāb tarkīb al-aflāk* by Sijzī) and another book, *Kitāb al-talwīhat* (Book on the Intimations), by the famous philosopher and founder of the philosophical school of Illumination, Shihāb al-Dīn al-Hakīm, known as al-Maqtūl, with the date 578/1191—the year that he was put to death by the Ayyubid sultan Ṣalāḥ al-Dīn's son al-Malik al-Ẓāhir, the viceroy of Aleppo. This suggests that the manuscript was completed sometime before the late twelfth century. It is highly unusual for a map to be included in an astronomical manuscript. This is in addition to the fact that the text of the main manuscript of Sijzī's *Tarkīb al-aflāk* is in Persian but the map is annotated in Arabic. Little is known about Sijzī other than that he was an astronomer especially interested in the problems associated with the measurement of spheres. Pinna reprints this map and devotes a page of discussion to it but does not consider the possible Macrobian connections. Pinna, *Il Mediterraneo*, 1:41–42. Pinna incorrectly lists the date of Leiden's Sijzī manuscript as 1248 CE based on a misreading of one of the dates on the opening folio.

86. The term "Islamo-Christian" comes from another one of Richard Bulliet's seminal contributions to the field: *Islamo-Christian Civilization.*

87. This map was first studied in the context of medieval European cartography in 1965 by Pidal, in *Mozarabes y asturianos*, and introduced two years later to the Arabic-speaking public by Mu'nis in *Tarīkh al-jughrāfiya*, fig. 10. Then it was forgotten for half a century. I first stumbled upon this map through the 2010 Guggenheim fellowship announcement for Simone Pinet's forthcoming research and book *The Task of the Cleric: Three Studies on the "Libro de Alexandre"*(in progress) (http://www.gf.org/fellows/all-fellows/simone-pinet/, accessed May 22, 2015). This map is a truly exciting renewed find that is destined to open up the discourse on medieval Christian and Muslim cartographic connections as is evidenced by the recent outpouring of interest and writing on the piece: Chekin, *Northern Eurasia*, 59–61; Aillet, *Les Mozarabes*, 167–70; and Schröder, "Kartographische Entwürfe," 268–76.

88. This finding runs counter to Dalché's skepticism in "Géographie Arabe et Géographie Latine" regarding connections between the two traditions. Aillet places the manuscript within the Mozarabic culture of Andalus, and Schröder follows suit. Both argue that the annotations were done by Arabic-speaking Christians who were trying to introduce Latin texts into the Ibero-Islamic environment of Andalus. Aillet, *Les Mozarabes*, 167–73; Schröder, "Kartographische Entwürfe," 274–76. Having examined the marginalia notations of place-names and other comments along with the colophon of this manuscript, I think there is more to the author of the Arabic notations than the average Mozarab trying to fit in with the new Arabo-Islamic elite of southern Spain.

89. Detailed templates translating and transcribing the Arabic are in preparation and forthcoming in the author's separate essay on this map, titled "A 9th Century Isidorian T-O Map Labeled in Arabic." *Farsakh* is a Persian word of Parthian origin referring to the measure of distance based on time. Originally akin to the "marching mile," the distance has been deteremined to be approximately 6 km. We should keep in mind that this was a guesstimate of the marching mile that must have varied over time, especially with changing material conditions. See *EI2*, s.v. "Farsakh," by Hinz.

90. Chekin rectifies this half-century hiatus and brings this map back to the attention of historians of medieval European cartography in his *Northern Eurasia*, 59–61.

91. Further collaborative research is underway with Visigothic Latin specialists. This is part of a larger book project on connections between medieval European and Middle Eastern maps tentatively titled *Islamo-Christian Cartographic Connections.*

92. Jacob, "Will There Be Histories of Cartography after the *History of Cartography* Project?"

93. For more on this, see, in particular, Eliade, "Waters and Water Symbolism," in *Patterns*, 188–213; and Eliade, "Symbols and History," in *Images and Symbols*, 151–78.

Chapter 7

1. Karamustafa, "Cosmographical Diagrams," 71–72.

2. *The Qur'an*, trans. with an introduction by Tarif Khalidi (London: Penguin, 2008). References are to this translation unless otherwise noted. Khalidi has striven to present a transla-

tion that gets away from the antiquated approach of "Lord" and "Thou" in order to capture the cadence of the language a la Seamus Heaney's approach to translation. For another translated version of this passage, see *Al-Qur'ān*, trans. Ahmed Ali, 5th ed. (Princeton, NJ: Princeton University Press, 1994).

3. *EI3*, s.v. "Barzakh," by Christian Lange, accessed April 2, 2014, http://referenceworks.brillonline.com/entries/encyclopaedia-of-islam-3/barzakh-COM_23704.

4. Ahmed Ali provides a variant and questionable version that translates *maraja al-baḥrayn* as "two bodies of water flow side by side." Confusion over whether *maraja* means "merged" or "side by side" may explain why some geographers interpreted the "two merged seas" as the Indian Ocean merging with the Encircling Ocean and others interpreted it as a description of the dual nature of the Encircling Ocean. Those medieval geographers who interpreted this verse as a reference to the Encircling Ocean asserted that its sweet and brine waters ran side by side and didn't mix. This made it possible to divide the Encircling Ocean into two separate seas: Baḥr al-Ẓulūmāt (Sea of Darkness) and Baḥr al-Akhdar (Green Sea).

5. Arberry's translation resembles that of Khalidi's with a few variations: "He let forth the two seas that meet together/between them a barrier they do not overpass." *The Koran Interpreted*, trans. A. J. Arberry (New York: Collier Books, 1955). Once again, Ali's translation presents a variation. In Ali's version the two seas are "in motion," and once again Ali uses "side by side" for *maraja*: "He has set two seas in motion that flow side by side together/With an interstice between them which they cannot cross."

6. "Remember when Moses said to his youthful attendant: 'I shall not pause until I reach the place where two seas meet, even if I journey for years to come'" (18:60). For similar references, see 25:53, 27:61, and 55:19–20. Among the more extreme interpretations of these two seas and their meeting is a meeting of the prophets Mūsā and Khaḍir, who are cast as two seas of wisdom. See *EI1*, s.v. "al-Ḵh̲aḍir," by A. J. Wensinck.

7. Burnham, "Edges of the Earth," 264–65.

8. This latter verse presents a unique Qur'ānic occurence of the word *ḥājiz* where one would typically expect to find the word *barzakh*.

9. *Al-Qur'ān*, trans. Ali.

10. Bashier, *Ibn al-ʿArabī's Barzakh*.

11. See, for example, Bakrī, *Masālik*, 1:304.

12. *Al-Qur'ān*, trans. Ali. See also *Koran Interpreted*, trans. Arberry, 338.

13. Muslim, *Ṣaḥīḥ*, 4:1243; Mālik, *Muwaṭṭa'*, 15:21–22.

14. Muslim, *Ṣaḥīḥ*, 19:4394.

15. Ibid., 20:4705–8.

16. Ibid., 20:4699–4701.

17. Mālik, *Muwaṭṭa'*, 25:12.

18. Muslim, *Ṣaḥīḥ*, 41:7028 (contains the most extensive version of this ḥadīth), 7029–31. For a brief but excellent discussion on Dajjāl, see *EI3*, s.v. "Dajjāl," by David. B. Cook, April 17, 2014, http://referenceworks.brillonline.com/entries/encyclopaedia-of-islam-3/dajjal-COM_25826.

19. Muslim, *Ṣaḥīḥ*, 41:6924, 6979.

20. *Al-Qur'ān*, trans. Ali; translator's parentheses.

21. Miquel refers to this imperative often in his *La géographie humaine*. See, for instance, his comment in connection with Jahiz (2:252).

22. Tirmidhī, *Sunan*, 39:2.

23. For details on the *ṭalab al-ʿilm* phenomenon and their profound impact on Islamic history and the development of the ḥadīth, see Bulliet, *View*, chaps. 3–6, and p. 53 in particular; Gellens, "Search for Knowledge," 53–63; and Touati, *Islam and Travel*, 79–100, for a flavor of the trials that these *ṭalab al-ʿilm* faced in their quest for knowledge.

24. Touati, *Islam and Travel*, 1.

25. Gellens, "Search for Knowledge," 51.

26. Touati, *Islam and Travel*, 3, 47–77, 80–93, 246–64.

27. Ibn Mājah, *Sunan*, 15:6779.

28. To this day the Pacific and Atlantic Oceans are not fully surveyed. In the medieval period sailors were afraid of venturing too far out into their turbulent waters and they could only be understood through hearsay and myth. I say this in spite of Picard's excellent exposé, *L'océan atlantique musulman*, on Muslim navigation of the Atlantic from the tenth and eleventh centuries onward. This is because although navigation increased, it was a localized phenomenon restricted to the coasts of Iberia and Morocco. The vast majority of the Atlantic remained unexplored and therefore daunting and mysterious.

29. As early as 1971, Dunlop recognized some of these influences and discussed, albeit briefly, Indic and Chinese influences along with Greek. Dunlop, *Arab Civilization*, 150–60. Ditto with Brinner in his introduction to Thaʿlabī's *Lives of the Prophets* (xxxi).

30. Andalusī, *Science in the Medieval World*. Recently, Hossein Kamaly updated the aforementioned translation with a new translation, submitted as a master's essay to Columbia University in 2000.

31. Masʿūdī, *Livre de l'avertissement*, 1:vi–vii.

32. Khalidi, *Islamic Historiography*, 2.

33. Touati, *Islam and Travel*, 129.

34. Yaʿqūbī, *Les Pays*, viii.

35. *EI2*, s.v. "Iṣṭakhrī," by A. Miquel.

36. *EI2*, s.v. "Ibn Ḥawḳal," by A. Miquel.

37. Touati, *Islam and Travel*, 135.

38. The elimination of the world map from Muqaddasī manuscripts is an unusual phenomenon that may be explained by the breakdown of the concept of the *mamlakat al-Islām* after the year 1000 CE with the advent of the Turks. See Miquel, *La géographie humaine*, 1:99, 2:525–28, and 3:x–xi. The absence of a world map is not the only hurdle with assessing Muqaddasī's cartographic contribution. There is only one authentic Muqaddasī manuscript containing maps and even that is from the fourteenth century or later.

39. Touati, *Islam and Travel*, 118, 133, 136.

40. See notes 5 and 6 regarding Qurʾān 25:53 and 55:20.

41. Burnham, "Edges of the Earth," 229, 231–32.

42. Idrīsī, *Nuzhat*, 943, 956, and 962.

43. Jacques Bačić devoted a lifetime to the study of naming conventions involving color and the directionality implied in them. For a treasure trove of information on this topic, see Bačić, *Red Sea—Black Russia*.

44. Bakrī, *Masālik*, 1:306; Masʿūdī, *Murūj*, 1:258. One could perhaps ascribe the diametrically opposed nomenclatures for sections of the Encircling Ocean used by these authors to

the places they herald from. Bakrī was from al-Andalus and saw the part of the ocean closest to his homeland as "green." Whereas Mas'ūdī was Iraqi. For him the connections with the east were stronger and more important than those with the Maghrib. So the safe "green" part of the ocean was that which was closer to China, not the section closest to the Maghrib.

45. Zuhrī, *Kitāb al-ja'rāfiya*, 9, 79. For a discussion on the naming and transliteration issues with the title of Zuhrī's work, see Pinto, "*Searchin' his eyes*," 74–78.

46. Readers interested in exploring in greater depth the different naming conventions of the Baḥr al-Muḥīṭ in medieval Arabic texts should refer to the excellent discussion in Burnham, "Edges of the Earth," 233–47. For a novel discussion on the grammatical underpinnings of the naming of seas in Arabic, see Zagórski, "Sea Names."

47. Qudāma, *Kharāj*, 230–31. For an excellent overview on Qudāma and his encyclopedic contribution to geography and other areas, see Heck, *Construction of Knowledge*.

48. Ibn Khaldūn, *Muqaddimah*, 1:118; translator's brackets.

49. *Ḥudūd al-'ālam*, 51.

50. Idrīsī, *Nuzhat*, 55.

51. For an erudite understanding of Mas'ūdī, in which Khalidi proves that Mas'ūdī is much more than just "a delightful raconteur," see Khalidi, *Islamic Historiography*.

52. Mas'ūdī, *Murūj*, 1:105.

53. Miquel, *La géographie humaine*, 2:234.

54. Mas'ūdī, *Tanbīh*, 108.

55. Ibn Khaldūn, *Muqaddimah*, 1:118; translator's parentheses. Ibn Khaldūn is known for his essentialist approach to history in which he provides theories and explanations for many social and geographical phenomena. Among his most famous theories is the relationship of the rise and fall of dynasties to "group feeling" ('aṣabiyya).

56. Ibn al-Wardī, *Kharīdat*, 105, 103.

57. One is tempted to think that this may be a reference to whales.

58. Miquel, *La géographie humaine*, 2:234.

59. For more details on Ibn Zunbul, refer to the work of Benjamin Lellouch and Robert Irwin.

60. For a fascinating account of the depictions of the monster Tinnīn in *'ajā'ib* manuscripts, see Berlekamp, *Wonder*, 81–85, 175.

61. Miquel, *La géographie humaine*, 3:259–60.

62. Ibn al-Wardī, *Kharīdat*, 116–17.

63. Mas'ūdī, *Murūj*, 105.

64. Mas'ūdī, *Tanbīh*, 101–2.

65. Abraha was the king of Yemen associated with the story of the Year of the Elephant when Mecca was attacked by an army from Yemen circa 570. The Prophet Muḥammad is said to have been born in this year. Why does the origin of these mythical statues change from Hercules to Abraha between the tenth and the fifteenth centuries? And, why are they attributed to the strong men of the classical and Islamic traditions. These are questions that belong to an in-depth examination of the pillars and their association with the mythical island of Jabal al-Qilāl that I intend to explore in a chapter devoted to Jabal al-Qilāl in my next book project *The Mediterranean in the Islamic Cartographic Imagination*.

66. Ibn al-Wardī, *Kharīdat*, 113–14.

67. There is an old theory by Reinaud about Jabal al-Qilāl representing Fraxinetum. But Reinaud's understanding is based exclusively on the geographical texts. He does not take into account the maps. Had he done so, he may have arrived at a different understanding of this enigmatic island. See Reinaud's translation of Abū'l-Fidā, *Géographie d'Aboulféda*. My next book *The Mediterranean in the Islamic Cartographic Imagination* studies this island in depth and suggests ways in which its presence on the Mediterranean maps can be interpreted.

68. For example, Ishtar going down to the fountain of the world in the netherworld, Vishnu sleeping on the primeval Ocean at the time of Creation, and numerous Greek myths involving Kronos.

69. The Qur'ān itself is conflicted on the matter of whether to view the sea as a positive or negative force. For example, compare sura 17, verses 66–70, which open this chapter, or sura 7, verse 136 ("Therefore We inflicted retribution on them and drowned them in the sea because they rejected Our signs and were heedless to them"), with sura 16, verse 14 ("And it is He Who has made the sea subservient that you may eat fresh flesh from it and bring forth from it ornaments which you wear, and you see the ships cleaving through it, and that you might seek of His bounty and that you may give thanks"). With the ḥadīth, the matter is less clear, although from my initial survey of ḥadīth containing a reference to the word *baḥr* (sea), it would seem that the ḥadīth presents, overall, a view of the sea as a negative force. After all, it is quite clearly expressed that at the end of the world, sea, fire, and Dajjal will rise to bring about mayhem and chaos. There are numerous ḥadīth references to this scenario of the apocalypse. Ḥadīth instruct us to limit sailing on the sea to when one is going to perform hajj, umrah, or jihād. Abū Dāwūd, *Sunan*, 14:2483.

70. This quote is from the Ali translation. For a similar verse, see 24:45 where it is explicitly said that "God created every moving thing from water."

71. Ibn Khaldūn, *Muqaddimah*, 1:95; translator's parentheses.

72. Qazwīnī, *'Ajā'ib*, 101.

73. Famous Qur'ānic Throne verse 11:7 discussed at the outset of this chapter.

74. Ṭabarī, *Tafsir*, 1:409.

75. Nuwayrī quoted in Wensinck, "Ocean," 7.

76. Indeed, on the basis of Qur'ān 18:109 (Ali translation), it may even be considered the ink for God's pen.

77. This representation of a pliable and obedient primeval Ocean stands in sharp contrast to the earlier traditions in which the primeval Ocean is presented as a sign of chaos.

78. What is so unusual about this textual color scheme is that it does *not* match the maps. There is no color variation between the Encircling Ocean and the other seas as represented on the maps. If the Encircling Ocean is painted in with dark green so are the other seas. If the other seas are blue with elaborate patterns, the same color and pattern scheme is applied to the Encircling Ocean as well.

79. The paradisiacal origin of certain rivers is, as André Miquel tells us, nothing but a variation on "ancient oriental belief," that is, Mesopotamian, Iranian, Egyptian beliefs. Miquel, *La géographie humaine*, 3:117–18. It is yet another confirmation for viewing the Encircling Ocean as a transcultural cosmographic motif that finds its way into Islamic and biblical texts. Some

geographers even suggest that the four rivers issue from below the throne of the pharoah in Memphis.

80. Muqaddasī, *Best Divisions,* 23; translator's brackets. Versions of this story are found in Mas'ūdī's writing as well. It is tempting to read possible knowledge, albeit vague, of the American continent into this location of Paradise on the other side of the waters of the Encircling Ocean—which is sometimes referred to as the emerald land. It could be postulated that the reference to the green Jabal Qāf that ring some of the later medieval Islamic maps is a further manifestation of the existence of another continent on the farthest horizon. Manfred Kropp has written on the possibility of the premodern discovery of the Americas by Muslim navigators. See Kropp, "*Kitāb al-bad'*" and "*Kitāb ǧuġrāfiyā*."

81. Mas'ūdī, *Tanbīh,* 103.

82. Muqaddasī, *Best Divisions,* 13. Qazwīnī relates a similar story in *'Ajā'ib,* 102.

83. For earlier and later variants of this image, see Berlekamp, *Wonder,* 159.

84. See ibid. for an example.

85. Mas'ūdī, *Murūj,* 1:256.

86. *Ḥudūd al-'ālam,* 8.

87. Bakrī, *Geography of Al-Andalus,* 24; my brackets.

88. *EI1,* s.v. "al-Khālidāt," by P. Schwarz, suggests that this could be a reference to the Celts since Maqqarī includes Britain as part of the Jazā'ir al-Sa'āda.

89. Qazwīnī, *'Ajā'ib,* 106.

90. Minorsky provides some intriguing explanations for these islands, including a possible linguistic relation to Amazānūs (Amazons) and a Finnish tribe associated with the myth of Amazons who named the island of Nargen "Naissare," meaning "the island of women." *Ḥudūd al-'ālam,* 191.

91. Ibid., 58–59.

92. Ibid., 56.

93. See chapter 3 for an overview of the Islamic mapping tradition including that of Idrīsī.

94. Jacob, *L'empire des cartes,* 267. See also my discussion on the act of naming or not naming at the end of chapter 9.

95. For a basic explanatory diagram, see "Circular World map," under bk. 2, chap. 5, of the *Book of Curiosities* online (http://cosmos.bodley.ox.ac.uk, accessed May 2, 2015).

96. Miquel, *La géographie humaine,* 3:270–75. See also *EI2,* s.v. "Ḳāf," by M. Streck and A. Miquel. Wendell, "Baghdad," 121–22, concurs and discusses specific emerald-related connections between the two traditions. For more on the pre-Islamic Iranian traditions, see chapter 5.

97. 'Aṭṭār, *Conference,* 33–34.

98. Ibid., 48.

99. M. S. Simpson, *Arab and Persian Painting,* 99.

100. Similarly in the Mesopotamian traditions, Ishtar has to descend to the bottom of the netherworld in order to acquire the water of life.

101. Miquel, *La géographie humaine,* 3:117.

102. Iṣṭakhrī, *Masālik,* 14; Ibn Ḥawqal, *Ṣūrat,* 12.

103. See note 37.

104. Miquel, *La géographie humaine,* 3:235–37.

105. Masʿūdī, *Murūj*, 192–292.

106. For an excellent historiography of the Qiṣaṣ al-anbiyā' tradition including the issue of *isrā'īliyyāt* (borrowings from the Jewish traditions), see Brinner's introduction (xi–xxxiii) to his translation of Thaʿlabī's *'Arā'is al-majālis, Lives of the Prophets.*

107. Al-Kisā'ī, *Qiṣaṣ al-anbiyā'*, 5; Kisā'ī, *Tales*, 8–9. In *Tales*, Thackston translates these names as Baytush, Asamm, Qaynas, Sakin, Mughalib, Muannis, Baki (without transliteration), but a mid-eighteenth-century manuscript (copied 1164/1751) housed at the University of Michigan lists the names differently (http://catalog.hathitrust.org/Record/006805276, accessed March 18, 2015). I have relied on the names in this manuscript copy and transliterated them accordingly. The names that Kisā'ī gives to the seven seas have puzzled many a medieval Islamic geographer who has tried to find an equivalent earthly sea name. Although Thaʿlabī, author of one of the earliest extant versions from the late tenth/early eleventh century does not name the seas specifically, he does allude to them in a number of places. Thaʿlabī, *Arā'is al-majālis*, 21, 31–32, 35.

108. Ṭabarī, *From the Creation to the Flood*, 207–8; translator's parentheses.

109. Abū Dāwūd, *Kitāb al-Jihād*, 14:2483.

110. Ibid., 40:4705. Emphasis my own; original parentheses.

111. Ibn al-Wardī quoted in Wensinck, "Ocean," 43.

112. Hamdhānī, *Kitāb al-buldān*, 7.

113. Qazwīnī, *'Ajā'ib*, 104. Kaʿb al-Aḥbār is said to have been a Yemenite Jewish convert to Islam around 17/638 (although Prof. Halivni suggests that he may be of Mesopotamian origin). According to tradition, he came to Medina during the caliphate of ʿUmar ibn al-Khaṭṭāb, was apparently on intimate terms with the caliph, and even predicted his death. He was also a vigorous supporter of ʿUthmān. Both Yāqūt and Ibn Battuta mention that his tomb is located in Damascus where, apparently, a gravestone bearing his name is still extant. Kaʿb is considered to have possessed knowledge of the Hebraic and southern Arabian traditions and was sometimes accused of introducing Jewish elements into Islam. Although Prof. Halivni has confirmed that this specific mention of multiple Encircling Oceans does not have a parallel in the documented early midrash sources, it is possible that it was introduced into Hebrew lore at a later stage by Arabs; but Prof. Halivni was not able to confirm this. (I would like to thank Prof. Halivni [emeritus] of the Dept. of Religion at Columbia University for his assistance in this matter.) See also *EI2*, s.v. "Kaʿb al-Aḥbār," by M. Schmitz.

114. Carboni, "Constellations, Giants and Angels," 91.

115. Note the parallel between the pattern of the background cloud decoration of this image and the decoration of the seas in the Safavid world map shown in fig. 4.14.

116. Miquel, *La géographie humaine*, 2:275.

117. Picard prefers a more technical, less mythological approach to the Atlantic end of the Encircling Ocean. In *L'océan atlantique musulman*, he explores the question of Muslim navigation of the Atlantic. During the course of his exposé, he argues that prior to the tenth century the accounts of the western ocean were dark and dismal but that from the tenth and eleventh centuries onward, especially during the Almohad and Almoravid Berber kingdoms' unification of the African and European continents, Muslim navigation picks up. But, as Picard himself reveals, this was a navigation that hugged the coasts and had upper and lower limits of Lisbon

and Nūl Lamta. In other words, the vast body of the Encircling Ocean was still unexplored and terribly mysterious to the majority of Muslims. I don't doubt that at a local level the increased Muslim navigation of the Atlantic coasts had a major effect on the economy, technology, and even the law of the region. But it is hard to see how this isolated understanding of the Encircling Ocean limited to the world of southern Iberia and coastal North Africa would have affected the vast majority of literature (geographic, cartographic, and wondrous) being written thousands of miles away. Thus we see that in the Islamic domains east of the Levant, Anatolia, Syria, Iraq, Iran, and even as late as Mughal India, mythical notions of the Encircling Ocean continued to hold sway as indicated in the accounts of Qazwīnī and Ibn al-Wardī, to mention but a few. It must be noted that there were exceptions, such as the work of the well-travelled compiler of the first alphabetical geographical encyclopedia, Yāqūt, who provides a more nuanced reading of the Encircling Ocean that fits with the changing approach that Picard points to.

118. The context of the epigraph is René Magritte's painting of the pipes, *Les deux mystéres*.

119. In just this one form lies the key to the medieval Muslim vision of the world, fitting right in with one of the central principles of Islam: *tawḥīd*, that is, "unity." Since it is one of the central concepts of Islam, it is not surprising, therefore, to find it as the essential, all-encompassing foundational piece in the structure of the world, all exclusive and all inclusive at the same time, an impermeable cage within which the believers are forcibly encased. *Tawḥīd* implies the unity of God, the indivisible, and absolute. This principle of unity arises from the very first step of becoming a Muslim, the *shahada*, namely, the loud and firm declaration "There is no god except Allah."

120. Kufic is a specialized Arabic script used frequently in the early medieval period to adorn objects and manuscripts. Some of the earliest Qur'āns were written in elaborate Kufic. For details on the plate, see von Folsach, *The David Collection*, cat. no. 117.

121. The examples are so numerous that it is impossible to present a definitive catalogue here. Instead I recommend that the reader pick up any book on Islamic art and architecture to see a multitude of additional examples. Fortunately, I am not alone in seeing these visual parallels. In his analysis of Islamic mosaics, in particular those of the Dome of the Rock, Nees spots similar encircling ocean motifs underlying Roman and Byzantine mosaics and reveals reflections of these motifs. Nees, "Blue behind Gold," 159–65. In a fascinating and insightful article on medieval Islamic flags, Yuka Kadoi argues that the "advanced stylistic treatment" of the Las Navas de Tolosa banner from Burgos "demonstrates that the banner was no longer a mere military tool, but it is more likely to have been treated as a work of art at that time." Kadoi, "Timurid Flag," 145–46.

122. For a discussion of this image, see Robinson, *Fifteenth-Century*, 69; and Gruber, "Between Logos," 246–47. For a discussion on the illustrated *miʿrāj* tradition, see other works by Gruber, including, *Timurid Book*, *Ilkhanid Book*, and *The Prophet's Ascension* (coedited with Colby).

123. For an excellent discussion on the issues surrounding the depiction of the Prophet and other human figures, see Elias, *Aisha's Cushion*; and the older Arnold, *Painting in Islam*.

124. For an insider's view on the history and meaning of the *hilye*, see Zakariya, "Hilye of the Prophet Muhammad."

125. Creswell, *Early Muslim Architecture*, 1:18–21; Creswell, *Short Account*, 170–73; Lassner,

Topography, 132–34; Hillenbrand, *Islamic Art*, 40. Some of these pre-Islamic round cities are also discussed in chapter 5.

126. Ya'qūbī, *Kitāb al-buldān*, 233.

127. Wendell goes so far as to suggest that "the plan of Baghdad is really nothing less than an Islamic *mandala* worked out on the huge scale of urban architecture." Wendell, "Baghdad," 122. His arguments in this article on Indic, Central Asian, and Iranian influences on Islamic visual vocabulary fit with my theory about the metaiconographic roots of the encircling ocean.

128. For details of reconstruction, see Creswell, *Short Account*, 165. In 1980, Jacob Lassner updated Creswell's diagram to indicate a double ring—inner and outer—encircling the palace and the location of the police headquarters (Dār al-Qaṭṭān) and the guard's quarters. Lassner still places the caliph's palace at the navel of the image. Lassner, *Shaping*, 190.

129. David King is the expert on this subject. His publications on the subject are too numerous to list. See, for instance, his article with Lorch, "Qibla Charts, Qibla Maps, and Related Instruments." See also chapter 3 for a brief discussion on Qibla charts and maps.

130. Krauss, *Optical Unconscious*, 9.

Chapter 8

1. Globes present a paradox for mapping. They appear to be three-dimensional and yet they too fall into the constraints of two-dimensional space. They are not flat, but they are constrained by the amount of surface area that is available to indicate places and spaces. In order to avoid this, one would have to create a 1:1 map of the world—that is, an exact replica of the world, in shape as well as size. And we already know from Borges about the fate of such idealistic maps, when he says in his novella "On Exactitude in Science": "In that Empire, the craft of Cartography attained such Perfection that the Map of a Single province covered the space of an entire City, and the Map of the Empire itself an entire Province. In the course of Time, these Extensive maps were found somehow wanting, and so the College of Cartographers evolved a Map of the Empire that was of the same Scale as the Empire and that coincided with it point for point. Less attentive to the Study of Cartography, succeeding Generations came to judge a map of such Magnitude cumbersome, and, not without Irreverence, they abandoned it to the Rigours of sun and Rain. In the western Deserts, tattered fragments of the Map are still to be found, Sheltering an occasional Beast or Beggar; in the whole Nation, no other relic is left of the Discipline of Geography." Borges, *Universal History*, 131.

2. Monmonier, *Lie with Maps*, 1.

3. As W. J. T. Mitchell puts it in *Language of Images*: "The fact is that spatial form is the perceptual basis of our notion of time, that we literally cannot 'tell time' without the mediation of space. All our temporal language is contaminated with spatial imagery: we speak of 'long' and 'short' times, of 'intervals' (literally, 'space between'), of 'before' and 'after'—all implicit metaphors which depend on a mental picture of time as a continuum" (274).

4. Wood, *Power of Maps*, 130.

5. "Reproductions" are replays of past experiences of events of a remembered or fantasied reconstructed "now" in the past; "retentions" are based on past phases of experience from the

vantage of the "now" moment which slips forward. See Gell's *Anthropology of Time*, 223–25, which is based on Husserl's *Internal Time Consciousness.*

6. It is my intention to explore the representation of other places on KMMS maps in a separate book on place in the context of time and space.

7. Even the use of an ethnonym as opposed to the customary toponym format is unconventional for medieval Islamic maps.

8. Gell, *Anthropology of Time*, 223–25.

9. Heinz Halm suggests the word is an Arabicized rendition of the Berber "Vaga." Halm also mentions a region called "Baja / Beja" in northern Tunisia. But this region is quite different from the one in East Africa that is the subject of this chapter. Halm, *Empire of the Mahdi*, 99.

10. Mostly Tunisia and eastern Algeria.

11. The Hilālī and Sulaym tribes are ignominiously credited with devastating and "bedouinizing" wide areas of central North Africa from the mid-eleventh century onward.

12. Endress, *Islam*, for instance, mentions neither West nor East Africa.

13. Takrūr was important for its export of gold and slaves to North Africa, as well as for its commitment to Islamic jihād.

14. Timbuktu was an important center for Islamic scholarship and learning during the late medieval and early modern periods.

15. Lapidus, *Islamic Societies*, 524.

16. Abyssinia captured and dominated the Muslim imagination as one of the ultimate frontiers of Africa. Nobody knew what lay beyond the lands of Abyssinia. Hence it is often cited as extending all the way to the Encircling Ocean in the west and, sometimes, even as far as the Maghrib. For more detail on the mythical extension of the mysterious and impenetrable Abyssinia in the Muslim imagination, see Miquel, *La géographie humaine*, 2:132–33.

17. The Zanj rebelled in 868–883 CE in the swamps south of Basra. The black slaves set up their own enclave, which was not suppressed by the Abbasids for twenty years.

18. To the credit of the editors of *EI2*, there is a four-paragraph entry on the Beja, but half of it focuses on the Beja from the fourteenth century onward and their relationship with the kingdom of the Funj and later Sudan (s.v. "Bed̲ja").

19. Lapidus, *Islamic Societies*, 524–40.

20. Ibid., 524.

21. Note other well-known scholars of Islamic history do not mention the Beja at all. For instance, Dominique Sourdel, *Medieval Islam.*

22. Shaban, *Islamic History*, 78–79, 110.

23. Lewis, *Race and Slavery*, 48; Lewis, *Race and Color*, 52. Barring the slight change in name, the texts of these two books are virtually identical.

24. But even this is limited to little more than a page. See Spaulding, "Precolonial Islam," 117–18.

25. Hiskett, *Islam in Africa*, 67.

26. On the Persian maps these deserts are referred to as "Biyābān." Some maps abbreviate the entry to just "al-Mafāza" or "Biyābān."

27. Some Idrīsī maps contain three markings. See Miller, *Mappae Arabicae*, vol. 3, pt. 2, sec. 5.

28. In figure 8.3 the area is named in Persian "Zamīn biyābān-e Būja" (Land of the Beja Desert).

29. To the same question, Antonio Palmisano says there is only one "tautological answer": "the Beja are those people who have a strong Beja identity." Palmisano, *Beja as Representation*, 7.

30. Miquel, *La géographie humaine*, 2:163; Trimingham, *Islam in West Africa*, 104.

31. *EI2*, s.v. "Bed̲ja," by Holt. Andrew Paul disagrees and devotes a chapter to the discussion of the origins of the Beja. Paul, *Beja Tribes*, 20–26.

32. Paul, *Beja Tribes*, 20–63; and Arkell, *History of the Sudan*, 41–42, 179.

33. See, for instance, Abū'l-Fidā, *Taqwīm*, 2:157.

34. See, for instance, Mas'ūdī, *Tanbīh*, 1:226; and Ibn al-Wardī, *Kharīdat*, 65.

35. Ibn Ḥawqal, *Configuration*, 1:48.

36. Miquel, *La géographie humaine*, 2:163.

37. Today there are about two million Beja spread out between Egypt, Sudan, and Eritrea. "The Beja Project" of the Oakland Baptist Church (Virginia) is on a mission to convert the tribes. The Joshua Project 2000 reports a similar mission in the region. According to Andrew Paul and other writers, the Beja are not known for holding hard and fast to religious mores, and this means that they are perceived as good candidates for conversion. See Paul, *Beja Tribes*, 37, 78–79; Palmisano, *Beja as Representation*, 67–69; and Clark, "Northern Beja."

38. For more on the history of the Beja tribes and their migration and occupation of this western flank of the Red Sea, see Paul, *Beja Tribes*, 67–68. It is interesting to note that the sixteenth- and seventeenth-century European cartographers seem to have had very limited knowledge of the Beja and their location. If there is any truth to the association of the Beja to Greek and Roman myths of Troglodytes (Cave Dwellers), it could be argued that the Beja did appear on medieval European mappamundi as one of the monstrous races represented along the coast of Africa in maps, such as, the Psalter map (see fig. 6.9). Herodotus (ca. 450 BCE) speaks of the Ichthyophagi (Fish-Eaters), who lived between Aswān and the Red Sea. Later writers, such as Diodorus Siculus (ca. 140 BCE), distinguish between the Troglodytes and the Ichthyophagi. The Agriophagi (Flesh-Eaters or Locust-Eaters) are also mentioned in ancient accounts as living in approximately the same area. For details, see Paul, *Beja Tribes*, 35–36, 54. It is possible that medieval European cartographers acquired their concepts of the strange and bizarre monstrous races inhabiting Africa through these ancient accounts.

39. *EI2*, s.v. "Ibn Ḥawḳal," by A. Miquel.

40. Hasan suggests that Muslim contacts date from even earlier. He mentions that the first caliph, Abū Bakr (632–634 CE), banished a group of Arabs to the region of 'Aydhāb in Beja country. Unfortunately, Hasan does not cite a source for this. See Hasan, "Eastern Sudan," 116. Andrew Paul also discusses these early Arab incursions in *Beja Tribes*, 78–79.

41. Cited in Spaulding, "Precolonial Islam," 117, although he does not specify his original source for this information.

42. *EI2*, s.v. "Bed̲ja," by Holt. Ibn Ḥawqal presents a different version of this event, noting that Abd Allah went on from the Nubian defeat to conquer the city of Aswān in the year AH 31/652 CE. In the process of this conquest, Ibn Ḥawqal tells us that Abd Allah also "subdued the Būja and other lords." Ibn Ḥawqal also gives the name of the commander of Arab forces as 'Abd Allāh ibn Abī Sarḥ. See Ibn Ḥawqal, *Configuration*, 1:48.

43. Once again Ibn Ḥawqal's version is quite different. Instead of exacting tribute, he says that the Beja were actively converted to Islam by force and outwardly adopted certain obligations. However, this would not have permitted the Rabi'a tribes to subjugate the Beja and use them as slaves in the mines, because conversion to Islam should have meant that they could not be enslaved. Perhaps Ibn Ḥawqal was conflating a later phenomenon present during his time with the earlier Muslim forays in the region. Ibn Ḥawqal, *Configuration*, 1:48.

44. *EI2*, s.v. "Bedja," by Holt.

45. The mines in the Beja deserts are mentioned by most of the geographers: see, for instance, Iṣṭakhrī, *Masālik*, 31–32; Masʿūdī, *Murūj*, 331, 334, 336; Ibn Ḥawqal, *Configuration*, 1:48–49; and Ibn al-Wardī, *Kharīdat*, 65.

46. See, for instance, discussion of active trade in the region in Yaʿqūbī, *Kitāb al-buldān*, 336–37. Key for pegging a ninth-century date to full-blown mine exploitation activity is the occurrence of the Beja uprisings, which began around the first quarter of the ninth century. From this, one would have to presume that the penetration and takeover of the mines in Beja territory by the Rābiʿa tribes from the Arabian mainland must have begun sometime in the late eighth century even though exact dates of the incursion are not mentioned in the texts. S. Hillelson and C. E. Bosworth, in their joint work on the second part of the "Nubia"entry in *EI2*, say that by the tenth century the Rābiʿa had gained control of the mines at al-ʿAllāqī and had imposed their rule on the Beja with whom they had allied themselves through marriage (90). Hasan notes that proof for extensive Muslim contact lies in the fact that two Arabic treaties were translated in their language, Bujāwī. See Hasan, "Eastern Sudan," 117.

47. Ibn Ḥawqal, *Configuration*, 1:48.

48. Spaulding, "Precolonial Islam," 118; Miquel, *La géographie humaine*, 2:163–64. The geographers cited above in note 44 also mention the use of Beja slaves for mining.

49. See, for example, Ibn Ḥawqal, *Configuration*, 1:48; also, Miquel, *La géographie humaine*, 2:161–62. Accounts of the Beja before and after the Arab incursions from Pharaonic and Roman times to the Funj and the Mahdist period tell the same story of the Beja's independent streak and their consistently defiant resistance to external domination. See Paul, *Beja Tribes*, 30–31, 52–53, 56–63, 91–119. Their tough fighting skills captured British colonial attention, and they have been immortalized in early twentieth-century British literature through the epithet "Fuzzy-Wuzzy" and Rudyard Kipling's poem extolling their brave resistance of the British invading forces as "first-class fighting men." Palmisano, *Beja as Representation*, 8, 11–13, 29–30.

50. The Beja raids and uprisings and refusal to comply with the requirements for tribute imposed on them are cited in numerous sources. See, for instance, Arkell, *History of the Sudan*, 188; Spaulding, "Precolonial Islam," 118; *EI2*, s.v. "Bedja," by Holt; and Miquel, *La géographie humaine*, 2:163; as well as Ibn Ḥawqal, *Configuration*, 1:48–53.

51. Spaulding, "Precolonial Islam," 118. Spaulding does not cite his source. He only makes a vague reference to a comment by an unnamed Iranian in the eleventh century.

52. There is one notable exception to this: Ghana, Kanem, Kuga, Awdaghost, and a whole host of unidentifiable territories in western Africa are marked on the earliest extant world map of 479/1086. Sijilmāsa is never located on the world maps. It does, however, figure prominently in the regional maps of the Maghrib. The Maghrib maps only show the coast of North Africa and Muslim Spain. They cut off short of the Beja territory. The gold mines of Nubia, on the other hand, are never indicated on any map.

53. Andalusī, *Science in the Medieval World*, 8; my brackets. Bernard Lewis presents a more severe reading of this passage: "The only people who diverge from this human order and depart from this rational association are some dwellers in the steppes and inhabitants of the deserts and wilderness, such as the rabble of Bujja, the savages of Ghana, the scum of Zanj, and their like." Lewis, *Race and Slavery*, 48.

54. Iṣṭakhrī, *Masālik*, 16; Iṣṭakhrī, *Viae regnorum*, 4–5.

55. Iṣṭakhrī, *Masālik*, 16; Iṣṭakhrī, *Viae regnorum*, 5.

56. *Marāḥīl* is literally the plural of "stage" (marḥala) and refers to the stages of a journey. One *marḥala* is the distance a traveler can cover in one day and varies greatly depending on terrain.

57. Iṣṭakhrī, *Masālik*, 31–32; Iṣṭakhrī, *Viae regnorum*, 35. Ibn Ḥawqal begins his description of the Beja similarly.

58. Iṣṭakhrī, *Masālik*, 42; Iṣṭakhrī, *Viae regnorum*, 54. I am grateful to Richard Bulliet for his clarification of Iṣṭakhrī's confusing use of *najab* and *asīr*.

59. An example of how the geographers often copied verbatim from each other.

60. Ibn Ḥawqal, *Configuration*, 1:48–53 (brackets in the French); Ibn Ḥawqal, *Ṣūrat*, 55.

61. Maqdīsī, *Kitāb al-bad'*, 4:69–70. This passage is cited by both Miquel, *La géographie humaine*, 2:161, and Lewis, *Race and Slavery*, 52.

62. From the ancient period (Pharaonic, Greek, and Roman) accounts to twentieth-century studies, the depiction of the Beja as "extreme other" is a consistent feature. Clark's "Northern Beja" is possibly the most egregious of the modern twentieth-century accounts on the Beja, presenting them as the most freakish type of "other" (4–6). Even though Andrew Paul has taken the time to write a history on them, he has clearly done so out of fascination for their extreme "otherness," and he has made it clear that they were seen this way from the earliest recorded histories of them. He doesn't mince his words when he says, "Rude, wild, bestial, call them what you will, of unpleasant and unhygienic habits, their hair clotted with mutton fat, their bodies reeking of oil, sweat, and wood smoke, the Beja, for those whose knowledge of them goes beyond externals, will ever be a fascinating and rewarding study. . . . Primitive and bloodthirsty desert tough. . . . Of Beja indolence it is almost superfluous to speak. . . . Strolling aloof and unconcerned among them all, may be found the unchanging 'Fuzzy' with his easy nomad stride and his 'hay-rick head of hair,' his camel stick aslant across his shoulders, and an occasional goat, his familiar spirit, trotting dog-like at his heels." Paul, *Beja Tribes*, 2–11. Antonio Palmisano provides a summary of texts depicting this "extreme other" representation of the Beja across the span of recorded history from the pharoahs to the British colonials and Rudyard Kipling's immortalization of them as the "Fuzzy-Wuzzy." See Palmisano, *Beja as Representation*, 11–33. Even Edward Gibbon makes passing mention of their "extreme otherness" when he notes, "These barbarians whom antiquity, shocked by the deformity of their figure, has almost excluded from the human species, presumed to rank themselves among the enemies of Rome." Gibbon, *Decline and Fall*, 6:138. It is only with anthropological and linguistic studies from the late 1980s onward that there has been a concerted effort among scholars to see them as something other than a bizarre and monstrous group untouched by civilization. See, for instance, Ausenda, "Leisurely Nomads"; Jacobsen, *Sickness and Misfortune*; Nawata, "Rashāyda and the Beja"; and Vanhove, "Beja Language."

63. Ya'qūbī, *Kitāb al-buldān*, 336. Thanks go out to Paul Cobb, who is preparing an English

translation of Yaʿqūbī's geography, for his assistance with my reading of this passage. Thanks also to Matthew Gordon for bringing Paul Cobb's ongoing work on this text to my attention. The big question here is what exactly does *falak* mean? Normally it has a sense of "round" as in "as round as a celestial sphere," which is also called *falak*. It is also used to describe the ideal female breasts, round like orbs. But no Arabic dictionary records a meaning of "nipple." Yet it is the only meaning that seems to fit in the context, since the removal of the "roundness" of the male breast makes no sense. If this reading is accurate, it is likely the first recorded mention of the removal of male nipples! Presently, the Beja are not known for this practice, but, it seems, some Beja tribes in modern Sudan are still known for removing their front teeth so that they do not resemble asses. Ancient Greek writers used to refer to them as the *Colobi*, meaning the "Mutilated People," because the women practiced infibulation and the men had the habit of removing their right testicle. There is no mention of any of these practices among the Beja today, although they continue to practice female genital mutilation according to the "Pharaonic way." Jacobsen, *Sickness and Misfortune*, 27. Andrew Paul notes that the appearance of the Beja excited much astonishment among the Romans. If the Blemmyes are in fact the same as the Beja, then Pliny's references to the headless Blemmye, with eyes and ears sunk below the level of their shoulders, could be read perhaps as reference to their practice of self-mutilation. Paul, *Beja Tribes*, 34–35, 58–59. Gibbon, *Decline and Fall*, 6:138, also makes note of their deformities.

64. Miquel, *La géographie humaine*, 2:161.

65. Ibid., 2:164.

66. In his pathbreaking book, Houari Touati brings "the voyage" forward as the glue that bound Islamic society together from the eighth to the twelfth century, the fruits of which created the bedrock of *ʿilm* (knowledge) that would guide the community for centuries after. Out of this background of the centrality of travel emerges its raison d'être: the "effort to create sameness." All this was spearheaded by the efforts of the earliest linguists in the eighth century who returned to the desert and lived among the Bedouin (in what could be argued is the earliest manifestation of anthropologists) in order to collect and document the purest Arabic before it was corrupted by the mixed populations of the urban centers. Touati, *Islam and Travel*, 2–3, and chapter 2, "The School of the Desert," 45–77.

67. Miquel, *La géographie humaine*, 2:164.

68. Hamdhānī, *Kitāb al-buldān*, 252.

Chapter 9

1. Ṭabarī, *Incipient Decline*, 142.

2. The phrase "four hundred heads per year" also occurs in precisely this formula in other texts. See, for instance, Hamdhānī, *Kitāb al-buldān*, 76.

3. Some versions of this manuscript read "uppermost Nile area of Egypt."

4. Ibn Khurradādhbih, *Kitāb al-masālik*, 83. The text does not specify "heads" of what. Cattle, sheep, camels, or men?

5. In a recent book, *Islam and Travel*, Houari Touati argues that the geographers were active travelers, not the armchair scholars of earlier reputation. Still I doubt that they made it

as far upstream on the Nile as the Beja or that they personally witnessed any of the battles between Muslim forces and the Beja. Touati, *Islam and Travel*, 119–55.

6. Ṭabarī, *Incipient Decline*, 142; translator's brackets.

7. Although the sources don't specify, one must presume that it is Samarra because this is where the caliph Mutawakkil was based.

8. Ṭabarī, *Incipient Decline*, 141–42; translator's brackets.

9. Ibn Ḥawqal, *Configuration*, 1:49.

10. These stories are discussed at length in ibid., 49–52.

11. For more detail on Mutawakkil and his troubles with the Turks, see Gordon, *Breaking of a Thousand Swords*.

12. Ibn Ḥawqal, *Configuration*, 1:51.

13. Ṭabarī, *Incipient Decline*, 142.

14. Ibn Ḥawqal, *Configuration*, 1:51. The absurdly large numbers cited by Ibn Ḥawqal reflect the fantasy that he was projecting onto this event.

15. Both Ṭabarī and Ibn Hawqal report this but their descriptions of the plan vary.

16. Ibn Ḥawqal, *Configuration*, 1:51.

17. Ibid., 51–52.

18. Ṭabarī, *Incipient Decline*, 144; translator's brackets.

19. Ibid; translator's brackets. Procopius, in *History of the Wars*, reports that the Blemmyes worshipped Isis, Osiris, and Priapus, and all the Greek gods. In particular, Procopius notes that the Blemmye worshipped the sun in one specific manifestation—Mandulis—to whom they offered human sacrifices (bk. 1, xix). Paul, *Beja Tribes*, 37. Could Mandulis have been the stone god that ʿAlī Baba is reported worshipping?

20. The latter certainly does seem to be a bit of hyperbole.

21. In Ṭabarī's version of the story, after the audience with the caliph, ʿAlī Baba, accompanied by Qummī, returned to East Africa. Ibn Ḥawqal mentions the return of Qummī to the region but not ʿAlī Baba.

22. Ibn Ḥawqal, *Configuration*,1:52.

23. There is a great deal of dispute over the authenticity of *Alf Layla wa-Layla* (The Thousand and One Nights). Only the first 271 lesser-known stories are confirmed by extant fragments. The remainder are now considered later innovations, possibly by the original French translator, Antoine Galland (1646–1715). He played such an important part in composing and popularizing the tales that some call him the "real author." Not only did Galland freely embellish his translation; one of the four Arabic manuscripts he used is no longer extant, and it is theorized that he may have employed a second set of manuscripts that is no longer extant. Many erudite Orientalists have worked to identify the "true" *Nights*, including Duncan Black Macdonald and, more recently, Muhsin Mahdi. It is Mahdi who proposed in his new edition that only the first 271 tales should be considered authentic based on a thirteenth-century Syrian manuscript. Robert Irwin, however, disagrees. He suggests that "the *Nights* are really more like the New Testament, where one cannot assume a single manuscript source, nor can one posit an original fixed cannon. Stories may have been added and dropped in each generation. Mahdi's stemma suggests that there were very few thirteenth-century manuscripts of the *Nights*; for, in the end, the stemma narrows down to one single manuscript source. The references

in the Geniza and in al-Maqrīzī's topography of Cairo suggest, however, that the work was quite well known in the eleventh and twelfth centuries. Is it conceivable, then, that only one thirteenth-century manuscript served as the basis of all subsequent copies?" Irwin, *Arabian Nights*, 59–60.

24. Why is the chief of the Beja named ʿAlī Baba? It sounds like an unlikely name for a pagan African chief.

25. Refer to chapters 10, 11, and 12 for a detailed discussion of the Ottoman cluster.

26. What is particularly noticeable in this focus on the Beja is the way in which the Ottoman renditions of the KMMS maps blow up the size of Africa, implying virgin, yet to be "conquered" empty space available. This is discussed in chapter 11.

27. The maps retain the Arabic naming convention.

28. If this reading of the Beja is correct, it presents a significant challenge to the idea that it was Abyssinia exclusively that dominated the medieval Muslim imagination.

29. Hasan, "Eastern Sudan," 117.

30. Ibid. cites the translation of Arab treaties into Bujāwī as proof of this.

31. Miquel, *La géographie humaine*, 2:166.

32. Hasan, "Eastern Sudan," 119.

33. It is hard to prove the provenance of the maps in this manuscript because they have been retouched a number of times over the centuries.

34. The scribal errors in this manuscript raise questions about its provenance. The maps in this manuscript resemble those of Ahmet 3012, which is also of questionable origin. See fig. 9.10.

35. The Safavids ruled in Iran from the sixteenth to the eighteenth century.

36. The identification of the Beja is missing from the Bodleian's online translation of this map: see "Circular World map," under bk. 2, chap. 5, of the *Book of Curiosities* online (http://cosmos.bodley.ox.ac.uk, accessed May 2, 2015).

37. Refer to chapter 2 for a discussion of the Ibn al-Wardī manuscripts. This particular Ibn al-Wardī–type KMMS world map occurs oddly in a KMMS manuscript based on the earliest extant Ibn Ḥawqal model of Ahmet 3346.

38. The Funj are a nomadic cattle-herding people from the region of northern Sudan who set up a major sultanate that ruled over a substantial portion of northeast Africa between the sixteenth and nineteenth centuries.

39. Jacob, *L'empire des cartes*, 267.

Chapter 10

1. Barber, "England I," 27.

2. Barber, "England II," 77.

3. Delano-Smith and Cain, *English Maps*, 1.

4. Pedley, *Commerce of Cartography*, 26–29, 161–62.

5. Edney, *Mapping an Empire*, 325.

6. Harley, "Maps, Knowledge, and Power," 282. Other examples include Boud, "Carto-

graphic Patronage"; Edney, "Patronage of Science"; Edney, "Mathematical Cosmography"; Doel, Levin, and Marker, "Extending Modern Cartography."

7. Adıvar mentions references to these earlier translations in the opening chapter of his *İlim*. *Ṭabaqāt* (biographical dictionaries) were an important element of medieval Islamic culture. Readers interested in learning more about the phenomenon of biographical dictionaries in medieval Islamic societies may begin by consulting the many works of Richard W. Bulliet that are based on the extensive use of *ṭabaqāt*. See, for example, Bulliet, *Conversion*.

8. Readers interested in knowing more should consult Gutas's *Greek Thought*. Note that I also reference Gutas's work in chapter 5 on the "Encircling Ocean."

9. Brentjes mentions two carto-geographical manuscripts in passing but at least one of them is incorrectly attributed. Although, in "Courtly Patronage," Brentjes omits mention of the manuscript number, it is clear from the discussion that the reference to the British Library's Iṣṭakhrī manuscript is Ms. Or. 3101 and is therefore not Mamluk but Ottoman, as I prove in the following chapter. This glitch should in no way take away Brentjes's overall contribution to the subject of patronage, and one can only hope that she will write a monograph on courtly patronage and princely education in the Islamic world where she can have the space to fully unfold her ideas.

10. Ramaswamy's article addresses patronage only in the specific Mughal context, not in its Middle Eastern orbit. This is the subject of her forthcoming work on a digital *muraqqaʿ* album.

11. Another cause of this "cloud of controversy" is the dearth of comprehensive secondary scholarship on Meḥmed and the period. To date, the most comprehensive source is Babinger's *Mehmed*.

12. Julian Raby, "Gran Turco," xv–xvi, provides an unmistakable flavor of Meḥmed's eclecticism; see also Raby, "Sultan of Paradox." For Meḥmed's eclecticism, see Babinger, *Mehmed*, in particular the final chapter, "The Personality and Empire of Mehmed the Conqueror," 409–508. For primary source descriptions of Meḥmed and the period, see Kritovoulos's *History* and Ṭursun Beg's *Ebü'l-feth*. Necipoğlu reaffirms this when she asserts that Meḥmed "deliberately negotiated the expanding Western and Eastern cultural horizons of his empire through visual cosmopolitanism and creative translation" ("Visual Cosmopolitanism," 1).

13. Çigdem Kafescioğlu, "Ottoman Capital," 4, notes that "he [Meḥmed] patronized the art and *sciences* of the *Iranian* world and of Italy" (emphasis my own). See also Çigdem Kafescioğlu, *Constantinopolis/Istanbul*, 62. Raby, "Gran Turco," xi, says that it was his intention at the outset to present "a more balanced picture of Mehmed's contribution," but that he was forced by the lack of sources to put his "'Eastern brief' to one side." Even Babinger, a strong proponent of the "Meḥmed-the-eclectic" view, devotes many more pages to Meḥmed's Western face than his Eastern one in his book *Mehmed*. See also Brotton, *Trading Territories*, 87–103, who mentions possible Eastern scholarly influences but focuses on the impact of European cartographic models.

14. A good example of early scholarship on the period that exclusively saw the "Western side" is F. R. Martin's claim that "Gentile Bellini was the father of Turkish painting"! E. Simms, "Turks," 747.

15. Among the exceptions to this is the recent contribution by Ahmet Karamustafa to the subject of Ottoman cartography where he notes that the "terrestrial maps contained in geo-

graphical and historical works that derive from earlier Islamic texts are invariably world maps," and that "these maps are derivative of previous Islamic geographical traditions and can be viewed as further examples of the different trends of mapping the world that existed in the Islamic Middle Ages." Karamustafa, "Scholarly Maps and Plans," 221. Taeschner, in "Literatur der Osmanen" and "Ottoman Geographers" (in *EI2*, s.v. "Djughrāfiyā," by S. Maqbul Ahmed and F. Taeschner), mentions the translations of some earlier medieval Islamic geographical works into Turkish (in particular Qazwini's *ʿAjāʾib al-makhlūqāt* as well as Abū'l-Fida's *Taqwīm al-buldān*). Adıvar also mentions references to these earlier translations in the opening chapter of his *İlim* (13–15), discussing among others the works of Yazıcı-zâde Ahmed Bican. None of them, however, discuss Meḥmed's interest in and sponsorship of medieval Islamic cartographical works. In all the cases (Adıvar, Babinger, Brotton, Karamustafa, Raby, Taeschner, Adıvar, Kafescioğlu, etc.), when discussing Meḥmed II, the focus is on his interest in Western cartography.

16. Any use of "Istanbul" in this paper and indeed in this period must be seen for what it is: purely anachronistic. When exactly "Istanbul" as the name for the former Byzantine capital comes into regular use is the subject of much debate among Ottoman scholars. According to the seventeenth-century historian/chronicler Evliya Çelebi, by his time "Istanbul" was regarded by the "educated classes" as "the Ottoman" way of referring to the city. However, it was not officially decreed as the name of the city until the *farmān* of AH 1174/1760 CE, issued by Mustafa III (r. 1171–1187/1757–1774), when the mint name on the coins was changed. Up until then "Qusṭanṭīniyya," the Arabic version of Constantinople, continued. *EI2*, s.v. "Istanbul," by H. Inalçık, provides a reference to a *wāqf* (pious foundation) document from the fifteenth century that uses Meḥmed's term. The colophons of the Ottoman cluster manuscripts that form the basis of this paper, to be discussed later, use "Bilad al-Qusṭanṭīniyya." I have discovered a series of geographical manuscripts of the KMMS tradition, in which the maps have been omitted, dating from the early sixteenth century, which refer specifically in their colophons to "Istanbul." These may be among the earliest known colophonic examples of the use of the name "Istanbul."

17. Raby, "Sultan of Paradox," 6, makes note of "the very private character" of Meḥmed's patronage of Western art, especially Christian works, such as the depiction of Madonna and Child that he commissioned from Bellini.

18. Meḥmed's interest in commissioning portraits of himself and his importation of Italian portrait artists for this purpose, in particular the famous Venetian Gentile Bellini, who spent two years in Istanbul (1479–1481) and to whom is attributed a number of portraits of Meḥmed and other paintings, is probably one of the best-known aspects of painting during Meḥmed II's period. Countless articles and books discussing Meḥmed II and early Ottoman painting mention this, including Babinger, *Mehmed*, 201; Raby, "Sultan of Paradox"; Atıl, "Ottoman Miniature"; Andaloro, "Costanzo da Ferrara"; Raby, "Portrait Meda"; Necipoğlu, "Süleyman"; Grube, "Ottoman Painting"; Jardine and Brotton, *Global Interests*, 32–42; and, most recently, Necipoğlu, "Visual Cosmopolitanism," 33–34.

19. For the latest general summary and detailed references on this, see Karamustafa, "Scholarly Maps and Plans." See also numerous references to Meḥmed and Italian maps found in Babinger, *Meḥmed,* 118, 201, 248, 425, 495, 505; Babinger, "Italian Map"; and Adıvar, *İlim*, 19–43.

20. There are a multitude of references to this, some previously cited above. See, for instance, Babinger's recording of Meḥmed's request of the Lord of Rimini, Sigismondo Pandolfo Malatesta (1417–1468), that one of his court painters, specifically Matteo de' Pasti, be dispatched to Istanbul to paint Meḥmed's portrait. Babinger, *Mehmed*, 201. See also the introductory chapter to Raby's doctoral thesis, "Gran Turco"; and Necipoğlu, "Visual Cosmopolitanism," 16, 21, and 35.

21. Raby is quick to point out that the literature has focused far too much on Bellini and too little on others. Bellini came into Meḥmed's life late, in 1479, just a few years before he died. Raby tells us to look earlier for the European influence and fascination, particularly to Meḥmed's schoolbook with sketches and portraits of figures in a "naturalistic" proto-European Renaissance style. Aside from the most crucial question of where and how Meḥmed was exposed to European portraiture technique as a child (raising of course the distinct possibility that his father, Murād II, was also familiar with this type of art form), the schoolbook points to Meḥmed's early fascination with visual material. As Raby puts it: "Throughout his life Mehmed showed a consistent interest in three developing areas of European art—cartography, painted portraits and medallions—and each of these interests Bellini was called upon to satisfy." Raby, "Gran Turco," 82. Despite the plethora of references to Bellini and other Italian portrait painters, there is also a considerable dispute about them. Esin Atıl argues that the presence of Bellini in Istanbul cannot be confirmed. Many of the portraits and paintings attributed to Bellini are also disputed: for example, the so-called Bellini portrait of Meḥmed the Conqueror housed in the National Gallery, London. Whether these paintings are Bellini's does not change the overall evaluation that Meḥmed was interested in having his portrait painted. Atıl, "Ottoman." For more on this, see the articles in Campbell and Chong, *Bellini and the East.*

22. This suggestion is made in passing by Ahmet Karamustafa, "Scholarly Maps and Plans," 210, in what is probably the best article to date on the subject of Ottoman mapping, and certainly warrants investigation. That the map contains transliterations of place-names in late fifteenth- and early sixteenth-century Cyrillic alongside the place-name markings in Latin also needs to be explained. For details on this latter bit of crucial information, see the discussion by Gallo, "*Terraferma*," who on the basis of certain landmark anomalies on the map argues convincingly for a pre-1496 date for the map. On the basis of the style of illustration of the map, Gallo asserts that "we are not dealing with the work of a professional cartographer who had in mind to combine in one single map the various maps of the provinces of the Venetian territory"; rather the *Terraferma* map, as it is called, appears, according to Gallo, to have been "a makeshift map drawn up for specific military purposes, and sketched from memory on the basis of the personal knowledge of the designer." Gallo, "*Terraferma*," 56–57. If the strange anomaly of the presence of Cyrillic on the map can be accounted for, these conclusions would seem to favor Karamustafa's hypothesis. Gallo, *Terraferma*, 57. The reference to Meḥmed's request of Bellini to make him a map comes from Babinger, "Italian Map," 12. See also *PDH*, 51.

23. Babinger, *Mehmed*, 295. See Raby, "Gran Turco," 194.

24. J. R. M. Jones, *Siege*, 127; also discussed in Babinger, *Mehmed*, 112; and Necipoğlu, "Visual Cosmopolitanism," 53.

25. Santini's tract occupies the first half of the manuscript. For details on this, see Antonio Marsand, *Bibliotheca Parigina*, 2:1–5.

26. Babinger, "Italian Map." See also Banfi, "Two Italian Maps," in which he challenges Babinger's conclusions. Banfi closely studies the places on the map and gives the map an earlier date of 1443, which would fit with the theory that Meḥmed used this map during his conquest of Constantinople. Banfi argues that it probably fell into the hands of the Turks during the sack of the Corvinian library at Buda in 1541. It is possible that this *is* the illustrated military treatise in which Ducas said Meḥmed immersed himself in the "feverish" last nights of planning right before the conquest.

27. Brotton mentions this treatise but does not indicate the presence of Bāyezīd II's stamps. Brotton, *Trading Territories*, 102–3.

28. Babinger, *Mehmed*, 81.

29. From Jakob Unrest's *Chronicon Austriacum* (1472), quoted in Babinger, *Mehmed*, 295.

30. I use "attributed" to indicate the problems regarding the attribution of extant manuscripts to Ptolemy since the earliest extant copy of his work dates from the thirteenth century. None of these can be firmly attributed to Ptolemy. For the debate, see Polascek, "Ptolemy's *Geography*"; Bagrow, "Origin"; and Bagrow, "Wilczek-Brown."

31. This story about Meḥmed and his interest in the Ptolemaic manusript during the summer of 1465 is cited by numerous sources: most recently, Karamustafa, "Military," 210; Adıvar, *İlim*, 18–21; Raby, "Sultan of Paradox," 6; and Deissmann, *Forschungen*, 32–35.

32. Exactly which Ptolemy-attributed manuscript of the two presently housed at the Topkapı Saray library Meḥmed examined is a matter of some debate. Deissmann, *Forschungen*, 32–35, suggests that Meḥmed must have taken a look at the later fourteenth-century one (G.I. 27), which contains the signature of the priest Joseph Bryennios and a 1421 date of deposit in the Aya Sofya church, because it does not contain a world map. Deissmann presumes that if it had contained a world map, Meḥmed would not have asked Amirutzes to construct one. It is also possible that Meḥmed recognized the discrepancy between the standard Ptolemaic world map and the detail of the individual regional ones, and that he wanted to see the image that the sum of the smaller individual maps added up to.

33. Amirutzes's enjoyed a position of considerable influence at Meḥmed's court from 1461 until his death in 1475, both for his academic abilities and for his relationship with Meḥmed's grand vizir, Maḥmūd Paşa. There is a rather elaborate tale told about how he originally tried to persuade the sultan to convert to Christianity but was won over by Meḥmed and converted to Islam instead! Adıvar, *İlim*, 19. Raby suggests that George of Treibzond and George Amirtuzes were two different personalities. Raby, "Gran Turco," 27. Necipoğlu presents him as one in "Visual Cosmopolitanism," 11–12, and 15.

34. The problem with this request is that Meḥmed was not considered proficient in Arabic. If he was planning on reading Ptolemy's text, why would he have had a translation commissioned in Arabic?

35. Although the map prepared by Amirutzes is no longer extant a number of elaborate descriptions and guesses abound about this map. Raby suggests that "Amirutzes was commissioned by the Sultan to produce a *wall map*." Raby, "Gran Turco," 20; emphasis my own. Given that the map was no longer extant when he conducted his research at Topkapı Saray, Deissmann, *Forschungen*, 32–35, provides a surprisingly elaborate description of the map, its size, and bright colors! In his 1954 summer trip to examine the maps in Topkapı Saray, Leo Bagrow

makes a most unusual suggestion: namely, that the Amirutzes's world map was woven into a carpet. Leo Bagrow, "Bosphorus," 25. Bagrow also suggests that the Ptolemaic-type world map contained in the so-called Benincasa Atlas may be the missing Amirutzes map. Karamustafa, "Scholarly Maps and Plans," 210, gives up the map for lost, while Brotton, *Trading Territories*, 100–102, avers that the manuscript and the maps survive in the Aya Sofya library, Istanbul. Brotton does not provide any details to back up his claim.

36. Kritovolous, *History*, 209–10, confirms Meḥmed's commissioning of the map and the translation of Ptolemy, but the only details he provides are as follows: "He [i.e., Amirtuzes] also put down on the chart the names of the countries and places and cities, writing them in Arabic, using as an interpreter his son, who was expert in the languages of the Arabs and of the Greeks."

37. These are Sülemaniye's Ayasofya 2596 and 2610. As with everything else, there is considerable debate on this matter, but one cannot help wondering whether at least one of these was the one prepared by Amirutzes.

38. Raby, "Gran Turco," 27. Raby's reference is, however, extremely confusing. He does not specify who the "George" is that he is refering to. At one point he says, "Mehmed's interest in Ptolemy and the crucial role of Amirtuzes' recur, because George of Treibzond acknowledges Amirutzes' help with his Greek." This would seem to suggest that George of Treibzond and Amirtuzes are two different personalities, where as in every other reference to the person, and indeed, in other parts of Raby's dissertation, they are one personality.

39. For the debate on this atlas, see Salinari, "Atlas of the 15th Century," who gives the atlas its "Benincasa" attribution. Destombes, "Venetian Nautical Atlas," argues against Salinari's attribution. See also Bagrow, "Bosphorus," 27.

40. TSMK G.I. 84. For further details on this map, see *PDH*, 52–53; Deissmann, *Forschungen*, 84; Babinger, *Mehmed*, 506; Karamustafa, "Scholarly Maps and Plans," 209–10; Brotton, *Trading Territories*, 90, 93–96; and Roberts, "Francesco Berlinghieri's *Geographia*," 145–60.

41. For a general summary with detailed references, see Karamustafa, "Scholarly Maps and Plans," 209; Goodrich, "Old Maps," 120–33; and Ebel, "Ottoman Town Views," 2. For older references, see Babinger, *Meḥmed*, 118, 201, 248, 425, 495, 505; Deissmann, *Forschungen*; Bagrow, "Bosphorus," 25; Adıvar, *İlim*, 19–43; and Babinger, "Italian Map."

42. Raby, "Gran Turco," 190.

43. Ibid., 195–98. Raby, "Sultan of Paradox," 3–4, includes sample images from Meḥmed's sketchbook. Portraits and sculpture are noticeably absent from traditional Islamic arts. Their absence is a matter of considerable debate among Islamic scholars. The ban is related to a ḥadīth linked to the Prophet Muhammad banning graven images. See Arnold, *Painting in Islam*; and Grabar, *Formation*, 72–98. Thus, the appearance of portraits in Ottoman arts is automatically identified as "European" influence. See Pamuk, *Red*, for an entertaining and insightful fictional account regarding this matter.

44. Babinger, *Mehmed*, 81. Emphasis my own.

45. Necipoğlu, *Topkapı Palace*, 8, 15.

46. Hamdhānī, *Kitāb al-buldān*, 5:283. It is said that a plan of the round city of Baghdad was prepared in advance of building the city for the Abbasid caliph Manṣūr (r. 136–158/754–775). Yaʿqūbī, *Kitāb al-buldān*, 7:238.

Chapter 11

1. Kafescioğlu skillfully problematizes the issue of the slash between Constantinople / Istanbul: "A symbolic locus embodying and representing myriad meanings, the political center of the eastern Mediterranean and one of the largest urban centers of the world, Constantinople / Istanbul has been the site of large-scale urban and architectural interventions several times in the course of its history." Kafescioğlu, *Constantinopolis/Istanbul*, 1.

2. Due to political turmoil in Egypt, I was unable to secure high resolution images or permission to reprint images from the Ottoman cluster manuscripts in Cairo.

3. Two additional manuscripts appear to be related to this Ottoman cluster series: Biblioteca Universitaria Bologna, Cod. 3521; and Berlin, Staatsbibliothek Preussischer Kulturbesitz, Orientabteilung, Sprenger 1. Although there are similarities between the images and texts, I have excluded them from the Ottoman cluster because the maps are a touch different, suggesting that they have a different provenance and that they were not produced in the same time and place as the Ottoman cluster. They appear instead to be nineteenth-century copies of the Ottoman cluster produced in South Asia, possibly at the behest of Orientalist collectors. For this reason, I have decided not to include them in this discussion although their similarity to the Ottoman cluster cannot be denied.

4. The Dār al-Kutub library in Cairo has misplaced Jughrāfiyā 256. Based on Gerald R. Tibbetts's description in appendix 5.1, "Select List of Manuscripts," in *HC 2.1*, 130, and the reprinted maps in Ḥīnī's 1961 Arabic edition of Iṣṭakhrī's text, J. 256 can be identified as part of the Ottoman cluster set.

5. I am grateful to Cemal Kafadar for his insights on this colophon.

6. For reasons unknown to us, the original copyist of Aya Sofya 2971a chose to not insert a colophon of his own, but a later copyist of Aya Sofya 2971a, namely, Ibrāhīm ibn Aḥmad al-Sinābbī, did (gratefully from our perspective) choose to insert his name and the date, which was then carried forward in subsequent copies, with some later copyists eliminating his name but keeping the original date.

7. None of the Ottoman cluster manuscripts that I examined contain original covers. One wonders if this series went through heavy use and if this is the reason that none of the original covers have survived. In the case of the Dār al-Kutub manuscripts in Cairo, I relied on a microfilm, so I am unable to comment on the covers of Jughrāfiyā 257.

8. Erünsal, *Türk Kütüphaneleri*, 283–84, where he says: "The reign of the bibliophile sultan Maḥmūd I (1730–1754) can be considered the golden age of the Ottoman libraries, a period which saw the establishment of the libraries in every part of [the] empire, even in the border fortresses." See also Ṭursun Beg, *Tārīḫ-i Ebü'l-fetḥ*, fols. 1b–2a, 27.

9. I cannot confirm the stamps for Jughrāfiyā 256 because the manuscript has been missing from Dār al-Kutub for at least two decades. All that is available of Jughrāfiyā 256 are the photographs of the maps that Ḥīnī used for his edition of Iṣṭakhrī's *Kitāb al-masālik wa-al-mamālik* in 1961. These match the Ottoman cluster.

10. The lack of a Bāyezīd II stamp in Jughrāfiyā 257 may mean that this manuscript found its way to Cairo by the late fifteenth / early sixteenth century.

11. Note the order of the maps remains the same and matches earlier versions of the KMMS series.

12. For a detailed discussion of the various shapes of the Mediterranean in medieval Islamic maps, see my contribution to *Eastern Mediterranean Cartographies*, "Surat Bahr al-Rum," and 2013–2014 NEH project, "The Mediterranean in the Islamic Cartographic Imagination."

13. The Muslim cartographers designed their world maps with south on top. This is confirmed by the direction of the geographical text that accompanies these maps. Not all the maps have south on top—such as the Mediterranean and Maghrib maps, which face west.

14. Oxford, Bodleian Library, Ouseley 133. See also Çağman and Tanındı, *Topkapı Saray Museum*, 184.

15. Topkapı Saray, Revan 989. See Çağman, "Kulliyat-i Katibi."

16. Bibliothèque nationale de France, Paris, Supp. turc 693. Reprinted as a color facsimile by Uzel in Sabuncuoğlu, *Cerrāḥiyyetü'l-Ḥāniyye*. It is intriguing to postulate a connection between the illustration schools of Amasya and Sinop. The colophons in four of the Ottoman cluster KMMS manuscripts tell us that the copyist was originally from Sinop, which is about 162 kilometres from Amasya.

17. For details on this manuscript, refer to Çağman, "Kulliyat-i Katibi."

18. TSMK, Hazine, 799, in particular fol. 179a, "Bahram Gur with the Tatar princess in the Green Pavilion," reproduced in Çağman and Tanındı, *Topkapı Saray Museum*, 192, pl. 129.

19. For more examples, see Evans and Wixom, *Glory of Byzantium*, cat. nos. 42, 43, 57, 63, 143.

20. Grube, "Notes on Ottoman Painting," 445–46.

21. Atıl, "Ottoman Miniature," 107–8. Emphasis my own. Note I stress Atıl's earlier evaluation rather than her inexplicable change of opinion a decade later in Atıl, *Turkish Art*.

22. It is also possible that the *Kulliyat-i Kātibī*, which does not contain a colophon but resembles the *Dilsizname*, was done in Meḥmed's Istanbul atelier. The *Dilsizname*, the *Kulliyat-i Kātibī*, and the *Cerrāḥiyyetü'l-Ḥāniyye* could in this way also be seen as representative of the synthesis and further refinement that took place with the move of the Ottoman miniature painters from their court in Edirne to their new capital in Istanbul.

23. It is true that the colophons only say "Bilād Qusṭanṭīniyya" (a reference to Istanbul), which means that they could have been locally produced outside of Meḥmed's atelier. But the chances of this are highly unlikely. For starters, the mother manuscript that generated the Istanbul-produced Ottoman cluster, which is still housed in the library of Topkapı Saray, appears to have been a gift to the sultan. (This manuscript is discussed in great detail in the next chapter.) Whoever made the first copy made it on the basis of the one at Topkapı Saray, and it stands to reason that Meḥmed would not necessarily have sent it out of the palace for copying. The early Ottomans are not known for their manuscript illustration skills. What little exists not only displays an unsophisticated hand but also emerged in the context of palace ateliers. There is no reason to presume that these KMMS manuscripts from Istanbul, linked as they are to a manuscript in the sultan's possession, would not have been produced in the sultan's atelier. In the devastation and rebuilding of the immediate postconquest period, if anybody is likely to have had an atelier, it is Meḥmed.

24. I fully explain this reasoning in the following chapter with my discussion of Ahmet 2830, which I hold to be the mother manuscript that spawned the Ottoman cluster.

25. The *Taqwīm-i ta'rīkhī, taqwīm fī al-aḥkām* is dated to the reign of Murād II by the mention of the conquest of Salonica: 5 Rajab 833/30 March 1430. The calendar opens with

chronological tables (fols. 1b–3a), which give the number of years elapsed from Adam's descent to Sultan Murād II's enthronement.

Chapter 12

1. It should be noted that the illumination of the world map is superior to the rest. TSMK A2830 displays signs of a variety of different hands.

2. The copyist has incorrectly attributed this manuscript to Abū Zayd Aḥmad ibn Sahl Balkhī—a common mistake of the period. Even the title of the manuscript, *Kitāb ṣūrat al-aqālīm al-sab'a*, is different, even though the text is definitely that of Iṣṭakhrī's *Kitāb al-masālik wa-al-mamālik*. See Raby and Tanındı's comment on this at the start of their entry on this atlas in *Turkish Bookbinding* (132). For a good discussion on the range of different titles given to Iṣṭakhrī manuscripts, see Ducène, "Quel est le titre veritable."

3. Karatay, *Arpaça Yazmalar*, 3:580, vaguely says, "Fâtih Sultan Mehmed için istinsah edilmiştir" (It was copied perhaps for Fâtih Sultan Meḥmed).

4. Öz, *Fâtih Sultan*, 20–23.

5. Reproduced in Lentz and Lowry, *Timur and the Princely Vision*, 110–11.

6. For detailed information on this miniature, refer to ibid., 130, 338.

7. For detailed information on this miniature, refer to ibid., 108, 335.

8. The date of Tīmûr's birth is disputed. The reigning expert on Tīmûr, Beatrice Manz, believes that it was an invention from the time of his successor, Shāh Rukh, who chose it for its astrological meaning and coincidence with the death of the last Ilkhanid ruler. Manz, "Tamerlane," 113–14.

9. Kadoi, "Timurid Flag," 153.

10. Another example of this is the medallion decoration opening a 1330 copy of Ghazālī's *Iḥyā' 'ūlūm al-dīn* in the Saltykov-Schedrin State Public Library, Leningrad: Dorn. 255, fol. 1A. For a reproduction, see *ABC*, 37.

11. On the Morgan's *Manāfi'*, see Schmitz, *Pierpont Morgan Library*, 9–24.

12. Robinson, *Persian Painting*, 23, 39–40.

13. See, for instance, TSMK Revan 1021, fol. 86b. This same manuscript also contains an example of the *islimī-i bargi* and *band-i rūmī* found on the opening folios of TSMK A2830. Reproduced and discussed in Robinson, *Persian Painting*, 31–33.

14. By 1455, Uzun Ḥasan's predecessor, Jahānshāh, chief of the Qaraquyunlu, had conquered Tabriz, Baghdad, Isfahan, Shiraz, and Kirman, and even occupied Herat for a short period. When Uzun Ḥasan caught Jahānshāh unawares on a hunting expedition, he not only acquired Qaraquyunlu territory but Jahānshāh's Timurid artists as well.

15. Robinson, *Persian Painting*, 21.

16. There is much debate about the sixty-odd manuscripts like TSMK A2830 in the collection at TSMK. See Öz, *Fâtih Sultan*, 20–23.

17. Raby and Tanındı identify this manuscript and its covers as originating in Meḥmed's mythically advanced atelier. I disagree with their identification, just as I disagree with the identification of the other manuscripts, which I believe were part of the Aqqoyunlu ransom payment. Raby and Tanındı, *Turkish Bookbinding*, 132–33.

18. Woods, *Aqquyunlu*, 270 and 129, notes that the precise identity or name of Yusuf is uncertain (he is known by a variety of names). Confirmation of the kidnapping is also to be found in Tursun Beg, *Tārīḫ-i Ebü'l-fetḥ*, 59.

19. Raby, "Gran Turco," 159 (brackets in original), does not cite the source of this information, but Woods, *Aqquyunlu*, 270, notes that Yusuf was on the Ottoman ransom list. Obscure as he is, Yūsuf Mīrzā must have had considerable importance for Meḥmed II to presume that he could demand such a heavy ransom. Necipoğlu confirms Meḥmed's demand for a "blood money" ransom "to be accompanied by cultural currency that would especially please the sultan, 'namely, wondrous manuscripts and gifts of novelties such as albums,'" and clarifies that the ransom was for four Aqqoyunlu princes not just one. Necipoğlu, "Visual Cosmopolitanism," 43. What this confirms is that Meḥmed thought so highly of Aqqoyunlu miniature work as to place it in a ransom demand.

20. It is tempting to see these manuscripts as evidence of Uzun Ḥasan's attempt to please Meḥmed with books and calligraphy in order to distract him from his designs on Aqqoyunlu territory. I was permitted access only to cartographic manuscripts at the Topkapı Saray library, and for information about the other Aqqoyunlu manuscripts, I have had to rely on published sources such as Öz, *Fâtih Sultan*; Raby and Tanındı, *Turkish Bookbinding*; and Yoltar, "Ottoman Luxury Book Production." How many Aqqoyunlu manuscripts are in the collection is unclear, but the total could be as high as sixty. These manuscripts are registered at Topkapı Saray as definitive productions of Meḥmed's Istanbul atelier. Aqqoyunlu provenance is not easily broached. A number of senior scholars of Islamic art history, however, privately agree with my conclusions, having long identified these manuscripts as Aqqoyunlu in origin, but are hesitant to publicly say so. Necipoğlu reinforces the Ottoman claim of this exquisite heritage as theirs and not Aqqoyunlu in "Visual Cosmopolitanism," 43–44, this in spite of the fact that in note 179 (p. 78) Necipoğlu reveals the existence of "a letter addressed by Mehmed II to his son Prince Cem in 1473, in which he announced the capture of Uzun Hasan's personal belongings." This fits perfectly with my theory that TSMK A2830 arrived in Ottoman Constantinople sometime in 1473, a year before the earlier Ottoman Cluster copy.

21. Brentjes, "Courtly Patronage," 417, notes that "when produced for an Inju, Timurid, Mughal, Safavid or Ottoman ruler, prince, governor, eunuch, or other member of the courtly sphere, mathematical, astronomical, astrological, medical and philosophical texts were often carefully executed and finely illustrated with *shamses*, *sarlawhs*, *'unwans*, gilded frames, calligraphic writings and in some cases even illuminated with miniatures and other paintings."

22. Atıl, "Ottoman Miniature," 116.

23. Some believe that the famous Timurid miniature painter Bihzad was responsible for the Freer copy and suggest that it was executed in 1480–1481 when Bihzad visited Istanbul. Galerkina, "Bihzad," 128–29. For a book-length discussion on Bellini and his visit to Istanbul, see Campbell and Chong, *Bellini*. For a flavor of the strident debate on who painted the Freer Portrait, see Necipoğlu, "Visual Cosmopolitanism," 37–43.

24. Atıl, "Ottoman Miniature," 112–13, 115–17. I have personally examined the portrait (F1932.28) at the Freer and confirmed Atıl's findings.

25. Atıl, "Ottoman Miniature," 112–13, 115–17, pegs the "Freer Portrait" to around 1478–1481, that is, Meḥmed's final years. On the basis of this portrait's relationship with TSMK A2830

and TSMK A2830's relationship with the Ottoman cluster, we may be able to push back the date on the "Freer Portrait" to the early 1470s instead.

26. From reading Babinger's *Mehmed* and other accounts of Meḥmed, one gets the impression of a man very actively involved in daily affairs. He seems to have enjoyed mingling with his people in the local bazaars and strolling through the Hippodrome area; he relished debate and sat up nights and weeks listening to scholars argue points. He was even involved in the architectural design of Topkapı Saray. Necipoğlu, *Topkapı Palace*, 8, 15.

27. I suggest that TSMK A2830 is forgotten or misplaced because, unlike the other illuminated manuscripts, TSMK A2830 does not contain Bāyezīd's stamp—implying perhaps that Bāyezīd never saw it. Given that TSMK A2830 is still housed in the library of Topkapı Saray and that Bāyezīd II had a habit of stamping everything he perused, the only presumption that one can make is that it was misplaced and thus forgotten.

28. According to Kritovoulos's *History*, this study took place during the summer of 1465.

29. On the difference between Meḥmed's public and private patronage in connection with European artists, see Raby, "Sultan of Paradox," 6. Brentjes, "Courtly Patronage," 419, notes that a key factor in determining which manuscripts were copied "was the image a ruler or prince wished to portray."

30. At Samarqand, ʿĀlī Qūshjī was a member of the Ulugh Beg's observatory. Ulugh Beg, one of the last Timurid sultans, is noted in the annals of Islamic science for his active patronage of scholars and for building one of the earliest observatories in Samarqand. After Ulugh Beg's assassination, Qūshjī fled Samarqand and took up residence at Uzun Ḥasan's court. Adıvar, *İlim*, 33.

31. Brentjes uses ʿĀlī Qūshjī as an example of the way in which famous scholars, forced to flee their original patrons because of deteriorating political conditions, were patronized by other rulers as honored guests. She notes how Qūshjī was first welcomed by the Aqqoyunlu chief Uzun Ḥasan and later took up residence at the court of the Meḥmed II. She also cites examples of rededications of Qūshjī's works on arithmetic and mathematical cosmography from Ulugh Beg to Meḥmed, thus confirming that Qūshjī considered Meḥmed as important a patron as his former Timurid benefactor. Brentjes, "Courtly Patronage," 415, 428.

32. Adıvar, *İlim*, 32–35. "Fâtih Kütüphanesi ilk defa 1474 yılında Fâtih Camii içinde mihrabın iki tarafına yerlestirilen dolaplarda kurulmustur" (Two cupboards on either side of the *miḥrāb* established the Fâtih library for the first time in 1474). Bayraktar and Kut, *Vakıf Mühürler*, 17.

33. Erünsal, "Medieval Ottoman Libraries," 745; Erünsal, *Türk Kütüphaneleri*, 21–22.

34. None of the manuscripts contain any stamps for the Fâtih Kütüphanesi (Fâtih Library). It would seem that Meḥmed, unlike his successors, was not in the habit of stamping manuscripts. This we know from the rarity of the occurrence of his stamp. His son Bāyezīd II on the other hand, seems to have stamped everything that came his way.

35. Aslanapa, *Mimarisi*, 107. Erünsal notes that books were transferred to these Fâtih complex libraries from Aya Sofya and other libraries. Erünsal, "Development of Ottoman Libraries."

36. It may help the reader to think of these northern and southern *barārī* extremities as akin to the North and South Poles, although the Arab geographers are not referring specifically to the astronomical poles but rather to the wastelands that surround them.

37. The Sarīr are identified as a Turkic tribe and with the Avars of present-day upper Dagh-

estan, where the Avar language is still the lingua franca. *Ḥudūd al-ʿālam*, 447–50; *EI2*, s.v. "Avars," by H. Carreère-d'Encausse and A. Bennigsen; and *EI3*, s.v. "Avars," by Peter Golden, accessed May 2, 2015, http://referenceworks.brillonline.com/entries/encyclopaedia-of-islam-3/avars-COM_22906. The Khazar are a mysterious group variously identified as Turkic, Jewish, Christian, and Muslim. In the seventh and eight centuries, they were considered the undisputed masters of the Eurasian steppe, before they mysteriously melted back into the steppes from whence they came. Dunlop, *Jewish Khazars*; *EI2*, s.v. "Khazar," by W. Barthold and P. B. Golden. Some scholars consider the Burṭās a pagan Turkic tribe located in the area of the Itil (Volga) River related to modern-day Finns, whereas others identify them as relatives of the Ghuzz, who are considered the eponymous ancestors of the Ottomans. *EI2*, s.v. "Burṭās," by W. Barthold and Ch. Quelquejay; and *EI3*, s.v. "Burṭās," by Peter Golden, accessed May 3, 2015, http://referenceworks.brillonline.com/entries/encyclopaedia-of-islam-3/burtas-COM_23728. For debates on the origin and etymology of the term "Rūs," whence the people of modern Russia, Ukraine, and Belarus derive, see *EI2*, s.v. "Rūs," by P. B. Golden; and *Ḥudūd al-ʿālam*, 432–38. For an entertaining and informative tenth-century account of these tribes, see Ibn Fadlan, *Journey to Russia*.

38. Marquart, "Volkstum der Komanen," 25–238, identifies the "Inner Bulghār" and the Ṣaqāliba in KMMS manuscripts as the Danubian Bulghār. See also *Ḥudūd al-ʿālam*, 425–31, 438–40.

39. Findley, *Turks*, 21–92. Brummett, "Ottoman Space," 49, notes "that Ottomans, like some Europeans, also imagined for themselves a 'classical' past, including biblical times, the reign of Alexander, Turkic steppe antecedents, and the life of the Prophet Muhammad."

40. Enunciating this view, Necipoğlu, *Topkapı Palace*, 12, quotes George Trepuzuntios writing to Meḥmed in 1466: "No one doubts that you are emperor of the Romans. Whoever holds by right the center of the Empire is emperor and the center of the Empire is Constantinople." In similar vein, Brotton, *Trading Territories*, 92–93, points to medals cast by Constanzo de Ferrara for Meḥmed based on Roman imperial prototypes. In later years, Ottoman sultans, such as Süleymān the Magnificent, reinforced the concept of a neo-caesar through the commissioning of pseudo-Roman ceremonial helmets. Necipoğlu, "Süleyman," 410, 424, also notes that "both Mehmed II and Süleymān I, shared ambition to revive the Roman Empire by uniting Constantinople with Rome." See also Ebel, "Ottoman Town Views," 2.

41. Schapiro, *Selected Papers*, 78–79. Emphasis my own. Schapiro is referring in this passage to Weitzmann's *Fresco Cycle*.

42. For a masterful discussion of Ottoman maritime exploits in their age of exploration, see Casale, *Ottoman Age*.

43. The ḥadīth cited by Muqaddasī, *Best Divisions*, 17, tells us that Allah blessed the Indian Ocean / Persian Gulf for carrying his people and cursed the Mediterranean for drowning them. Pinto, "Bahr al-Rūm" (in *Eastern Mediterranean Cartographies*).

44. The same words are also routinely inscribed on the opening folios of numerous manuscripts below the stamps of Maḥmūd I (r. 1730–1754). Ruler of the two continents is also implied in Berlinghieri's scratched-out dedication on his atlas: "To Mehmed of the Ottomans, illustrious prince and lord of the throne of God. Emperor and merciful lord of all Asia and Greece, I dedicate this work." Quoted from Brotton, *Trading Territories*, 90–92. Necipoğlu makes note of the "three crowns of Byzantium, Trebizond, and Asia that were included in

painted or medallic portraits of Mehmed II by Gentile Bellini." Necipoğlu, "Süleyman," 412. "The Sultan measured his domains in terms of land, seas, reputation, and submission. He was the 'Lord of the two seas and two continents' and the 'Refuge of the World' (*alempenah*), both titles designating expansive power and authority." Brummett, "Ottoman Space," 47.

45. Kafadar, *Between Two Worlds*, 152.

46. For Meḥmed's Alexander worship, see Babinger, *Mehmed*; Brotton, *Trading Territories*, 92; and Brummett, "Ottoman Space," 49. Necipoğlu, "Süleyman," 411, 424–25, states, "Noting Mehmed's ambition to conquer Rome as early as 1453, contemporary European observers pointed out that the Sultan who took Alexander the Great as his model, was planning to join East and West by creating a world empire unified by a single faith and a single monarch."

47. The tradition of *Kızıl Elma* (Red Apple) is an important part of early Ottoman conquering mythology. Whichever was the most desired city was designated *Kızıl Elma* until it was conquered. Constantinople was the first one to receive this designation. Some say that the name is based on the statue of Justinian that guarded the Golden Horn and contained in its hand a golden orb. In early Turkish *kızıl* could also mean "gold." *EI2*, s.v. "Ḳizil-elma," by P. N. Boratav.

48. İnalcık, "Ottoman Empire," 41–42.

49. The debate over the role of the *ghazi* ethos in the motivation for the rise and spread of the Ottomans, deeply embedded in the master narrative of early Ottoman history, continues. It dates back to the 1938 Sorbonne lectures by Wittek that were published as both "De la défaite d'Ankara," 1–34, and *Rise of the Ottoman Empire*. Many scholars, including Rudi Lindner, Halil İnalcık, and Cemal Kafadar, have elaborated at length on the topic, but the seminal book that lays this debate to rest is Heath Lowry's *The Nature of the Early Ottoman State*. My analysis of the propagandistic use of the KMMS Ottoman cluster fits well with Lowry's reappraisal, namely, that the *ghazi* ethos is in fact an invention of the late fifteenth- and sixteenth-century chroniclers who probably invented this ideal and planted it in their state-sponsored historiographies at the behest of sultans such as Meḥmed II and his progeny.

50. Babinger, *Mehmed*, 411.

51. Islamic art historians do not distinguish between work for public versus private consumption. This is an important distinction that needs to be explored.

52. Portraits and sculpture are noticeably absent from traditional Islamic arts. Their absence is a matter of considerable debate among Islamic scholars. The ban is related to a ḥadīth by the Prophet Muhammad forbidding graven images. See Arnold, *Painting in Islam*; and Grabar, *Formation*, 72–98. Because of this, the appearance of portraits in Ottoman arts is automatically identified as related to "European" influence. Raby, "Gran Turco," 195–98. Necipoğlu, "Süleyman," 419, notes that the "Islamic prohibition of figural representation" undermined projects for the Ottoman sultan Süleymān.

53. Confirmed by Raby, "Sultan of Paradox," 6.

54. Raby, "Gran Turco," 410–11, 469–94.

Chapter 13

1. The Beja still live in the same region of East Africa, and they continue to capture the imagination of anthropologists and missionaries. Contemporary pictures show a life little

changed from Ibn Ḥawqal's tenth-century descriptions of them, with one major difference: the influence of Islam. As in the medieval period so today the identity of the Beja is under threat of being subsumed by the national interests of the countries within which they are divided. One of the most ancient peoples of recorded history—known since Pharaonic times (Twelfth Dynasty)—now fear extinction at the hands of the policies of the Sudanese government. See Young, "Beja," 7–10.

Bibliography

Abbreviations

ABC *The Arts of the Book in Central Asia*. Edited by Basil Gray. Paris: UNESCO; London: Serindia Publications, 1979.

AP *Ancient Perspectives: Maps and Their Place in Mesopotamia, Egypt, Greece, and Rome*. Edited by Richard J. A. Talbert. Chicago: University of Chicago Press, 2012.

BGA *Bibliotheca geographorum arabicorum*. Edited by Michael Jan de Goeje. 8 vols. Leiden: E. J. Brill, 1870–1894.

CAMA *Cartography in Antiquity and the Middle Ages: Fresh Perspectives, New Methods*. Edited by Richard J. A. Talbert and Richard W. Unger. Leiden: E. J. Brill, 2008.

CFT *Cartes et Figures de la Terre*. Paris: Centre Georges Pompidou, Centre de création industrielle, 1980.

EI1 *Encyclopaedia of Islam*. 1st ed. Edited by M. Th. Houtsma, T. W. Arnold, R. Basset, and R. Hartmann. Leiden: E. J. Brill, 1913–1936.

EI2 *Encyclopaedia of Islam*. 2nd. ed. Edited by P. Bearman, Th. Bianquis, C. E. Bosworth, E. van Donzel, and W. P. Heinrichs. Leiden: E. J. Brill, 1960–2005.

EI3 *Encyclopedia of Islam*. 3rd. ed. Edited by Kate Fleet, Gudrun Krämer, Denis Matringe, John Nawas, and Everett Rowson. Leiden: Brill, 2007–.

EIr *Encyclopaedia Iranica*. Edited by Ehsan Yarshater. London: Routledge and Kegan Paul, 1982–.

GE *Geography and Ethnography: Perceptions of the World in Pre-modern Societies*. Edited by Kurt A. Raaflaub and Richard J. A. Talbert. Chichester, West Sussex, UK: Wiley-Blackwell, 2010.

HC 1 *The History of Cartography*. Vol. 1, *Cartography in Prehistoric, Ancient, and Medieval Europe and the Mediterranean*. Edited by J. B. Harley and David Woodward. Chicago: University of Chicago Press, 1987.

HC 2.1 *The History of Cartography*. Vol. 2, bk. 1, *Cartography in the Traditional Islamic and South Asian Societies*. Edited by J. B. Harley and David Woodward. Chicago: University of Chicago Press, 1992.

HC 2.2 *The History of Cartography*. Vol. 2, bk. 2, *Cartography in the Traditional East and Southeast Asian Societies*. Edited by J. B. Harley and David Woodward. Chicago: University of Chicago Press, 1994.

HWM *The Hereford World Map: Medieval World Maps and Their Context*. Edited by P. D. A. Harvey. London: The British Library, 2006.

ICTA *International Congress of Turkish Art*

IJMES *International Journal of Middle East Studies*

IM *Imago Mundi*

JMISR *The Journey of Maps and Images on the Silk Road*. Edited by Phillipe Forêt and Andreas Kaplony. Leiden: Brill, 2008.

KMMS *Kitāb al-masālik wa-al-mamālik* ṣūrat (Iṣṭakhrī, Ibn Ḥawqal, and Muqaddasī *Book of Routes and Realms* map manuscripts)

MMM *Monarchs, Ministers, and Maps: The Emergence of Cartography as a Tool of Government in Early Modern Europe*. Edited by David Buisseret. Chicago: University of Chicago Press, 1992.

PDH *Istanbul Topkapı Sarayı Müzesi ve Venedik Correr Müzesi koleksiyonlarından XIV–XVIII Yüzyıl Portolan ve Deniz Haritaları/Portolani e Carte Nautiche XIV–XVIII Secolo*. Istanbul: Istanbul Italyan Kültür Merkezi, 1994.

Persian Art *A Survey of Persian Art from Prehistoric Times to the Present*. Edited by Arthur Upham Pope and Phyllis Ackerman. 16 vols. London: Oxford University Press, 1938–1977.

TSMK Topkapı Sarayı Müzesi Kütüphanesi, Istanbul.

Primary Sources

Abū'l-Fidā. *Géographie d'Aboulféda*. Translated by Joseph Toussaint Reinaud and S. Stanislas Guyard. 2 vols. Paris: Imprimerie Nationale, 1848–1883.

———. *Taqwīm al-buldān*. Edited by Joseph Toussaint Reinaud and Baron MacGluckin de Slane. Paris: Imprimerie Royale, 1840.

Akhbār al-Ṣin wa al-Ḥind. *'Aḫbār aṣ-Ṣīn wa l-Hind: Relation de la Chine et de l'Inde*. Edited and translated by Jean Sauvaget. Paris: Belles Lettres, 1948.

Alf layla wa layla: The Thousand and One Nights. Edited by Muhsin Mahdi. 2 vols. Leiden: E. J. Brill, 1984.

al-Andalusī, Ṣā'id. *Science in the Medieval World: "Book of the Categories of Nations"* [*Ṭabaqāt al-'umam*]. Translated and edited by Sema'an I. Salem and Alok Kumar. Austin: University of Texas Press, 1991. Reprint, 1996.

The Arabian Nights. Translated by Husain Haddawy. New York: Norton, 1990.

Aristophanes. *The Clouds.* Translated by William Arrowsmith. New York: New American Library, 1962.

Aristotle. *Meteorologica.* Translated by H. D. P. Lee. Loeb Classical Library. Cambridge, MA: Harvard University Press, 1952.

'Aṭṭār, Farīd al-Dīn. *The Conference of the Birds* [*Manṭiq al-ṭayr*]. Translated by Afkham Darbandi and Dick Davis. London: Penguin Books, 1984.

al-Balādhurī. *Futūḥ al-buldān. Liber expugnationis regionum.* Edited by Michael Jan de Goeje. Leiden: E. J. Brill, 1866.

al-Bakrī, Abū 'Ubayd. *Description de l'Afrique septentrionale par Abou-Obeid-El-Bekri.* Translated by M. de Slane. Paris: Imprimerie Impériale, 1911–1913.

———. *Geografía de España.* Translated by E. Vidal Beltrán. Zaragoza, Spain: Anubar Ediciones, 1982.

———. *The Geography of Al-Andalus and Europe from the Book "Al-Masalik wal-Mamalik"* [*The Routes and the Countries*]. Edited by Abdurrahman Ali El-Hajji. Beirut: Dar Al-Irshad, 1968.

———. *Kitāb al-masālik wa-al-mamālik.* Edited by Adrian van Leeuwen and Andre Ferre. Tunis: Al-Dār al-'Arabiya, 1992.

al-Bīrūnī. *Alberuni's India.* Translated by Edward Sachau. 2 vols. London: Kegan Paul, Trench, Trubner, 1888, 1910.

———. *The Book of Instruction in the Elements of the Art of Astrology.* Translated by Robert Ramsay Wright. London: Luzac, 1934.

———. *Kitāb al-qānūn al-Mas'ūdī fī al-hay'ah wa-al-nujūm. Biruni's Picture of the World.* Edited by Ahmed Zeki Velidi Togan. Memoirs of the Archaeological Survey of India, no. 53. Delhi, 1941.

———. *Kitāb al-tafhīm li-avā'il ṣinā'at al-tanjīm.* Edited by Jalal al-Din Huma'i. Tehran, 1974.

———. *Kitāb fī taḥqīq mā lil-Hind min maqūla. Book of the verification of what is said about India.* Hyderabad, India: Maṭba'a Majlis Dā'irat al-Ma'ārif al-'Uthmāniyya, 1958.

al-Bukhārī. *Ṣaḥīḥ.* Edited by Muḥammad Tawfīq 'Uwaydah. 7 vols. Cairo: Lajnat Iḥyā' Kutub al-Sunnah, 1966/67–1976/77.

al-Dimashqī. *Nukhbat al-dahr fī 'ajā'ib al-barr wa-al-baḥr. Manuel de la cosmographie du moyen age.* Translated by A. F. M. van Mehren. Copenhagen: C. A. Reitzel, 1874. Reprint, Amsterdam: Meridian, 1964.

An Eleventh-Century Egyptian Guide to the Universe: The Book of Curiosities. Edited and translated by Yossef Rapoport and Emilie Savage-Smith. Leiden: Brill, 2013.

"Enûma Elish." In *Alpha: The Myths of Creation.* Translated by Charles H. Long. New York: George Braziller, 1963.

Geminus. *Introduction to the Phenomena.* Edited and translated by Germaine Aujac. Paris: Belles Lettres, 1975.

al-Hamdhānī, Ibn al-Faqīh. *Kitāb al-buldān. Compendium libri kitâb al-boldân.* In *BGA*, vol. 5, 1885. Reprint, 1967.

al-Harawī, ʿAlī ibn Abī. *A Lonely Wayfarer's Guide to Pilgrimage: ʿAlī ibn Abī al-Harawī's Kitāb al-Ishārāt ilā Maʿrifat al-Ziyārāt.* Translated by Josef Meri. Princeton, NJ: Darwin Press, 2004.

Herodotus. *Histories.* Translated by George Rawlinson. 2 vols. London: Everyman's Library. Reprint, 1964.

Holy Bible: New Revised Standard Edition with Apocrypha. Oxford: Oxford University Press, 1977.

Homer. *Iliad.* Translated by Richmond Lattimore. Chicago: University of Chicago Press, 1951.

———. *Odyssey.* Translated by Richmond Lattimore. Chicago: University of Chicago Press, 1965.

Ḥudūd al-ʿālam: "The Regions of the World." Translated V. Minorsky. Preface by V. V. Barthold. London: Messrs. Luzac & Co., 1937. Reprint, Karachi: Indus, 1980.

Ibn al-Nadīm. *The Fihrist of al-Nadīm: A Tenth-Century Survey of Muslim Culture.* Translated by Bayard Dodge. 2 vols. New York: Columbia University Press, 1970.

———. *Kitāb al-fihrist.* Edited by Gustav Flügel. 2 vols. Leipzig: F. C. W. Vogel, 1871–1872.

Ibn al-Wardī. *Fragmentum libri Margarita mirabilium, auctore Ibn-el-Vardi.* Translation of fragment by Carolus Johannes Tornberg. Upsala, Sweden: Excudebant Regiæ Academiæ Typographi, 1839.

———. *Kharīdat al-ʿajāʾib wa-farīdat al-gharāʾib* [The Unbored Pearl Of Wonders And The Precious Gem Of Marvels]. Arabic edition. Cairo, 1863.

———. *Specimen Operis-Cosmographici Ibn el Vardi.* Translation of fragment by Andreas Hylander. Lunde: Litteris Berlinianis, 1792.

Ibn Fadlan. *Ibn Fadlan's Journey to Russia.* Translated with commentary by Richard Frye. Princeton, NJ: Marcus Wiener, 2005.

Ibn Ḥawqal. *Configuration de la Terre* (*Kitab Surat al-Ard*). Translated by J. H. Kramers and G. Wiet. 2 vols. Beirut: Commision Internationale pour la Traduction des Chefs-d' Œuvre; Paris: Éditions G.-P. Maisonneuve & Larose, 1964.

———. *Kitāb ṣūrat al-arḍ. Opus geographicum.* In *BGA*, vol. 2, 1873. Reedited by J. H. Kramers, 1938. Reprint, 1967.

Ibn Khaldūn. *The Muqaddimah: An Introduction to History.* Translated by Franz Rosenthal. 3 vols. New York: Pantheon Books, 1958.

Ibn Khurradādhbih. *Kitāb al-masālik wa-al-mamālik. Liber viarum et regnorum.* In *BGA*, vol. 6, 1889. Reprint, 1967.

Ibn Rusta. *Kitāb al-aʿlāq al-nafīsah.* In *BGA*, vol. 7, 1892. Reprint, 1967.

Ibn Saʿid. *Kitāb basṭ al-arḍ fī ṭūlihā wa-al-arḍ. Libro de la extension de la tierra en longitud y latitud.* Translated by Juan Vernet Ginés. Tetuan: Instituto Muley el-Hasan, 1958.

al-Idrīsī. *L'Afrique dans le Uns al-muhağ wa-rawḍ al-furağ d'al-Idrīsī.* Edited and translated with commentary by Jean-Charles Ducène. Leuven, Belgium: Peeters, 2010.

———. *Géographie d'Edrisi.* Translation by Pierre Amédée Emilien Probe Jaubert. 2 vols. Paris: Imprimerie Royale, 1836–1840.

———. *Kitāb nuzhat al-mushtāq fī ikhtirāq al-āfāq. Opus Geographicum; sive, "Liber ad eorum*

delectationem qui terras peragrare student." Edited by E. Cerulli. Naples: 1st Univ. Orientale di Napoli, 1970.
———. *Uns al-muhaj wa rawḍ al-furaj wa nuzhat al-muhaj. The Entertainment of Hearts, and Meadows of Contemplation.* Facsimile edition by Fuat Sezgin. Frankfurt: Institut für Geschichte der Arabisch-Islamischen Wissenschaften, 1984.
Isidore. *The Etymologies of Isidore of Seville.* Edited and translated by Stephen Barney, W. J. Lewis, J. A. Beach, and Oliver Berghof. Cambridge: Cambridge University Press, 2006.
al-Iṣṭakhrī. *Das Buch der Länder.* Translated by Andreas David Mordtmann with a forward by C. Ritter. Hamburg: Druck und Lithographie des Rauhen Hauses in Horn, 1845.
———. *Kitāb al-masālik wa-al-mamālik.* Edited by Muḥammad Jābir ʿAbd al-ʾĀl al-Ḥīnī. Cairo: Wazārat al-Thaqāfa, 1961.
———. *Liber climatum.* Edited by J. H. Moeller. Gotha: Libraria Beckeriana, 1839.
———. *The Oriental Geography of Ebn Haukal.* Translated by William Ouseley. London: Wilson for T. Cadell and W. Davies, 1800.
———. *Tarjamah-i masālik wa-al-mamālik.* Edited by Iraj Afshar. Tehran: Bungāh-i Tarjamah va Nashr-i Kitāb, 1961.
———. *Viae regnorum descriptio ditionis moslemicae.* In *BGA*, vol. 1, 1870. Reprint, 1927, 1967.
al-Kāshgarī. *Dīwān lughāt al-Turk. Divanü Lügat-it-Türk Tercemesi.* Translated by Besim Atalay. 4 vols. Ankara: Türk Tarih Kurumu Basimevi, 1992.
al-Khwārazmī. *Kitāb ṣūrat al-arḍ. Das Kitāb ṣūrat al-arḍ des Abū Ǧaʿfar Muḥammad Ibn Mūsā al-Ḫuwārizmī.* Edited by Hans von Mžik. Leipzig: Otto Harrassowitz, 1926.
al-Kisāʾī, Muḥammad ibn ʿAbd Allāh. *Qiṣaṣ al-anbiyāʾ.* Edited by I. Eisenberg. Leiden: E. J. Brill, 1922.
———. *Tales of the Prophets-Qisas al-anbiya.* Translated by Wheeler M. Thackston Jr. Chicago: Great Books of the Islamic World, 1997.
The Koran Interpreted. Translated by A. J. Arberry. New York: Collier Books, 1955.
Kritovoulos. *History of Mehmed the Conqueror.* Translated by Charles T. Riggs. Princeton, NJ: Princeton University Press, 1954.
Mālik ibn Anas. *Kitāb al-muwaṭṭaʾ.* Dubai: Majmūʿat al-Furqān al-Tijārīya, 2003.
al-Maqdīsī. *Kitāb al-badʾ wa-al-taʾrīkh.* Edited by C. Huart. 6 vols. Paris: Ernest Leroux, 1899–1919.
al-Maqrīzī. *Ittiʿāẓ al-Hunafāʾ.* Edited by Jamāl al-Dīn al-Shayyāl. Cairo: Dār al-Fikr al-ʿArabī, 1967.
al-Masʿūdī. *Kitāb al-tanbīh wa-al-ishrāf.* In *BGA*, vol. 8, 1894.
———. *Le livre de l'avertissement et de la revision.* Translated by B. Carra de Vaux. Paris: Imprimerie Nationale, 1896.
———. *Murūj al-dhahab wa-maʿādin al-jawhar. Les prairies d'or.* Translated by C. Barbier de Meynard and J. Pavet Courteille. 9 vols. Paris: Imprimerie Impériale, 1861–1877. Revised by Charles Pellat, 7 vols, Beirut: Manshūrāt al-Jāmiʿah al-Lubnānīyah, 1965–1979.
al-Muqaddasī. *Aḥsan al-taqāsīm fī maʿrifat al-aqālīm. Description imperii moslemici.* In *BGA*, vol. 3, 1877. Reprint, 1906, 1967.
———. *Aḥsan at-taqāsīm fī maʿrifat al-aqālīm.* Translated by André Miquel. Damascus: Institut Français de Damas, 1963.

———. *The Best Divisions for Knowledge of the Earth.* Translated by Basil Anthony Collins. London: Centre for Muslim Contribution to Civilization and Garnet Publishing, 1994.
Muslim ibn al-Ḥajjāj. *Ṣaḥīḥ Muslim.* 5 vols. Cairo: Dār al-Fikr al-ʿArabī, 1956.
Pīrī Reʾīs. *Kitāb-ı Baḥrīye.* Edited by Ertuğrul Zekai Ökte. Translated by Vahit Çabuk, Tülay Duran, and Robert Bragner. 4 vols. Ankara: Ministry of Culture and Tourism of the Turkish Republic, 1988.
Plato. *Timaeus.* Translated by B. Jowett. n.p. n.d. The Internet Classics Archive. http://classics.mit.edu/Plato/timaeus.html.
Procopius. *History of the Wars, Books I and II.* Translated by H. B. Dewing. London: William Heinemann, 1914. Reprint, Cambridge, MA: Harvard University Press, 1971; Project Gutenberg eBook, 2005.
Ptolemy. *Geography. Claudii Ptolemaei Geographia.* Edited by Karl Müller. 2 vols. Paris: Firmin-Didot, 1883–1901.
———. *Ptolemy's "Almagest."* Translated and annotated by G. J. Toomer. London: Duckworth, 1984.
Qāḍī Aḥmad. *Calligraphers and Painters. A Treatise by the Qāḍī Aḥmad, son of Mīr-Munshī (circa A.H. 1015/A.D. 1606).* Translated by V. Minorsky. Washington, DC: Freer Gallery of Art, 1959.
al-Qazwīnī. *Kitāb ʿajāʾib al-makhlūqāt wa-gharāʾib al-mawjūdāt.* Edited by Ferdinand Wüstenfeld. Göttingen: Druck und Verlag der Dieterichschen Buchhandlung, 1848–9.
Qudāma ibn Jaʿfar. *Kitāb al-kharāj wa-ṣināʿat al-kitāba.* In *BGA*, vol. 6, 1892.
al-Qurʾān. Translated by Ahmad Ali. 5th edition. Princeton, NJ: Princeton University Press, 1994.
The Qurʾan. Translated with an introduction by Tarif Khalidi. London: Penguin, 2008.
Sabuncuoğlu, Şerefeddin. *Cerrāḥiyyetüʾl-Ḫāniyye.* Edited by İlter Uzel. 2 vols. Ankara: Atatürk Kültür, Dil ve Tarih Yüksek Kurumu Yayınları, 1992.
Strabo. *Geography.* Translated by Horace Leonard Jones. 8 vols. Cambridge, MA: Harvard University Press, 1917–1932.
al-Ṭabarī. *General Introduction and From the Creation to the Flood.* Translated by Franz Rosenthal. Vol. 1 of *The History of al-Tabari*, edited by Ehsan Yarshater. Albany: State University of New York Press, 1989.
———. *Incipient Decline.* Translated by Joel L. Kraemer. Vol. 34 of *The History of al-Tabari*, edited by Ehsan Yarshater. Albany: State University of New York Press, 1989.
———. *Taʾrīkh al-rusul wa-al-mulūk. (History of prophets and kings). Annalesquos scripsit Abu Djafar Mohammed ibn Djarir at-Tabari.* Edited by Michael Jan de Goeje. 15 vols. Leiden: E. J. Brill, 1879–1901. Reprint, 1964–1965.
al-Thaʿlabī, Abū Isḥāq Aḥmad ibn Muḥammad ibn Ibrāhīm. *"Arāʾis al-majālis fī qiṣaṣ al-anbiyā" or "Lives of the Prophets."* Translated and annotated by William M. Brinner. Leiden: Brill, 2002.
al-Tirmidhī. *Jāmiʿ al-Tirmidhī.* 6 vols. Lahore: Dar-us-Salam, 2007.
Ṭursun Beg. *Tārīḫ-i Ebüʾl-feth. The History of Mehmed the Conqueror.* Translated by Halil İnalcık and Rhoads Murphey. Minneapolis: Bibliotheca Islamica, 1978.
al-Yaʿqūbī, Aḥmad ibn Abī Yaʿqūb. *Kitāb al-buldān.* In *BGA*, vol. 7, 1892. Reprint, 1967.

———. *Les pays.* Translated by Gaston Wiet. Cairo: Imprimerie de l'Institut français d'archéologie orientale, 1937.

Yāqūt. *The Introductory Chapters of Yāqūt's "Muʿjam al-buldān."* Translated and annotated by Wadie Jwaideh. Leiden: E. J. Brill, 1959. Reprint, 1987.

———. *Irshād al-arīb ilā maʿrifat al-adīb. Dictionary of Learned Men of Yaqut.* Edited by D. S. Margoliouth. 7 vols. Leiden: E. J. Brill, 1923–1931.

———. *Jacut's geographisches Wörterbuch.* 6 vols. Edited by Ferdinand Wüstenfeld. Leipzig: F. A. Brockhaus, 1866–1873.

———. *Muʿjam al-buldān.* Edited by Farīd ʿAbd al-ʿAzīz al-Jundī. 7 vols. Beirut: Dār al-Kutub al-ʿIlmiya, 1990.

al-Zuhrī, Muḥammad ibn Abū Bakr. *Kitāb al-jaʿrāfīya.* Edited by Maḥammad Hadj-Sadok, "Kitāb al-Djaʿrāfiyya: Mappemonde du calife al-Maʾmūn reproduite par Fazārī (IIIe/IX e s.) rééditée et commentée par Zuhrī (VIe/XIIe s.)," *Bulletin d'études orientales,* 21 (1968): 1–310.

Secondary Literature

Unpublished

Ausenda, Giorgio. "Leisurely Nomads: The Hadendowa (Beja) of the Gash Delta and Their Transition to Sedentary Village Life." PhD diss., Columbia University, 1987.

Burnham, Emily. "The Edges of the Earth: An Epistemology of the Unknown in Arabic Geographies from the 5/11th—7/13th Centuries." PhD diss., New York University, 2012.

Eryılmaz Arenas-Vives, Fatma Sinem. "The *Shehnamecis* of Sultan Süleyman: ʿArif and Eflatun and Their Dynastic Project." PhD diss., University of Chicago, 2010.

Jacob, Christian. "Will There Be Histories of Cartography after the *History of Cartography* Project?" Paper presented at the XVIIIth International Conference on the History of Cartography, Athens, Greece, July 11–16, 1999.

Kafescioğlu, Çiğdem. "The Ottoman Capital in the Making: The Reconstruction of Constantinople in the Fifteenth Century." PhD diss., Harvard University, 1996.

Kahalaoui, Tarek. "The Depiction of the Mediterranean in Islamic Cartography (11th–16th Centuries). The Ṣūras (Images) of the Mediterranean from the Bureacrats to the Sea Captains." PhD diss., University of Pennsylvania, 2008.

Ļaviņš, Imants. "Depiction of Scandanavia and Eastern Europe in Medieval Arabic and Persian Historical Sources." PhD diss., University of Latvia, 2013.

Pinto, Karen. "Fit for a Prince. A Possible 8th Century Umayyad Map Fresco." Under review.

———. "The Mediterranean in the Islamic Cartographic Imagination." National Endowment of the Humanities project, 2013–2014.

———. "A 9th Century Isidorian T-O Map Labeled in Arabic." In *Non-Muslim Contributions to Islamic Civilisation: A Selection of Colorful Stories,* edited by Myriam Wissa, Alasdair Watson, and Brian Catlos. Non-Muslim Contributions to Islamic Civilisation. Edinburgh: Edinburgh University Press, forthcoming.

———. "Surat Bahr al-Rum: The Mediterranean in the Muslim Cartographical Imagination." MA essay, Columbia University, 1992.

———. "Traces of the Diabolic and the Divine in Islamic Maps." Paper presented at the Fifth Biennial Virginia Garrett Lecture in the History of Cartography, University of Texas-Arlington, October 6–7, 2006.

———. "Ways of Seeing.3: Scenarios of the World in the Islamic Cartographic Imagination." PhD diss., Columbia University, 2002.

———. "What Is Islamic about Islamic Maps?" Paper Presented at Holy Places in Medieval Islam: Functions, Typologies and Narratives, University of Edinburgh, September 2–4, 2014.

Raby, Julian. "El Gran Turco: Mehmed the Conqueror as a Patron of the Arts of Christendom." PhD diss., Oxford University, 1980.

Renda, Günsel. "Üç Zübdet-üt Tevarih Yazmasının İncelenmesi." PhD diss., Hacettepe Üniversitesi, Ankara, 1969.

Roxburgh, David J. "'Our Works Point to Us': Album Making, Collecting, and Art (1427–1565) under the Timurids and Safavids." PhD diss., University of Pennsylavania, 1996.

von Bothmer, H.-D. "Die Illustrationen des Münchener Qazwini von 1280 A.D." PhD diss., University of Munich, 1976.

Yoltar, Ayşin. "The Role of Illustrated Manuscripts in Ottoman Luxury Book Production: 1413–1520." PhD diss., New York University, 2002.

Published

Ackerman, Phyllis. "The Art of the Parthian Silver and Goldsmiths." In *Persian Art*, 1:459–63.

———. "Symbol and Myth in Prehistoric Ceramic Ornament." In *Persian Art*, 14:2914–29.

Adams, C. E. P., and R. Laurence, eds. *Travel and Geography in the Roman Empire*. London: Routledge, 2001.

Adams, William Y. "Ibn Ḥawqal." In *The Coptic Encyclopedia*, edited by Aziz Suryal Atiya, 3:1266. New York: Macmillan, 1991.

Adamson, Peter, and Richard Taylor, eds. *The Cambridge Companion to Arabic Philosophy*. Cambridge: Cambridge University Press, 2005.

Adıvar, Adnan. *Osmanlı Türklerinde İlim*. 2nd ed. Istanbul: Maarif Matbaasi, 1943.

———. *A History of Arab-Islamic Geography*. Amman: Āl al-Bayt University, 1995.

Ahmad, Nafis. *Muslims and the Science of Geography*. Bangladesh: University Press Limited, 1980.

———. *Muslim Contribution to Geography*. Lahore: Sh. Muhammad Ashraf Kashmiri Bazar, 1972.

Aillet, Cyrille. *Les Mozarabes: Christianosme, Islamisation et Arabisation en Péninsule Ibérique (IXe-XIIe Siècle)*. Madrid: Casa de Velázquez, 2010.

Akalay, Zeren. "The Forerunners of Classical Turkish Miniature Painting." *ICTA* 5 (1978): 31–47.

Akerman, James, ed. *Maps: Finding Our Place in the World*. Chicago: University of Chicago Press, 2007.

Akimushkin, Oleg F., and Anatol A. Ivanov. "The Art of Illumination." In *ABC*, 35–57.

Aksoy, Şule, and Rachel Milstein. "A Collection of Thirteenth-Century Illustrated Hajj Certificates." In *Uğur Derman: 65 Yaş Armağanı*, edited by Irvin Cemil Schick, 101–34. Istanbul: Sabanci Üniversitesi, 2000.

Alai, Cyrus. *General Maps of Persia: 1477–1925*. Leiden: Brill, 2005.

Albers, Josef. *Interaction of Color*. New Haven, CT: Yale University Press, 1963.
Albu, Emily. “Imperial Geography and the Medieval Peutinger Map.” *IM* 57, no. 2 (2005): 136–48.
———. “Rethinking the Peutinger Map.” In *CAMA*, 111–19.
Allan, Sarah. *The Shape of the Turtle: Myth, Art, and Cosmos in Early China*. Albany: State University of New York Press, 1991.
Allen, James P. “The Egyptian Concept of the World.” In *Mysterious Lands*, edited by David O’Connor and Stephen Quirke, 23–30. London: Institute of Archaeology, University College, 2003.
Alpers, Svetlana. *The Art of Describing: Dutch Art in the Seventeenth Century*. Chicago: University of Chicago Press, 1983.
Andaloro, M. “Costanzo da Ferrara: Gli anni a Costantinopoli alla corte di Maometto II.” *Storia dell’arte* 40 (1980): 185–212.
Antrim, Zayde. *Routes and Realms: The Power of Place in the Early Islamic World*. Oxford: Oxford University Press, 2012.
Arkell, A. J. *A History of the Sudan: From the Earliest Times to 1821*. London: University of London, Athlone Press, 1966.
Arnheim, Rudolf. *New Essays on the Psychology of Art*. Berkeley: University of California Press, 1986.
———. *The Power of the Center: A Study of Composition in the Visual Arts*. Berkeley: University of California Press, 1988.
Arnold, Thomas W. *Painting in Islam: A Study of the Place of Pictorial Art in Muslim Culture*. 1928. Reprint, New York: Dover, 1965.
Arunachalam, B. “Technology of Indian Sea Navigation (c. 1200–c.1800).” *Medieval History Journal* 11 (2009): 187–226.
Aslanapa, Oktay. “The Art of Bookbinding.” In *ABC*, 58–91.
———. *Osmanlı Devri Mimarisi*. Istanbul: İnkilap Kitābevi, 1986.
Atıl, Esin. “Ottoman Miniature Painting under Sultan Mehmed II.” *Ars Orientalis* 9 (1973): 103–20.
———. *Renaissance of Islam: Art of the Mamluks*. Washington, DC: Smithsonian Institution Press, 1981.
———, ed. *Turkish Art*. Washington, DC: Smithsonian Institution Press, 1981.
Atıl, Esin, W. T. Chase, and Paul Jett, eds. *Islamic Metalwork in the Freer Gallery of Art*. Washington, DC: Smithsonian Institution Press, 1985.
Aujac, Germaine. *Claude Ptolémée astronome, astrologue, géographe: Connaissance et representation du monde habité*. Paris: CTHS, 1993.
———. “The Foundations of Theoretical Cartography in Archaic and Classical Greece.” In *HC 1*, 130–47.
Babinger, Franz. “An Italian Map of the Balkans, Presumably Owned by Mehmed II, the Conqueror (1452–53).” *IM* 8 (1951): 8–15.
———. *Mehmed the Conqueror and His Time*. Edited by William C. Hickman. Translated by Ralph Manheim. Princeton, NJ: Princeton University Press, 1978. Originally published as *Mehmed Der Eroberer und seine Zeit: Weltenstürmer einer Zeitenwende*. (Munich: Bruckmann Verlag, 1953).

Bacharach, Jere. *Islamic History through Coins: An Analysis and Catalogue of Tenth-Century Ikhshidid Coinage*. Cairo: The American University in Cairo Press, 2006.
Bačić, Jacques. *Red Sea—Black Russia: Prolegomena to the History of North Central Eurasia in Antiquity and the Middle Ages*. New York: East European Monographs, 1995.
Bağcı, Serpil, Filiz Çağman, Günsel Renda, and Zeren Tanındı. *Osmanlı Resim Sanatı*. Ankara: Republic of Turkey Ministry of Culture and Tourism, 2006. Translated by Ellen Yazar as *Ottoman Painting* (Istanbul: The Banks Association of Turkey, 2010).
Bagrow, Leo. *History of Cartography*. Revised and enlarged by R. A. Skelton. Translated by D. L. Paisey. Cambridge, MA: Harvard University Press, 1964. Reprint, Chicago: Precedent, 1985. Originally published as *Die Geschichte Der Kartographie* (Berlin: Safari-Verlag, 1951).
———. "The Origin of Ptolemy's Geographia." *Geografiska Annaler* 27 (1945): 318–87.
———. "A Tale from the Bosphorus." *IM* 12 (1955): 25–29.
———. "The Wilczek-Brown Codex." *IM* 12 (1955): 171–74.
Ball, John. *Egypt in the Classical Geographers*. Cairo: Government Press, Bulaq, 1942.
Banfi, Florio. "Two Italian Maps of the Balkan Peninsula." *IM* 11 (1954): 17–36.
Barber, Peter. "England I: Pageantry, Defense, and Government: Maps at Court to 1550." In *MMM*, 26–56.
———. "England II: Monarchs, Ministers, and Maps, 1550–1625." In *MMM*, 57–98.
———. *The Map Book*. New York: Walker, 2005.
———. *Medieval World Maps: An Exhibition at Hereford Cathedral, 29th June–1st October 1999*. Hereford: Reprodux Printers, 1999.
Barthes, Roland. *Image, Music, Text*. Translated by Stephen Heath. New York: Hill and Wang, 1977.
———. *Mythologies*. Translated from French by Annette Lavers. New York: Hill and Wang, 1976.
———. *The Responsibility of Forms: Critical Essays on Music, Art, and Representation*. Translated by Richard Howard. Berkeley: University of California Press, 1991.
Barthold, V. V. Preface to *Ḥudūd al-ʿālam: "The Regions of the World,"* translated V. Minorsky, 3–44. London: Messrs. Luzac & Co., 1937. Reprint, Karachi: Indus, 1980.
Bashier, Salman H. *Ibn al-ʿArabī's Barzakh: The Concept of the Limit and the Relationship between God and the World*. Albany: State University of New York Press, 2004.
Batchelor, Robert K. *London: The Selden Map and the Making of a Global City, 1549–1689*. Chicago: University of Chicago Press, 2014.
Baudrillard, Jean. *Cool Memories II*. Translated by Chris Turner. Durham, NC: Duke University Press, 1996.
Bautch, Kelley C. *A Study of Geography of 1 Enoch 17–19: No One Has Seen What I Have Seen*. Leiden: E. J. Brill, 2003.
Bayraktar, Nimet, and Günay Kut. *Yazma Eserlerde Vakif Mühürler*. Ankara: Basbakanlik Basimevi, 1984.
Beckingham, C. F. "Ibn Ḥauqal's Map of Italy." In *Iran and Islam*, edited by C. E. Bosworth, 73–78. Edinburgh: Edinburgh University Press, 1971.
Beer, Arthur. "Astronomical Dating of Works of Art." *Vistas in Astronomy* 9 (1967): 177–87.
Bekker-Nielsen, Tønnes. "Terra Incognita: The Subjective Geography of the Roman Empire."

In *Studies in Ancient History and Numismatics*, edited by Aksel Damsgaard-Madsen, 148–61. Aarhus, Denmark: Aarhus University Press.

Belenitksy, Aleksandr. *The Ancient Civilization of Central Asia*. London: Nagel, 1969.

Berger, John. *The Sense of Sight*. New York: Pantheon Books, 1986.

———. *Ways of Seeing*. London: British Broadcasting Company, 1972.

Berggren, J. L. "Al-Biruni on Plane Maps of the Sphere." *Journal of the History of Arab Science* 4 (1980): 47–112.

Berlekamp, Persis. *Wonder, Image, and Cosmos in Medieval Islam*. New Haven: Yale University Press, 2011.

Berthon, Simon, and Andrew Robinson. *The Shape of the World*. London: George Philip, 1991.

Birkholz, Daniel. *The King's Two Maps: Cartography and Culture in Thirteenth-Century England*. New York: Routledge, 2004.

Blachère, R., and H. Darmaun. *Extraits des Principaux Géographes Arabes du Moyen Age*. Paris: Librairie C. Klincksieck, 1957.

Blacker, Carmen, and Michael Lowe, eds. *Ancient Cosmologies*. London: George Allen & Unwin, 1975.

Bloom, Jonathan. *Paper before Print: The History and Impact of Paper in the Islamic World*. 2001. Reprint, New Haven, CT: Yale University Press, 2014.

Boardman, John, Jasper Griffin, and Oswyn Murray, eds. *The Roman World*. Oxford: Oxford University Press, 1990.

Borges, Jorge Luis. *A Universal History of Infamy*. Translated by N. T. di Giovanni. Harmondsworth: Penguin Books, 1975.

Boud, R. C. "Cartographic Patronage and the Highland and Agricultural Society: The Country Geological Premium Competitions, 1835–1847." *Cartographic Journal* 30, no. 1 (1993): 13–29.

Bouhdiba, Sofiane. "La tradition iconographique dans la cartographie arabe." In *Image et voyage: Representations iconographiques du voyage, de la Méditerranée aux Indes orientales et ocidentales, de la fin du Moyen Âge au XIX*, edited by Loïc P. Guyon and Sylvie Requemora-Gros, 13–19. Aix-en-Provence, Fr.: Presses de l'Université de Provence, 2012.

Bowersock, G. W. "The East-West Orientation of Mediterranean Studies and the Meaning of North and South in Antiquity." In *Rethinking the Mediterranean*, edited by W. V. Harris, 167–78. Oxford: Oxford University Press, 2005.

Boyce, Mary, ed. and trans. *Textual Sources for the Study of Zoroastrianism*. Chicago: University of Chicago Press, 1984.

Brauer, Ralph. *Boundaries and Frontiers in Medieval Muslim Geography*. Transactions of the American Philosophical Society, vol. 85, no. 6. Philadelphia: The American Philosophical Society, 1995.

Bray, Francesca, Vera Dorofeeva-Lichtmann, and Georges Métailié. *Graphics and Text in the Production of Technical Knowledge in China: The Warp and the Weft*. Leiden: Brill, 2007.

Brennan, Teresa, and Martin Jay. *Vision in Context: Historical and Contemporary Perspectives on Sight*. New York: Routledge, 1996.

Brentjes, Sonja. "Cartography in Islamic Societies." In *International Encyclopedia of Human Geography*, edited by R. Kitchin and N. Thrift, 1:414–27. Oxford: Elsevier, 2009.

———. "Courtly Patronage of the Ancient Sciences in Post-Classical Islamic Societies." *Al-Qantara* 29, no. 2 (2008): 403–36.
———. "Euclid's *Elements*, Courtly Patronage and Princely Education." *Iranian Studies* 41, no. 4 (2008): 441–63.
Brice, William C., ed. *An Historical Atlas of Islam*. Leiden: E. J. Brill, 1981.
Brockelmann, Carl. *Geschichte der Arabischen Litteratur*. 2d ed. 2 vols. and 3 suppl. vols. Leiden: E. J. Brill, 1937–1949.
Brodersen, Kai. *Terra Cognita: Studien zur römischen Raumerfassung*. Hildesheim: Olms, 1995.
Brodersen, Kai, and Jaś Elsner, eds. *Images and Texts on the "Artemidorus Papyrus": Working Papers on P.Artemid. (St. John's College Oxford, 2008)*. Stuttgart: Franz Steiner Verlag, 2009.
Brodersen, Kai, and R. J. A. Talbert, eds. *Space in the Roman World: Its Perception and Presentation*. Münster: LIT Verlag, 2004.
Brotton, Jerry. *A History of the World in Twelve Maps*. London: Allen Lane, 2012.
———. *Trading Territories: Mapping the Early Modern World*. Ithaca, NY: Cornell University Press, 1997.
Brummett, Palmira. "Imagining the Early Modern Ottoman Space." In *The Early Modern Ottomans: Remapping the Empire*, edited by Virginia Aksan and Daniel Goffman, 15–58. Cambridge: Cambridge University Press, 2007.
Brunet, J.-P., R. Nadal, and Cl. Vibert-Guigue. "The Fresco of the Cupola of Qusayr ʿAmra." *Centaurus*, 40 (1998): 97–123.
Bryson, Norman. *Vision and Painting: The Logic of the Gaze*. New Haven, CT: Yale University Press, 1983.
Buck-Morss, Susan. *The Dialectics of Seeing: Walter Benjamin and the Arcades Project*. Cambridge: MIT Press, 1989.
Bulliet, Richard. *The Camel and the Wheel*. Cambridge, MA: Harvard University Press, 1975. Reprint, Columbia University Press, 1990.
———. *The Case for Islamo-Christian Civilization*. New York: Columbia University Press, 2004.
———. *Conversion to Islam in the Medieval Period: An Essay in Quantitative History*. Cambridge, MA: Harvard University Press, 1979.
———. *Islam: The View from the Edge*. New York: Columbia University Press, 1994.
———. *The Wheel: Inventions and Reinventions*. New York: Columbia University Press, forthcoming.
Bunbury, Edward Herbert. *A History of Ancient Geography among the Greeks and Romans from the Earliest Ages till the Fall of the Roman Empire*. 2nd ed. 2 vols. London: John Murray, Albemarle Stree, 1879. Revised by W. H. Stahl. New York: Dover, 1959.
Burnett, Charles. "Hildegard and the Science of the Stars." In *Hildegard of Bingen: The Context of Her Thought and Art*, edited by Charles Burnett and Peter Dronke, 111–20. London: The Warburg Institute.
Bynum, Caroline. *Fragmentation and Redemption: Essays on Gender and the Human Body in Medieval Religion*. New York: Zone Books, 1991.
———. *The Resurrection of the Body in Western Christianity, 200–1336*. New York: Columbia University Press, 1995.
Çağman, Filiz. "The Miniatures of the Divan-i Hüseyni and the Influence of Their Style." *ICTA* 5 (1978): 231–59.

———. "Sultan Mehmed II Dönemine Ait Bir Minyatürlü Yazma: Kulliyat-i Katibi." *Sanat Tarihi Yıllıgı* 6 (1974–1975): 333–46.
Çağman, Filiz, and Nurhan Atasoy. *Turkish Miniature Painting.* Istanbul: Publications of the R. C. D. Cultural Institute, 1974.
Çağman, Filiz, and Zeren Tanındı. *The Topkapı Saray Museum: The Albums and Illustrated Manuscripts.* Translated, edited, and expanded by J. M Rogers. London: Thomas and Hudson, 1986.
Calvino, Italo. *Üç Deneme.* Translated by Bilge Karasu. Istanbul: Yapı Kredi Yayınları, 1993.
Campbell, Brian. *The Writings of the Roman Land Surveyors.* London: Society for the Promotion of Roman Studies, 2000.
Campbell, Caroline, and Alan Chong, eds. *Bellini and the East.* London: National Gallery; Boston: Isabella Stewart Gardner Museum, 2005.
Campbell, Mary B. *The Witness and the Other World: Exotic European Travel Writing, 400–1600.* Ithaca, NY: Cornell University Press, 1988.
Canby, Sheila. "Pilgrimage and Prayer." *Islamic Arts and Architecture*, April 4, 2012. http://islamic-arts.org/2012/pilgrimage-and-prayer/.
Carboni, Stefano. "Constellations, Giants and Angels from Al-Qazwini Manuscripts." In *Islamic Art in the Ashmolean Museum*, edited by James Allen, 83–97. Oxford: Oxford University Press, 1995.
Carder, J. N. *Art Historical Problems of a Roman Land Surveying Manuscript: The Codex Arcerianus A, Wolfenbüttel.* New York: Garland, 1978.
Carlstein, T. *Time Resources, Society, and Ecology.* London: George Allen & Unwin, 1982.
Casale, Giancarlo. *The Ottoman Age of Exploration.* Oxford: Oxford University Press, 2010.
Casey, Edward. *The Fate of Place: A Philosophical History.* Berkeley: University of California Press, 1997.
———. *Getting Back in Place: Toward a Renewed Understanding of the Place-World.* Bloomington: Indiana University Press, 1993.
———. *Representing Place: Landscape Painting and Maps.* Minneapolis: University of Minnesota Press, 2002.
Cattaneo, Angelo. *Fra Mauro's Mappa Mundi and Fifteenth-Century Venice.* Turnhout, Belgium: Brepols, 2011.
Caviness, Madeline H. "Hildegard as Designer of the Illustrations to Her Works." In *Hildegard of Bingen: The Context of Her Thought and Art*, edited by Charles Burnett and Peter Dronke, 29–62. London: The Warburg Institute, 1998.
Chang, Kwang-chih. *The Archaeology of Ancient China.* New Haven, CT: Yale University Press, 1977.
Chekin, Leonid S. *Northern Eurasia in Medieval Cartography: Inventory, Text, Translation, and Commentary.* Turnhout, Belgium: Brepols Publishers, 2006.
Chittick, William C. *Ibn al-'Arabi's Metaphysics of Imagination: The Sufi Path of Knowledge.* Albany: State University of New York Press, 1989.
Çıpa, H. Erdem, and Emine Fetvacı. *Writing History at the Ottoman Court: Editing the Past, Fashioning the Future.* Bloomington: Indiana University Press, 2013.
Clark, W. T. "Manners, Customs, and Beliefs of the Northern Beja." In *Sudan Notes and Records* 21, no. 1 (1938): 4–6.

Conley, Tom. *The Self-Made Map: Cartographic Writing in Early Modern France*. Minneapolis: University of Minnesota Press, 1996.
Contenau, G. "The Early Ceramic Art." In *Persian Art*, 1:171–94.
Cook, M. A., ed. *A History of the Ottoman Empire to 1730*. Cambridge: Cambridge University Press, 1976.
Corbin, Henry. *Avicenna and the Visionary Recital*. Princeton, NJ: Princeton University Press, 1988.
———. *Spiritual Body and Celestial Earth: From Mazdean Iran to Shi'ite Iran*. Translated by Nancy Pearson. Princeton, NJ: Princeton University Press, 1977.
Cortesão, Armando. *History of Portuguese Cartography*. 2 vols. Coimbra: Junta de Investigações do Ultramar-Lisboa, 1969–1971.
Cosgrove, Denis. *Apollo's Eye: A Cartographic Genealogy of the Earth in the Western Imagination*. Baltimore: Johns Hopkins University Press, 2001.
———. *Geography and Vision: Seeing, Imagining and Representing the World*. London: I. B. Tauris, 2008.
———, ed. *Mappings*. London: Reaktion Books, 1999.
Cosgrove, Denis, and Stephen Daniels, eds. *The Iconography of Landscape: Essays on the Symbolic Representation, Design, and Use of Past Environments*. Cambridge: Cambridge University Press, 1988.
Creswell, K. A. C. *Early Muslim Architecture*. 2 vols. Oxford: Clarendon Press, 1932–1940; Rev. ed., vol. 1, 1969.
———. *A Short Account of Early Muslim Architecture*. London: Penguin Books, 1958.
Crone, G. R. *Maps and Their Makers: An Introduction to the History of Cartography*. 1st ed., 1953. Reprint, Hamden, CT: Archon Books, 1978.
Curatola, Giovanni, and Gianroberto Scarcia. *The Art and Architecture of Persia*. Translated by Marguerite Shore. New York: Abbeville Press, 2004.
Curtis, Gregory. *The Cave Painters: Probing the Mysteries of the World's First Artists*. New York: Knopf, 2006.
Curtis, Vesta Sarkhosh. *Persian Myths*. London: British Museum Press, 1993.
Dalché, Gautier. "De la glose à la contemplation." *Testo e immagine nell'alto medioevo* 41 (1993): 693–764.
———. "Géographie Arabe et Géographie Latine au XIIe Siècle." *Medieval Encounters* 19, no. 4 (2013): 408–33.
———. "Maps in Words: The Descriptive Logic of Medieval Geography, from the Eighth to the Twelfth Century." In *HWM*, 223–42.
Dallal, Ahmad. *Islam, Science, and the Challenge of History*. New Haven, CT: Yale University Press, 2010.
Damisch, Hubert. "La grille comme volonté et comme représentation." In *CFT*, 30–40.
Däniken, Erich von. *Chariots of the Gods*. Translated by Michael Heron. New York: Berkley Books, 1980. Originally published as *Erinnerungen an die Zukunft: Ungelöste Rätsel der Vergangenheit* (Berlin: Econ-Verlag GMBH, 1968).
Darnton, Robert. *The Great Cat Massacre and Other Episodes in French Cultural History*. New York: Basic Books, 1984.

Daryaee, Touraj. *Šahrestānīha ī Ērānšahr: A Middle Persian Text on Late Antique Geography, Epic, and History*. Costa Mesa, CA: Mazda, 2002.

———. *Sassanian Iran (224–651 CE): Portrait of a Late Antique Empire*. Costa Mesa, CA: Mazda, 2008.

Davidson, Hilda Ellis. *The Lost Beliefs of Northern Europe*. London: Routledge, 1993.

de Goeje, Michael Jan. "Die Istakhrī-Balkhī Frage." *Zeitschrift der Deutschen Morgenländischen Gesellschaft* 25 (1871): 42–58.

Deissmann, Gustav Adolf. *Forschungen und Funde im Serai mit Einem Verzeichnis der Nichtislamischen Handschriften im Topkapu Serai zu Istanbul*. Berlin: Walter de Gruyter, 1933.

Delano-Smith, Catherine. "Cartography in the Prehistoric Period in the Old World: Europe, the Middle East, and North Africa." In *HC 1*, 54–101.

———. "*Imago Mundi*'s Logo: The Babylonian Map of the World." *IM* 48 (1996): 209–11.

———. "Milieus of Mobility: Itineraries, Route Maps, and Road Maps." In *Cartographies of Travel and Navigation*, edited by James R. Akerman, 16–68. Chicago: University of Chicago Press, 2006.

———. "Prehistoric Cartography in Asia." In *HC 2.2*, 1–22.

Delano-Smith, Catherine, and Roger P. Cain. *English Maps: A History*. Toronto: University of Toronto Press, 1999.

della Dora, Veronica. *Imagining Mount Athos: Visions of a Holy Place, from Homer to World War II*. Charlottesville: University of Virginia Press, 2011.

Derolez, Albert. *The Autograph Manuscript of the "Liber floridus": A Key to the Encyclopedia of Lambert of Saint-Omer*. Turnhout, Belgium: Brepols, 1998.

Derrida, Jacques. *Of Grammatology*. Translated by Gayatri Chakratvorty Spivak. Baltimore: Johns Hopkins University Press, 1976.

Destombes, Marcel. "A Venetian Nautical Atlas of the Late 15th Century." *IM* 12 (1955): 30.

Dilke, O. A. W. "Maps in the Service of the State: Roman Cartography to the End of the Augustan Era." In *HC 1*, 201–11.

———. "Roman Large-Scale Mapping." In *HC 1*, 212–33.

Doel, Ronald E., Tanya J. Levin, and Mason K. Marker. "Extending Modern Cartography to the Ocean Depths: Military Patronage, Cold War Priorities, and the Heezen-Tharp Mapping Project, 1952–1959." *Journal of Historical Geography* 32 (2006): 605–26.

Donini, Pier Giovanni. *Arab Travelers and Geographers*. London: Immel, 1991.

Ducène, Jean-Charles. "L'Afrique dans les mappemondes circulaires arabes médiévales: Typologie d'une representation." *Cartographier l'Afrique* 210 (December 2011): 19–35.

———. "Al-Bakrī et les Étymologies d'Isidore de Séville." *Journal Asiatique* 297, no. 2 (2009): 379–97.

———. "La carte circulaire du Kitāb dalā'il al-qibla d'Ibn al-Qāṣṣ: Représentation du monde et toponymie originales." *Folio Orientalia* 38 (2002): 115–46.

———. "Les coordonnées géographiques de la carte manuscrite d'al-Idrīsī." *Der Islam* 86, no. 2 (2009): 271–85.

———. "Le delta du Nil dans les cartes d'Ibn Ḥawqal." *Journal of Near Eastern Studies* 63, no. 4 (2004): 241–56.

———. "France in the Two Geographical Works of Al-Idrīsī (Sicily, 12th century)." In *Space in*

the Medieval West, Places, Territories, and Imagined Geographies, edited by M. Cohen and F. Madeline, 175–96. London: Ashgate, 2014.
———. "Les Îles de l'Océan Indien dans les sources Arabes Médievales Lente Découverte ou Imagination Galopante?" In *L'Île, Regards Orientaux: Varia Orientalia, Biblica et Antiqua. Hans Hauben in honorem*, edited by Christian Cannuyer, 125–34. Lille, Fr.: The Belgian Society of Oriental Studies, 2013.
———. "Un nouveau manuscrit du Ṣuwar al-aqālim d'al-Iṣṭaḫrī: Le ms. ʿAref Ḥakamt Ǧuġrāfiya 910/7 (Médine, Maktabat ʿAbd al-ʿAziz)." *Folia Orientalia* 40 (2004): 279–311.
———. "Un Nouvelle Description de Ṣuhār (ʿUmān) Extraite d'un Manuscrit Inexploité du Kitāb Ṣuwar al-Aqālīm d'al-Iṣṭaḫrī." *Arabica* 50, no. 1 (2003): 109–13.
———. "Quel est le titre veritable de l'ouvrage géographique d'al-Iṣṭaḫrī?" *Acta Orientalia Belgica* 19 (2006): 99–108.
———. "Soufisme et cosmographie musulmane aux XIIe et XIIIe siècles: Convergence ou influence à propos d'une conception commune du monde." In *Mystique: La passion de l'Un, de l'Antiquité à nos jours*, edited by A. Dierkens and B. Beyer de Ryke, 205–14. Brussels: Editions de l'Universite de Bruxelles, 2005.
Duncan, James S., and David Ley. *Place/Culture/Representation*. London: Routledge, 1993.
Dunlop, D. M. *Arab Civilization to A.D. 1500*. New York: Praefer, 1971.
———. *History of the Jewish Khazars*. New York: Schoken Books, 1967.
Eagleton, Terry. *Against the Grain: Essays, 1975–1985*. London: Verso Books, 1986.
———. *Literary Theory: An Introduction*. Oxford: Blackwell, 1983.
Ebel, Kathryn. "Representations of the Frontier in Ottoman Town Views of the Sixteenth Century." *IM* 60, no. 1 (2008): 1–22.
Edney, Matthew. *Mapping an Empire: The Geographical Construction of British India, 1765–1843*. Chicago: University of Chicago Press, 1997.
———. "Mathematical Cosmography and the Social Ideology of British Cartography." *IM* 46 (1994): 101–16.
———. "The Patronage of Science and the Creation of Imperial Space: The British Mapping of India, 1779–1843." *Cartographica* 30, no. 1 (1993): 61–67.
———. "Theory and the History of Cartography." *IM* 48 (1996): 185–91.
Edson, Evelyn. *Mapping Time and Space: How Medieval Mapmakers Viewed Their World*. London: The British Library, 1999.
———. "Maps in Context: Isidore, Orosius, and the Medieval Image of the World." In *CAMA*, 219–36.
———. *The World Map, 1300—1492: The Persistence of Tradition and Transformation*. Baltimore: Johns Hopkins University Press, 2007.
Edson, Evelyn, and E. Savage-Smith. *Medieval Views of the Cosmos: Picturing the Universe in the Christian and Islamic Middle Ages*. Oxford: Bodleian Library, 2004.
Eliade, Mircea. *A History of Religious Ideas*. Translated by Willard R. Trask. 3 vols. Chicago: University of Chicago Press, 1978.
———. *Images and Symbols: Studies in Religious Symbolism*. Translated by Philip Mairet. Princeton, NJ: Princeton University Press, 1978. Reprint, 1991.
———. *Patterns in Comparative Religions*. Translated by Rosemary Sheed. New York: Meridian Books, 1958.

Elias, Jamal J. *Aisha's Cushion: Religious Art, Perception, and Practice in Islam*. Cambridge, MA: Harvard University Press, 2012.
Elkins, James. *Why Are Our Pictures Puzzles? On the Modern Origins of Pictorial Complexity*. New York: Routledge, 1999.
Eminoğlu, Münevver, ed. *Ressam, sultan ve portresi*. Istanbul: Yapı Kredi Kultur Sanat Yayıncılık, 1999.
Emiralioğlu, Pınar. *Geographical Knowledge and Imperial Culture in the Early Modern Ottoman Empire*. Surrey, UK: Ashgate, 2014.
Endress, Gerhard. *An Introduction to Islam*. Translated by Carole Hillenbrand. Edinburgh: Edinburgh University Press, 1988.
Engin, Fatma Egemen. *Mühür ve Mühürcülük Sanatımız*. Istanbul: Arıitan Yayınevi, 1994.
Erünsal, Ismail E. "The Development of Ottoman Libraries from the Conquest of Istanbul to the Emergence of the Independent Library." *Belleten* 60 (1996): 93–125.
———. "Medieval Ottoman Libraries." *Erdem* 1, no. 3 (1985): 745–54.
———. *Türk Kütüphaneleri Tarihi II, Kurulştan Tanzimat'a Kadar Osmanli Vakif Kütüphaneleri*. Ankara: Atatürk Kültür Merkezi Yayini, 1988.
Ettinghausen, Richard. *Arab Painting*. New York: Rizzoli, 1977.
———. "Die bildliche Darstellung der Ka'ba im Islamischen Kulturkreis." *Zeitschrift der Deutschen Morgenländischen Gesellschaft* 87 (1934): 111–37.
Euben, Roxanne. *Journey to the Other Shore: Muslim and Western Travelers in Search of Knowledge*. Princeton, NJ: Princeton University Press, 2006.
Evans, Helen C., and William D. Wixom, eds. *The Glory of Byzantium: Art and Culture of the Middle Byzantine Era A.D. 843–1261*. New York: The Metropolitan Museum of Art, 1997.
Fernandez-Armesto, Felipe. *Civilizations: Culture, Ambition, and the Transformation of Nature*. New York: Free Press, 2002.
———. *1492: The Year the World Began*. New York: Harper Collins, 2009.
———. *Millennium*. London: Bantham Books, 1995.
———. *Near a Thousand Tables: A History of Food*. New York: Free Press, 2003.
———. *Pathfinders: A Global History of Exploration*. London: Norton, 2006.
Fetvacı, Emine. "The Office of the Ottoman Court Historian." In *Studies on Istanbul and Beyond*, edited by Robert G. Ousterhout, 6–21. Philadelphia: University of Pennsylvania, 2007.
———. *Picturing History at the Ottoman Court*. Bloomington: Indiana University Press, 2013.
Findley, Carter. *The Turks in World History*. New York: Oxford University Press, 2005.
Fiorani, Francesca. *The Marvel of Maps: Art, Cartography, and Politics in Renaissance Italy*. New Haven, CT: Yale University Press, 2005.
Fleischer, Cornell. *Bureaucrat and Intellectual in the Ottoman Empire: The Historian Mustafa Âli (1541–1600)*. Princeton, NJ: Princeton University Press, 1986.
———. "Shadows of Shadows: Prophecy in Politics in 1530s İstanbul." *Turkish Studies* 13 (2007): 51–62.
Flood, Finbarr Barry. *Objects of Translation: Material Culture and Medieval "Hindu-Muslim" Encounter*. Princeton, NJ: Princeton University Press, 2009.
Foucault, M. *The Archaeology of Knowledge and the Discourse on Language*. Translated by A. M. Sheridan Smith. New York: Pantheon Books, 1972. Originally published as *L'Archéologie du*

Savoir (Paris: Editions Gallimard, 1969) and *L'ordre du discours* (Paris: Editions Gallimard, 1971).
———. *The Order of Things: An Archaeology of the Human Sciences.* New York: Pantheon Books, 1971. Reprint, New York: Vintage Books, 1994. Originally published as *Les Mots et les choses* (Paris: Editions Gallimard, 1966).
Fowden, Garth. *Quṣayr ʿAmra: Art and the Umayyad Elite in Late Antique Syria.* Berkeley: University of California Press, 2004.
Galerkina, Olympiade. "On Some Miniatures Attributed to Bihzad." *Ars Orientalis* 8 (1970): 121–38.
Gallo, Rodolfo. "A Fifteenth Century Military Map of the Venetian Territory of *Terraferma*." *IM* 12 (1955): 55–57.
Gari, Lutfallah. "About al-Shayzarī and Ibn Bassām: Who Preceded the Other?" *Studies in Islam and the Middle East* 5, no. 1 (2008). http://majalla.org/papers/2008/article3.pdf.
Gell, Alfred. *The Anthropology of Time.* Oxford: Berg, 1996.
Gellens, Sam. "The Search for Knowledge in Medieval Muslim Societies: A Comparative Approach." In *Muslim Travellers: Pilgrimage, Migration, and the Religious Imagination*, edited by Dale F. Eickelman and James Piscatori, 50–65. Berkeley: University of California Press, 1990.
Genette, Gérard. The Aesthetic Relation. Ithaca: Cornell University Press, 1999.
———. *The Work of Art: Immanence and Transcendence.* Translated by G. M. Goshgarian. Ithaca: Cornell University Press, 1977.
Ghebreyesus, Abba Isaak. "Mereni's People, Origins of Geshinashim." *The Mirror*, October 22, 1997.
Ghirshman, Roman. *Persian Art: The Parthian and Sassanian Dynasties, 249 B.C.–A.D. 651.* New York: Golden Press, 1962.
al-Ghunaim, Abdullah Y. *al-Makhṭūṭāt al-jughrāfiya al-ʿarabiya fī al-mathaf al-briṭānī.* Kuwait: Al-ʾAmal Bookshop, 1999.
Gibbon, Edward. *The History of the Decline and Fall of the Roman Empire.* Edited J. B. Bury with an introduction by W. E. H. Lecky. 12 vols. New York: Fred de Fau, 1906. http://oll.libertyfund.org/titles/1681.
Ginzberg, Louis. *The Legends of the Jews.* 7 vols. Baltimore: Johns Hopkins University Press, 1998.
Gole, Susan. *Indian Maps and Plans: From Earliest Times to the Advent of European Surveys.* New Delhi: Manohar, 1989.
Gombrich, E. H. *Art and Illusion: A Study in the Psychology of Pictorial Representation.* The A. W. Mellon Lectures in the Fine Arts, 1956. Bollingen Series. Princeton, NJ: Princeton University Press, 1960.
———. *The Image and the Eye: Further Studies in the Psychology of Pictorial Representation.* Oxford: Phaidon, 1982.
———. *The Sense of Order: A Study in the Psychology of Decorative Art.* Ithaca, NY: Cornell University Press, 1979.
———. *Studies in the Art of the Renaissance.* 4 vols. Oxford: Phaidon, 1966–1977.
Goodrich, Thomas. "Old Maps in the Library of Topkapi Palace in Istanbul." *IM* (1993): 120–33.

———. *The Ottoman Turks and the New World: A Study of Tarih-i Hind-i Garbi and Sixteenth-Century Ottoman Americana*. Wiesbaden, Ger.: Otto Harrassowitz, 1990.
Gordon, Matthew S. *The Breaking of a Thousand Swords: A History of the Turkish Military of Samarra (200–275 A.H./815–889 C.E.)*. Albany: State University of New York Press, 2001.
Gould, Peter, and Rodney White. *Mental Maps*. Winchester, MA: Allen & Unwin, 1986.
Gow, Andrew. "Fra Mauro and the End of Authority: Legends and Empirical Evidence on the 'Last' Mappamundi." In *Mappa Mundi: The Hereford World Map*, 405–14. London: British Library, 2005.
———. "Gog and Magog on Mappaemundi and Early Printed Maps: Orientalizing Ethnography in the Apocalyptic Tradition." *Journal of Early Modern History* 2, no. 1 (1998): 1–28.
Grabar, Oleg. *The Formation of Islamic Art*. 2nd ed. New Haven, CT: Yale University Press, 1987.
———. *The Mediation of Ornament*. The A. W. Mellon Lectures in the Fine Arts, 1989. Bollingen Series. Princeton, NJ: Princeton University Press, 1992.
Grabar, Oleg, and Richard Ettinghausen. *The Art and Architecture of Islam, 650–1250*. London: Penguin Books, 1987. Reprint, 1991.
Gray, Basil, ed. *The Arts of the Book in Central Asia*. Geneva: UNESCO, 1979.
———. "History of Miniature Painting: The Fourteenth Century." In *ABC*, 92–120.
———. *Persian Painting*. 1st ed. Geneva: Editions d'Art Albert Skira S.A., 1961. Reprint, 1995.
———. "The School of Shiraz from 1392 to 1453." In *ABC*, 121–45.
———. *The World History of Rashid al-Din: A Study of the Royal Asiatic Society Manuscript*. London: Faber and Faber, 1978.
Gregory, Derek. *Geographical Imaginations*. Cambridge, MA: Blackwell, 1994.
Grube, Ernst. *The Classical Style in Islamic Painting: The Early School of Herat and Its Impact on Islamic Painting of the Later 15th, 16th and 17th Centuries*. Edizioni Oriens, 1968.
———. "Notes on Ottoman Painting in the 15th Century." In *Islamic Art and Architecture* (1981), 51–62. Reprint, in *Studies in Islamic Painting*, edited by Ernst Grube, 442–62. London: Pindar Press, 1995.
———. "The School of Herat from 1400 to 1450." In collaboration with Dr. Eleanor Sims. In *ABC*, 146–78.
Grube, Ernst, and Eleanor Sims, eds. *Between China and Iran: Paintings from Four Istanbul Albums*. Colloquies on Art & Archaeology in Asia. London: School of Oriental and African Studies, University of London, 1980.
Gruber, Christiane. "Between Logos (*Kalima*) and Light (*Nūr*): Representations of the Prophet Muhammad in Islamic Painting." *Muqarnas* 26 (2009): 229–62.
———. *The Ilkhanid Book of Ascension: A Persian-Sunni Devotional Tale*. London: I. B. Tauris, 2010.
———. *The Timurid Book of Ascension (Miʿrajnama): A Study of Text and Image in a Pan-Asian Context*. Valencia, Spain: Patrimonio Edciones, 2008.
Gruber, Christiane, and Frederick S. Colby, eds. *The Prophet's Ascension: Cross-Cultural Encounters with the Islamic Miʿrāj Tales*. Bloomington: Indiana University Press, 2010.
Gruzinski, Serge. *Painting the Conquest: The Mexican Indians and the European Renaissance*. Paris: Flammarion, 1992.

Gurevich, A. J. *Categories of Medieval Culture*. Translated by G. L. Campbell. London: Routledge Kegan & Paul, 1985.

Gurney, Q. R. *The Hittites*. London: Penguin Books, 1952. Reprint, 1990.

Gutas, Dimitri. *Greek Thought, Arabic Culture: The Graeco-Arabic Translation Movement in Baghdad and Early ʿAbbāsid Society (2nd–4th/8th–10th Centuries)*. Oxon, UK: Routledge, 1998.

Halm, Heinz. *The Empire of the Mahdi: The Rise of the Fatimids*. Translated by Michael Bonner. Leiden: E. J. Brill, 1996.

Hapgood, Charles H. *Maps of the Ancient Sea Kings: Evidence of Advanced Civilization in the Ice Age*. Philadelphia: Chilton; Toronto: Ambassador Books, 1966. Rev. ed., New York: E. P. Dutton, 1979.

Harbison, Robert. *Eccentric Spaces*. Boston: David R. Godine, 1988.

Hardie, P. R. "Imago Mundi: Cosmological and Ideological Aspects of the Shield of Achilles." *Journal of Hellenic Studies* 105 (1985): 11–31.

Harley, J. B. "Ancient Maps Waiting to Be Read." *Geographical Magazine* 53 (1981): 316–17.

———. "Deconstructing the Map." *Cartographica* 26, no. 2 (1989): 1–20. Rev. ed. in *Writing Worlds: Discourse, Text and Metaphor in the Representation of Landscape*, edited by Trevor Barnes and James S. Duncan, 231–47. London: Routledge, 1992.

———. "Historical Geography and the Cartographic Illusion." *Journal of Historical Geography* 15, no. 1 (1989): 80–91.

———. "The Iconology of Early Maps." In *Imago et Mensura Mundi: Atti del IX Congresso Internazionale di Storia della Cartographia*, edited by Carla Marzoli, 1:29–38. Rome: Istituto della Enciclopedia Italiana, 1985.

———. "Maps, Knowledge, and Power." In *Iconography of Landscape*, edited by Denis Cosgrove and Stephen Daniels, 277–312. Cambridge: Cambridge University Press, 1988.

———. "Meaning and Ambiguity in Tudor Cartography." In *English Map-Making, 1500–1650: Historical Essays*, edited by Sarah Tyacke, 22–45. London: The British Library Reference Division Publications, 1982.

———. "Monarchs, Ministers, and Maps." *Map Collector* 35 (1986): 42–43.

———. "Secrecy and Silences: The Hidden Agenda of Cartography in Early Modern Europe." *IM* 40 (1988): 111–30.

Harley, J. B., and M. J. Blackmore. "Concepts in the History of Cartography: A Review and Perspective." *Cartographica* 17, no. 4 (Winter 1980): 1–120.

Harley, J. B., and David Woodward, eds. *The History of Cartography*. 6 vols. Chicago: University of Chicago Press, 1987–.

———. "Why Cartography Needs Its History." *American Cartographer* 16, no. 1 (1989): 5–15.

Harley, J. B. (published posthumously), and Kees Zandvliet. "Art, Science, and Power in Sixteenth-Century Dutch Cartography." *Cartographica* 29, no. 2 (1992): 10–19.

Harper, Prudence Oliver. *The Royal Hunter: Art of the Sasanian Empire*. New York: The Asia Society, 1978.

Hartner, Willy. "Qusayr ʿAmra, Farnesina, Luther, Hesiod. Some Supplementary Notes to A. Beer's Contribution." *Vistas in Astronomy* 9 (1967): 225.

Harvey, David. *The Condition of Postmodernity*. Oxford: Blackwell, 1990.

Harvey, P. D. A. *Medieval Maps.* London: The British Library, 1991.

Harwood, Jeremy. *To the Ends of the Earth: 100 Maps That Changed the World.* Cincinnati: F+W Publications, 2006.

Hasan, Yusuf Fadl. "The Penetration of Islam in the Eastern Sudan." In *Islam in Tropical Africa,* edited by I. M. Lewis, 112–23. Bloomington: Indiana University Press, 1966. Reprint, 1980.

Heck, Paul L. *The Construction of Knowledge in Islamic Civilization: Qudāma ibn Ja'far and His "Kitāb al-kharāj wa ṣinā'at al-kitāba."* Leiden: Brill, 2002.

Heidel, Alexander. *The Babylonian Genesis: The Story of Creation.* 2nd ed. Chicago: University of Chicago Press, 1951.

Henderson, John B. "Chinese Cosmographical Thought: The High Intellectual Tradition," *HC* 2:2, 203–27.

———. "Nonary Cosmography in Ancient China." In *GE*, 64–73.

Hermes, Nizar F. *The [European] Other in Medieval Arabic Literature and Culture: Ninth–Twelfth Century AD.* New York: Palgrave Macmillan, 2012.

Herrera, Monica. "Granada en los atlas de al-Sharafi, e identificación de un modelo mallorquín para la carta de al Mursi." *Al Qantara* 30, no. 1 (2009), 221–35.

Hesiod. *"The Works and Days"; "Theogony"; "The Shield of Herakles."* Translated by Richmond Lattimore. Ann Arbor: University of Michigan Press, 1959. Reprint, 1991.

Hiatt, Alfred. "The Map of Macrobius before 1100." *IM* 59, no. 2 (2007): 149–76.

Higgins, Reynold. *Minoan and Mycenaen Art.* 2nd rev. ed. London: Thames and Hudson, 1997.

Hillenbrand, Carole. *The Crusades: Islamic Perspectives.* Edinburgh: Edinburgh University Press, 1999. Rev. ed., 2006.

Hillenbrand, Robert. *Islamic Art and Architecture.* London: Thames and Hudson, 1999.

Hiskett, Mervyn. *The Course of Islam in Africa.* Edinburgh: Edinburgh University Press, 1994.

Hodgson Marshall. *Venture of Islam: The Classical Age of Islam.* 3 vols. Chicago: University of Chicago Press, 1977.

Horowitz, Wayne. "The Babylonian Map of the World." *Iraq* 50 (1988): 147–65.

Hourani, George. *Arab Seafaring in the Indian Ocean in Ancient and Early Medieval Times.* Princeton, NJ: Princeton University Press, 1951. Revised by John Carswell. Princeton, NJ: Princeton University Press, 1995.

Hoyland, Robert G. *Arabia and the Arabs: From the Bronze Age to the Coming of Islam.* London: Routledge, 2001.

Hüe, Denis. "Tracé, écart: Le sens de la carte chez Opicinus de Canistris." In *Terres Medievales,* edited by Bernard Ribémont, 129–58. Editions Klincksieck, 1993.

Husserl, E. *The Phenomenology of Internal Time Consciousness.* Translated by John Barnett Brough. Dordrecht, Neth.: Kluwer Academic, 1991. Originally published as *Zur Phänomenologie des inneren Zeitbewusstseins (1893–1917)* (The Hague: Martinus Nijhoff, 1966).

Hsu, Hsin-Mei Agnes. "The Qin Maps: A Clue to Later Chinese Cartographic Development." *IM* 45 (1993): 90–100.

———. "Structured Perceptions of Real and Imagined Landscapes in Early China." In *GE*, 43–63.

İnalcık, H. "The Rise of the Ottoman Empire." In *History of the Ottoman Empire to 1730,* edited by M. A. Cook, 41–42. Cambridge: Cambridge University Press, 1976.

Irby, Georgia L. "Mapping the World: Greek Initiatives from Homer to Eratosthenes." In *AP*, 81–107.
Irwin, Robert. *The Arabian Nights: A Companion.* London: Penguin Press, 1994.
———. "Ibn Zunbul and the Romance of History." In *Writing and Representation in Medieval Islam,* edited by Julia Bray, 3–15. London: Routledge, 2006.
Ivanov, Anatol A., and Oleg Akimushkin. "The Art of Illumination." In *ABC,* 30–47.
Ivins, William M., Jr. *Art & Geometry: A Study of Space Intuitions.* New York: Dover, 1946.
Jacob, Christian. "Alexandre et la maitrise de l'espace." *Quaderni di storia* 34 (July–December 1991): 5–40.
———. "La carte du monde: De la clôture visuelle à l'expansion des savoirs." *Fini & Infini* 24/25 (1992): 241–58.
———. "De la terre à la lune: Les débuts de la sélénographie au XVII Siècle." In *Cartographiques,* edited by Marie-Ange Brayer, 9–43. Paris: Réunion des Musées, 1996.
———. *La description de la terre habitée de Denys d'alexandrie ou la leçon de géographie.* Paris: Albin Michel, 1990.
———. "Écritures du monde." In *CFT,* 104–19.
———. *L'Empire des cartes: Approche théorique de la cartographie à travers l'histoire* (Paris: Albin Michel, 1992).
———. "Géométrie, graphisme, figuration: Pour une esthétique des cartes anciennes." *Encyclopaedia Universalis* 1 (1995): 296–332.
———. "Il faut qu'une carte soit ouverte ou fermée: Le tracé conjectural." *Revue de la Bibliothéque Nationale* 45 (1992): 35–41.
———. "L'Inde imaginaire des géographes alexandrins." In *Inde, Grèce ancienne: Regards croisés en anthropologie de l'espace,* edited by Jean-Claude Carrière, 61–80. Paris: Presses Univ. Franche-Comté, 1995.
———. "Inscrire la terre habitée sur une tablette: Réflecions sur la fonction des cartes géographiques en grece ancienne." In *Les Savoirs de l'écriture en grèce ancienne,* edited by Marcel Détienne, 273–304. Lille, Fr.: Presses Universitaires, 1988.
———. "Mapping in the Mind: The Earth from Ancient Alexandria." In *Mappings,* edited by Denis Cosgrove, 24–49. London: Reaktion Books, 1999.
———. "Quand les cartes réfléchissent." *Espaces Temps* 62–63 (1996): 36–49.
———. *The Sovereign Map: Theoretical Approaches in Cartography throughout History.* Translated by Tom Conley. Chicago: University of Chicago Press, 2006. Originally published as *L'Empire des cartes: Approche théorique de la cartographie à travers l'histoire* (Paris: Albin Michel, 1992).
———. "Toward a Cultural History of Cartography." *IM* 48 (1996): 191–97.
Jacobs, Rabbi Louis. "Jewish Cosmology." In Blacker and Lowe, *Ancient Cosmologies,* 66–86.
Jacobsen, Frode F. *Theories of Sickness and Misfortune among the Hedandowa Beja of the Sudan: Narratives as Points of Entry into Beja Cultural Knowledge.* London: Kegan Paul International, 1998.
Jameson, Frederic. *The Political Unconscious: Narrative as a Socially Symbolic Act.* Ithaca, NY: Cornell University Press, 1981.
———. *Signatures of the Visible.* New York: Routledge, 1992.

———. *Valences of the Dialectic*. London: Verso, 2009.

Jardine, Lisa, and Jerry Brotton. *Global Interests: Renaissance Art between East and West*. Ithaca, NY: Cornell University Press, 2000.

Jay, Martin. *The Dialectical Imagination: A History of the Frankfurt School and the Institute for Social Research, 1923–1950*. Berkeley: University of California Press, 1973.

Jettmar, K. *Art of the World of the Steppes*. New York: Crown, 1964.

Johns, Jeremy, and Emilie Savage-Smith. "*The Book of Curiosities*: A Newly Discovered Series of Islamic Maps." *IM* 55 (2003): 7–24.

Jones, Alexander. "Ptolemy's *Geography*: Mapmaking and the Scientific Enterprise." In *AP*, 109–28.

Jones, J. R. Melville. *The Siege of Constantinople 1453: Seven Contemporary Accounts*. Amsterdam: Hakkert, 1972.

Jung, C. G. *Mandala Symbolism*. Bollingen Series. Princeton, NJ: Princeton University Press, 1959. Reprint, 1969.

Kadoi, Yuka. "On the Timurid Flag." In *Beiträge zur Islamischen Kunst und Archäologie*, edited by Markus Ritter and Lorenz Korn, 2:143–62. Wiesbaden, Ger.: Dr. Ludwig Reichert Verlag, 2010.

Kafadar, Cemal. *Between Two Worlds: The Construction of the Ottoman State*. Berkeley: University of California Press, 1995.

Kafescioğlu, Çigdem. *Constantinopolis/Istanbul: Cultural Encounter, Imperial Vision, and the Construction of the Ottoman Capital*. University Park, PA: Penn State Press, 2009.

Kahalaoui, Tarek. "Towards Reconstructing the Muqaddima following Ibn Khaldun's Reading of the Idrisian World Map." In "The Worlds of Ibn Khaldun," special issue, *Journal of North African Studies* 13, no. 3 (2008): 293–306.

Kamal, Youssouf. *Monumenta Cartographica Africae et Aegypti*. 5 vols. In 16 pts. Cairo: private edition (printed and bound by E. J. Brill), 1926–1951. Facsimile reprint. 6 vols. Edited by Fuat Sezgin. Frankfurt: Institut für Geschichte der Arabisch-Islamischen Wissenschaften, 1987.

Kaplony, Andreas. "Appendix: List of Geographical Nomenclature in Al-Kāshgarī's Text and Map." In *JMISR*, 209–25.

———. "Comparing al-Kāshgarī's Map to His Text: On the Visual Language, Purpose, and Transmission of Arabic-Islamic Maps." In *JMISR*, 137–53.

———. "Die fünf Teile Europas der arabischen Geographen. Die Berichte von Ibn Rusta, Ibn Ḥawqal und Abū Ḥāmid al-Ġarnāṭī." *Archiv orientální* 71 (2003): 485–98.

———. "Ist Europa eine Insel? Europa auf der rechteckigen Weltkarte des arabischen, 'Book of Curiosities' (Kitāb Ġarā'ib al-funūn)." In *Europa im Weltbild des Mittelalters: Kartographische Konzepte*, 143–56. Berlin: Akademie Verlag GmbH, 2008.

———. "Das Verkehrsnetz Zentralasiens: Die Raumgliederung der arabischen Geographen al-Muqaddasī und Ibn Ḥawqal." In *Strassen- und Verkehrswesen im hohen und späten Mittelalter*, edited by Rainer Christoph Schwinges, 353–64. Sigmaringen, Ger.: Thorbecke, 2007.

Karamustafa, Ahmet. "Cosmographical Diagrams," *HC 2.1*, 71–89.

———. "Introduction to Islamic Maps." In *HC 2.1*, 3–11.

———. "Maps and Mapmaking: Islamic Terrestrial Maps." In *Encyclopedia of the History of Science, Technology, and Medicine in Non-Western Cultures*, 1303–7. 2nd ed. Berlin: Springer-Verlag, 2008.

———. "Military, Administrative, and Scholarly Maps and Plans." In *HC 2.1*, 209–27.

Karatay, Fehmi Edhem. *Topkapı Saray Müzesi Kütüphanesi: Arpaça yazmalar katalogu*. 3 vols. Istanbul: Topkapı Saray Müzesi, 1962–1966.

———. *Topkapı Saray Müzesi Kütüphanesi: Farsça yazmalar katalogu*. 2 vols. Istanbul: Topkapı Saray Müzesi, 1961.

———. *Topkapı Saray Müzesi Kütüphanesi: Türkçe yazmalar katalogu*. 2 vols. Istanbul: Topkapı Saray Müzesi, 1961.

Kashani-Sabet, Firoozeh. *Frontier Fictions: Shaping the Iranian Nation, 1804–1946*. Princeton, NJ: Princeton University Press, 2000.

Keightly, David N. *The Ancestral Landscape: Time, Space, and Community in Late Shang China (ca. 1200–1045 BC)*. Berkeley: University of California Press, 2000.

Kennedy, Philip F., ed. *On Fiction and "Adab" in Medieval Arabic Literature*. Wiesbaden, Ger.: Harrassowitz, 2005.

Khalidi, Tarif. *Arabic Historical Thought in the Classical Period*. Cambridge: Cambridge University Press, 1994.

———. *Islamic Historiography: The Histories of Mas'udi*. Albany: State University Press, 1975.

Kimble, George H. T. *Geography in the Middle Ages*. London: Metheun, 1938.

King, David. "Architecture and Astronomy: The Ventilators of Medieval Cairo and Their Secret Secrets." *Journal of the American Oriental Society* 104 (1984): 97–133.

———. *Islamic Astronomy and Geography*. Aldershot: Ashgate-Variorum, 2012.

———. "The Sacred Direction in Islam: A Study of the Interaction of Religion and Science in the Middle Ages." *Interdisciplinary Science Reviews* 10 (1985): 315–28.

———. "Science in the Service of Religion: The Case of Islam." *Impact of Science on Society*, (UNESCO) 159 (1991): 245–62.

———. *Some Medieval Qibla Maps: Examples of Tradition and Innovation in Islamic Science*. Preprint Series no. 11. Frankfurt: Johann Wolfgang Goethe Universität, Institut für Geschichte der Naturwissenschaften, 1989.

———. "Two Iranian World Maps for Finding the Direction and Distance to Mecca." *IM* 49 (1997): 62–82.

———. *World-Maps for Finding the Direction and Distance to Mecca: Innovation and Tradition in Islamic Science*. Leiden: Brill, 1999.

King, David, and Richard P. Lorch. "Qibla Charts, Qibla Maps, and Related Instruments," In *HC 2.1*, 189–205.

King, Geoff. *Mapping Reality: An Exploration of Cultural Cartographies*. New York: St. Martin's Press, 1996.

Kline, Naomi Reed. *Maps of Medieval Thought: The Hereford Paradigm*. Rochester, NY: Boydell Press, 2001.

Kloetzli, W. Randolph. "Buddhist Cosmology." In *The Encyclopaedia of Religion*, edited by Mircea Eliade, 114. New York: Macmillan, 1987.

Knapp, Bernard A. *The History and Culture of Ancient Western Asia and Egypt*. Dorsey Press, 1988.

Kosslyn, Stephen Michael. *Image and Mind*. Cambridge, MA: Harvard University Press, 1980.
Kowalska, M. "The Sources of al-Qazwīnī's *Āthār al-Bilād*." In *Folia Orientalia*, 8 (1966): 41-88.
Krachkovsky, I. J. *Arabskaya geograficheskaya literatura*. Moscow / Leningrad: Akademii Nauk SSSR, 1957.
Kramer, Baerbel, and C. Gallazzi. "Artemidor im Zeichensaal. Eine Papyrusrolle mit Text, Landkarte und skizzenbüchern aus späthellenistischer Zeit." *Archiv fuer Papyrusforschung* 44, no. 22 (1998): 189–208.
Kramers, J. H. *Analecta Orientalia: Posthumous Writings and Selected Minor Works of J. H. Kramers*. 2 vols. Leiden: E. J. Brill, 1954–1956.
———. "Geography and Commerce." In *The Legacy of Islam*, edited by Thomas Arnold and Alfred Guillaume, 78–107. Oxford: Oxford University Press, 1931.
———. "L'influence de la tradition iranienne dans la géographie arabe." In Kramers, *Analecta Orientalia*, 147–56.
———. "La question Balhī-Istahrī-Ibn Hawkal et l'Atlas de l'Islam." *Acta Orientalia* 10 (1932): 9–30.
Krauss, Rosalind. *The Optical Unconscious*. Cambridge, MA: MIT Press, 1993.
Kropp, Manfred. "*Kitāb al-bad' wa-t-ta'riḫ* von Abü l-Ḥasan 'Alï ibn Aḥmad ibn 'Alï ibn Aḥmad Aš-Šāwï al-Fāsï und sein Verhältnis zu dem *Kitāb al-ğa'rāfiyya'* von az-Zuhrī." In *Proceedings of the Ninth Congress of the Union Européenne des Arabisants et Islamisants*, edited by Rudolph Peters, 153–68. Leiden: E. J. Brill, 1981.
———. "*Kitāb ğuġrāfiyā* des Ibn Fāṭima: Eine unbekannte Quelle des Ibn Sa'īd oder 'Neues' von al-Idrīsī?" In *Un Ricordo che non si Spegne*, 163–79. Napoli: Istituto Universitario Orientale di Napoli, 1955.
Kühnel, Bianca. *From the Earthly to the Heavenly Jerusalem: Representations of the Holy City in Christian Art of the First Millennium*. Freiburg, Ger.: Herder, 1987.
———, ed. *Real and Ideal Jerusalem in Jewish, Christian, and Islamic Art*. Jerusalem: Center of Jewish Art, 1998.
Kuntizsch, Paul. "Celestial Maps and Illustrations in Arabic-Islamic Astronomy." In *JMISR*, 175–80.
Kuhn, Thomas S. *The Structure of Scientific Revolutions*. Chicago: University of Chicago Press, 1962.
Kuhns, Richard: *Psychoanalytic Theory of Art: A Philosophy of Art on Developmental Principles*. New York: Columbia University Press, 1983.
Lacoste, Yves. "Les objets géographiques." In *CFT*, 16–23.
Lanman, J. T. "The Religious Symbolism of the T in T-O Maps," *Cartographica* 18 (1981), 18–22.
Lapidus, Ira M. *A History of Islamic Societies*. Cambridge: Cambridge University Press, 1988.
Lassner, Jacob. *The Shaping of 'Abbāsid Rule*. Princeton, NJ: Princeton University Press, 1980.
———. *The Topography of Baghdad in the Early Middle Ages*. Detroit: Wayne State University Press, 1970.
Ļaviņš, Imants. "Al Bīrūnī's ''Kitāb Al Qānūn al-Mas'ūdī' as the Database for the World Map Reconstruction." In *Proceedings of 5th International Workshop on Digital Approaches in Cartographic Heritage*, 272–85. Vienna: Vienna University of Technology (2010).
———. "Cosmographic Vision in Nizāmī's 'Haft Paykar.'" *Materials of International Sympo-*

sium "Mythological Thinking, Folklore and Literary Discourse: European and Caucasian Experience." Tbilisi: Shota Rustaveli Institute of Georgian Literature, 2012, 65–75.

———. "Islamic Geographical Tables and Their Place in Cartographic Heritage." In *Proceedings of the 8th International Workshop on Digital Approaches in Cartographic Heritage*. Rome: ICA, Societa Geographica Italiana, 2013.

Lawrence, Bruce B. "Al-Biruni: Against the Grain." *Critical Muslim* 12 (October–December 2014): 61–71.

Laxton, Paul, ed. *The New Nature of Maps: Essays in the History of Cartography*. Baltimore: Johns Hopkins University Press, 2002.

Ledyard, Gari. "Cartography in Korea." In *HC* 2.2, 235–345.

Lefebvre, Henri. *The Production of Space*. Translated by Donald Nicholson-Smith. Oxford: Blackwell, 1992.

Le Goff, Jacques. *The Medieval Imagination*. Translated by Arthur Goldhammer. Chicago: University of Chicago Press, 1988.

Leitner, Wilhelm. "Abu Ga'far Muhammad Ibn Musa al-Khwarazmi, His "Map of the World" as Intellectual Bridge between Orient and Occident." In *Proceedings of the 27th International Geographical Congress*, 20–26. Commission on the History of Geographical Thought, Mary Washington College, Fredericksburg, Virginia, August 4–7, 1992.

Lelewel, Joachim. *Géographie du Moyen Age*. 4 vols. Brussels: J. Piliet, 1852–1857.

Lentz, Thomas W., and Glenn D. Lowry. *Timur and the Princely Vision: Persian Art and Culture in the Fifteenth Century*. Los Angeles: Los Angeles County Museum of Art and Arthur M. Sackler Gallery; Washington, DC: Smithsonian Institution, 1989.

Lellouch, Benjamin. "Ibn Zunbul." In *Historians of the Ottoman Empire*, edited by C. Kafadar and C. Flescher. https://ottomanhistorians.uchicago.edu/en.

———. "Ibn Zunbul, un Égyptien face àl'universalisme ottoman (siezième siècle)." *Studia Islamica* 79 (1994): 143–55.

Le Strange, Guy. *The Lands of the Eastern Caliphate: Mesopotamia, Persia, Central Asia from the Moslem Conquest to the Time of Timur*. London: Cambridge University Press, 1930.

———. *Palestine under the Moslems: A Description of Syria and the Holy Land from A.D. 650 to 1500*. London: Committee of the Palestine Exploration Fund, 1890.

Lestringant, Frank. "Suivre la guide." In *CFT*, 424–35.

Levtzion, Nehemia, and J. F. P. Hopkins. *Corpus of Early Arabic Sources for West African History*. Cambridge: Cambridge University Press, 1981.

Levtzion, Nehemia, and Randall L. Powels. *The History of Islam in Africa*. Athens: Ohio University Press, 2000.

Lewis, Archibald. "Maritime Skills in the Indian Ocean, 1368–1500." *Journal of the Economic and Social History of the Orient* 16 (1972): 238–64.

Lewis, Bernard. *Race and Color in Islam*. New York: Harper and Row, 1970. Reprint, 1971.

———. *Race and Slavery in the Middle East*. Oxford: Oxford University Press, 1990.

Lewis, Mark Edward. *The Construction of Space in Early China*. Albany: State University of New York Press, 2006.

Lewis, Martin W., and Kären E. Wigen. *The Myth of Continents: A Critique of Metageography*. Berkeley: University of California Press, 1997.

Lilley, Keith. *City and Cosmos: The Medieval World in Urban Form.* London: Reaktion Books, 2009.

———, ed. *Mapping Medieval Geographies.* Cambridge: Cambridge University Press, 2013.

Loewe, Michael. "Knowledge of Other Cultures in China's Early Empires." In *GE*, 74–88.

Lohuizen-Mulder, Marc von. "Frescoes in the Muslim Residence and Bathhouse Qusayr ʿAmra: Representations, Some of the Dionysiac Cycle, Made by Christian Painters from Egypt." *BaBesch* 73 (1998): 133.

Lowry, Heath. *The Nature of the Early Ottoman State.* Albany: State University of New York Press, 2003.

Lozovsky, Natalia. "The Uses of Classical History and Geography in Medieval St. Gall." In *Mapping Medieval Geographies*, edited by Keith Lilley, 65–82. Cambridge: Cambridge University Press, 2013.

Lukens-Swietochowski, Marie. "The School of Herat from 1450–1506." In *ABC*, 179–214.

Manz, Beatrice. "Tamerlane and the Symbolism of Sovereignty." *Iranian Studies* 21, no. 1–2 (1988): 105–22.

Marquart, J. "Über das Volkstum der Komanen." In *Osttürkische Dialektstudien*, by W. Bang and J. Marquart, 13, no. 1 (1914): 25–238.

Marsand, Antonio. *I Manoscritti Italiani della Regia Biblioteca Parigina.* 2 vols. Paris: Dalla Stamperia Reale, 1835–1838.

Maspero, G. *Histoire ancienne des peuples de l'Orient classique.* 3 vols. Paris, 1895.

McIntosh, Greg. *The Piri Reis Map of 1513.* Athens: University of Georgia Press, 2000.

Mecquenem, R. de. "The Achaemenid and Later Remains at Susa." In *Persian Art*, 10:321–23.

Mercier, Raymond. "Astronomical Tables in the Twelfth Century." In *Adelard of Bath: An English Scientist and Arabist of the Early Twelfth Century*, edited by Charles Burnett, 87–118. London: The Warburg Institute, 1987.

———. "Geodesy." In *HC 2.1*, 175–88.

———. "The Meridians of Reference of Indian Astronomical Canons." In *History of Oriental Astronomy*, edited by G. Swarup, A. K. Bag, and K. S. Shukla, 97–107. Cambridge: Cambridge University Press, 1987.

Michalowski, Piotr. "Masters of the Four Corners of the Heavens: View of the Universe in Early Mesopotamian Writings." In *GE*, 147–68.

Micheau, Françoise. "The Scientific Institutions in the Medieval Near East." In *Encyclopedia of the History of Arabic Science*, edited by Rashed Roshdi, 985–1007. London: Routledge, 1996.

Mignolo, Walter D. *The Darker Side of the Renaissance: Literacy, Territoriality, and Colonization.* Ann Arbor: University of Michigan Press, 1995.

Millard, A. R. "Cartography in the Ancient Near East." In *HC 1*, 107–16.

Miller, Konrad. *Mappae Arabicae: Arabische Welt- und Länderkarten des 9–13. Jahrunderts.* 6 vols. Stuttgart, 1926–1931. Facsimile reprint, 2 vols. Edited by Fuat Sezgin. Frankfurt: Institut für Geschichte der Arabisch-Islamischen Wissenschaften, 1994.

Milstein, Rachel. "Futuh-i Haramayn: Sixteenth-Century Illustrations of the Hajj Route." In *Mamluks and Ottoman: Studies in Honour of Michael Winter*, edited by David J. Wasserstein and Ami Ayalon, 166–94. London: Routledge, 2006.

Minkowski, Christopher. "Where the Black Antelope Roam: Dharma and Human Geography in India." In *GE*, 9–31.
Minorsky, V. *Chester Beatty Library: A Catalogue of the Turkish Manuscripts and Miniatures.* Dublin: Hodges Figgis, 1958.
———. "A False Jayhani." *Bulletin of the School of Oriental and African Studies* 13 (1949–1951): 89–96.
Miquel, André. "Cartographes Arabes." In *CFT*, 55–60.
———. *La géographie humaine du monde musulman jusqu'au milieu du 11e siècle.* 4 vols. Paris: Mouton, 1967–1988.
Miri, Negin. *Sassanian Pārs: Historical Geography and Administrative Organization.* Costa Mesa, CA: Mazda, 2012.
Mitchell, W. J. T., ed. *Iconology: Image, Text, Ideology.* Chicago: University of Chicago Press, 1986.
———. *The Language of Images.* Chicago: University of Chicago Press, 1980.
———. *Picture Theory: Essays on Verbal and Visual Representation.* Chicago: University of Chicago Press, 1994.
Moers, Gerald. "The World and the Geography of Otherness in Pharaonic Egypt." In *GE*, 169–81.
Monmonier, Mark. *Air Apparent: How Meteorologists Learned to Map, Predict, and Dramatize Weather.* Chicago: University of Chicago Press, 1999.
———. *Drawing the Line: Tales of Maps and Cartocontroversy.* New York: H. Holt, 1995.
———. *From Squaw Tit to Whorehouse Meadow: How Maps Name, Claim, and Inflame.* Chicago: University of Chicago Press, 2007.
———. *How to Lie with Maps.* Chicago: University of Chicago Press, 1991.
———. *No Dig, No Fly, No Go: How Maps Restrict and Control.* Chicago: University of Chicago Press, 2010.
Morewedge, Parviz, ed. *Neoplatonism and Islamic Thought.* Albany: State University of New York Press, 1992.
Mottahedeh, Roy P. *Loyalty and Leadership in an Early Islamic Society.* Princeton, NJ: Princeton University Press, 1980.
Muhanna, Elias. "Why Was the Fourteenth Century a Century of Arabic Encyclopaedism?" In *Encyclopaedism from Antiquity to the Renaissance*, edited by Jason König and Greg Woolf, 343–56. Cambridge: Cambridge University Press, 2013.
Mu'nis, Husayn. *Tarīkh al-jughrāfiya wa-l-jughrāfiyyin.* Madrid: Institute of Islamic Studies, 1967.
Mžik, Hans von, ed. *Beitrage zur historischen Geographie, Kulturgeographie, Ethnographie und Kartographie, vornehmlich des Orients.* Leipzig: Franz Deuticke, 1929.
———. *Al-Iṣṭaḫrī und seine Landkarten im Buch "Ṣuwar al-aḳālīm," nach der pers. Handscrift Cod. Mixt. 344 der Österreichischen Nationalbibliothek.* Vienna: Georg Prachner, 1965.
———. "Ptolemaeus und die Karten der Arabischen Geographen." *Mitteilungen der Kaiserlich-Königlichen Geographischen Gesellschaft in Wien* 58 (1915): 152–76.
Nasr, Syed Hossein. *Islamic Science: An Illustrated Study.* Istanbul: Insan Yayinlari, 1989.
Naumann, Rudolf. "Takht-i Suleiman and Zindan-i-Suleiman." In *Persian Art*, 14:3050–60.

Nawata, Hiroshi. "Historical Scoio-economic Relationships between the Rashāyda and the Beja in the Eastern Sudan: The Production of Racing Camels and Trade Networks across the Red Sea." *Senri Ethnological Studies* 69 (2005): 187–213.

Nazmi, Ahmad. *The Muslim Geographical Image of the World in the Middle Ages: A Source Study.* Warsaw: Wydawnictwo Akademickie DIALOG, 2007.

Nebes, Norbert, Tilman Seidensticker, Hans-Caspar Graf von Bothmer, Karin Rührdanz, and Gottfried Hagen. *Orientalische Buchkunst in Gotha.* Ausstellung Zum 350 Jahrigen Jubliaum Der Forschungs- Und Landesbibliothek Gotha. Gotha, Ger.: Druck und Verpackung, 1997.

Necipoğlu, Gülru. *Architecture, Ceremonial, and Power: The Topkapı Palace in the Fifteenth and Sixteenth Centuries.* New York: The Architectural History Foundation, 1991.

———. "Plans and Models in 15th- and 16th-Century Ottoman Architectural Practice." *Journal of the Society of Architectural Historians* 45 (1986): 224–43.

———. "Süleyman the Magnificent and the Representation of Power in the Context of Ottoman-Hapsburg-Papal Rivalry." *Art Bulletin* 71 (1989): 401–27.

———. "Visual Cosmopolitanism and Creative Translation: Artistic Conversations with Renaissance Italy in Mehmed II's Constantinople." *Muqarnas* 29 (2012): 1–81.

Needham, Joseph. *The Shorter Science and Civilization in China: An Abridgement of Joseph Needham's Original Text.* 3 vols. Edited by Colin A. Ronan. Cambridge: Cambridge University Press, 1981.

Nees, Lawrence. "Blue behind Gold: The Inscriptions of the Dome of the Rock and Its Relatives." In *And Diverse Are Their Hues: Color in Islamic Art and Culture,* edited by Jonathan Bloom and Sheila Blair, 152–73. New Haven, CT: Yale University Press, 2011.

Netton, Ian. *Muslim Neoplatonists: An Introduction to the Thought of the Brethern of Purity (Ikhwān al-Ṣafā').* Oxon, UK: RoutledgeCurzon, 2002.

Nicolet, Claude. *Space, Geography, and Politics in the Early Roman Empire.* Ann Arbor: University of Michigan Press, 1991.

Nuti, Lucia. "Mapping Places: Chorography and Vision in the Renaissance." In *Mappings,* edited by Denis Cosgrove, 90–108. London: Reaktion Books, 1999.

O'Connor, David, and Stephen Quirke, eds. *Mysterious Lands.* London: Institute of Archaeology, University College, 2003.

O'Leary, De Lacy. *How Greek Science Passed to the Arabs.* 1949. Reprint, London: Routledge and Kegan Paul, 1964.

Olsson, J. T. "The World in Arab Eyes: A Reassessment of the Climes in Medieval Islamic Scholarship." *Bulletin of the School of Oriental and African Studies* 77 (2014): 487–508.

Oppenheim, A. Leo. *Ancient Mesopotamia: Portrait of a Dead Civilization.* Edited by Erica Reiner. 1964. Rev. ed., University of Chicago Press, 967.

Öz, Tahsin. *Fâtih Sultan Mehmet II'ye Ait Eserler.* Ankara: Türk Tarih Kurumu Basimevi, 1953.

Özgen, Engin, and Ilknur Özgen, eds. *Antalya Museum.* Ankara: T. C. Kültür Bakanligi, 1988. Rev., 1992.

Padron, Ricardo. *The Spacious Word: Cartography, Literature, and Empire in Early Modern Spain.* Chicago: University of Chicago Press, 2004.

Palmisano, Antonio. *Ethnicity: The Beja as Representation.* Berlin: Freie Universität, 1991.

Pamuk, Orhan. *My Name Is Red*. Translated by Erdağ M. Göknar. New York: Alfred A. Knopf, 2001. Reprint, New York: Vintage, 2002.

Pancaroğlu, Oya. "Concepts of Image and Boundary in a Medieval Persian Cosmography." *RES* 43 (2003): 31–41.

Panofsky, Erwin. *Meaning in the Visual Arts: Papers in and on Art History*. New York: Doubleday Anchor Books, 1955.

———. *Perspective as Symbolic Form*. New York: Zone Books, 1997.

———. *Studies in Iconology: Humanistic Themes in the Art of the Renaissance*. New York: Oxford University Press, 1939. Reprint, New York: Harper Torch Books, 1962.

Park, Hyunhee. *Mapping the Chinese and Islamic Worlds: Cross-Cultural Exchange in Premodern Asia*. New York: Cambridge University Press, 2012.

Patai, R. *Man and Temple*. London, 1967.

Paul, Andrew. *A History of the Beja Tribes of the Sudan*. Cambridge: Cambridge University Press, 1954.

Pedersen, Johannes. *The Arabic Book*. Translated by Geoffrey French. Princeton, NJ: Princeton Unviersity Press, 1984.

Pedley, Mary. *The Commerce of Cartography: Making and Marketing Maps in Eighteenth-Century France and England*. Chicago: University of Chicago Press, 2005.

Petto, Christine Marie. *When France Was King of Cartography*. Lanham, MD: Lexington Books, 2007.

Picard, Christophe. "La Mèditerranèe Musselmane, Un Hèritage Omeyyade." In *Umayyad Legacies: Medieval Memories from Syria to Spain*, edited by Antoine Borrut and Paul M. Cobb, 365–402. Leiden: E. J. Brill, 2010.

———. *L'océan atlantique musulman. De la conquête arabe à l'époque almohade. Navigation et mise en valeur des côtes d'al-Andalus et du Maghreb occidental (Portugal-Espagne-Maroc)*. Paris: Maisonneuve & Larose / Éditions UNESCO, 1997.

Pidal, Gonzalo Menendez. "Mozarabes y asturianos en la cultura de la alta edad media." *Boletin de la Real Academia de la Historia* 134 (1954): 169–71.

Pinder-Wilson, R. H. "An Islamic Ewer in Sassanian Style." In *Persian Art* 14:3061–63.

Pingree, David. *The Thousands of Abū Ma'shar*. London: The Warburg Institute, 1968.

Pinna, M. *Il Mediterraneo e la Sardegna nella Cartografia Musulmana*. 2 vols. Nuoro, Sardegna, Italy: Istituto Superiore, 1996.

Pinto, Karen. "Capturing Imagination: The Buja and Medieval Islamic Mappa Mundi." In *Views from the Edge: Essays in Honor of Richard W. Bulliet*, edited by Neguin Yavari, Lawrence G. Potter, and Jean-Marc Oppenheim, 154–83. New York: Columbia University Press for The Middle East Institute, Columbia University, 2004.

———. "The Maps Are the Message: Meḥmet II's Patronage of an 'Ottoman Cluster.'" *IM* 63, no. 2 (2011): 155–79.

———. "Middle East." In *Encyclopedia of the Modern Middle East*, edited by Reeva S. Simon, Philip Mattar, and Richard W. Bulliet, 1522–23. New York: Columbia University Press, 1996.

———. "Passion and Conflict: Medieval Islamic Views of the West." In *Mapping Medieval Geographies: Geographical Encounters in the Latin West and Beyond, 300–1600*, edited by Keith Lilley, 201–24. Cambridge: Cambridge University Press, 2013.

———. "*Searchin' his eyes, lookin' for traces*: Piri Reis' World Map of 1513 & Its Islamic Iconographic Connections (A Reading Through Bağdat 334 and Proust)." *Osmanlı Aratışrmaları/Journal of Ottoman Studies* 39, no. 1 (2012): 63–94.

———. "Surat Bahr al-Rum [Picture of the Sea of Byzantium]: Possible Meanings Underlying the Forms." In *Eastern Mediterranean Cartographies: The Cartography of the Mediterranean World*, edited by George Tolias and Dimitris Loupis, 234–41. Athens: Institute for Neohellenic Research, 2004.

Plofker, Kim. "Humans, Demons, Gods, and Their Worlds: The Sacred and Scientific Cosmologies of India." In *GE*, 32–42.

Plumey, J. M. "The Cosmology of Ancient Egypt." In Blacker and Lowe, *Ancient Cosmologies*, 17–41.

Polaschek, Erich. "Ptolemy's *Geography* in a New Light." *IM* 14 (1959): 17–37.

Polcaro, Andrea. "The Tuleilat al-Ghassul Star Painting: A Hypothesis Regarding a Solar Calendar from the Fourth Millennium B.C." In *Time and History in the Ancient Near East*, edited by L. Feliu, J. Llop, A. Millet Albà, and J. Sanmartín, 273–84. Winona Lake, IN: Eisenbrauns, 2013.

Pope, Arthur Upham. "The Relation between Geography and Art in Iran." In *Persian Art* 1: 106–28.

Porada, Edith. *The Art of Ancient Iran: Pre-Islamic Cultures*. New York: Greystone Press, 1969.

Rabbat, Nasser. *Mamluk History through Architecture: Monuments, Culture, and Politics in Medieval Egypt and Syria*. New York: I. B. Tauris, 2010.

Raby, Julian. "Mehmed II Fatih and the Fatih Album." In *Between China and Iran: Paintings from Four Istanbul Albums*, edited by Ernst Grube and Eleanor Simms, 42–68. London: Percival David Foundation, 1985.

———. "Pride and Prejudice: Mehmed the Conqueror and the Portrait Medal." In *Italian Medals*, edited by J. G. Pollard, 171–94. Washington, DC: National Gallery of Art, 1987.

———. "A Sultan of Paradox: Mahmed the Conqueror as a Patron of the Arts." *Oxford Art Journal* 5, no. 1 (1982): 3–8.

Raby, Julian, and Teresa Fitzherbert, eds. *The Court of the Il-khans, 1290–1340*. Studies in Islamic Art. London: Oxford University Press, 1996.

Raby, Julian, and Zeren Tanındı. *Turkish Bookbinding in the 15th Century: The Foundation of an Ottoman Court Style*. London: Azimuth Editions, 1993.

Ramaswamy, Sumathi. "Catastrophic Cartographies: Mapping the Lost Continent of Lemuria." *Representations* 67 (Summer 1999): 92–129.

———. "Conceit of the Globe in Mughal Visual Practice." *Comparative Studies in Society and History* 49, no. 4 (2007): 751–82.

———. *The Goddess and the Nation: Mapping Mother India*. Durham, NC: Duke University Press, 2010.

———. *The Lost Land of Lemuria: Fabulous Geographies, Catastrophic Histories, 1864–1981*. Berkeley: University of California, 2004.

Rapoport, Yossef. "The *Book of Curiosities*: A Medieval Islamic View of the East." In *JMISR*, 155–71.

———. "Reflections of Fatimid Power in the Maps of Island cities in the 'Book of Curiosities.'"

In *Herrschaft verorten: Politische Kartographie des Mittelalters und der frühen Neuzeit*, edited by Martina Stercken and Ingrid Baumgärtner, 183–210. Zürich: Chronos, 2012.

———. "The View from the South: The Maps of the *Book of Curiosities* and the Commercial Revolution of the Eleventh Century." In *Histories of the Middle East: Studies in Middle Eastern Society, Economy, and Law in Honor of A. L. Udovitch*, edited by R. Margariti, A. Sabra, and P. Sijpesteijn, 183–212. Leiden: Brill, 2011.

Rapoport, Yossef, and Emilie Savage-Smith. "Medieval Islamic View of the Cosmos: The Newly Discovered *Book of Curiosities*." *Cartographic Journal*, 41 (2004): 253–59.

———. "The *Book of Curiosities* and a Unique Map of the World." In *CAMA*, 121–38.

Reuther, Oscar. "Parthian Architecture." In *Persian Art*, 1:411–578.

Rice, David Talbot. *Islamic Art*. Singapore: Thames and Hudson, 1993.

Richer, Jean. *Sacred Geography of the Ancient Greeks: Astrological Symbolism in Art, Architecture, and Landscape*. Translated by Christine Rhone. Albany: State University of New York Press, 1994.

Ringbom, Lars Ivar. "Three Sasanian Bronze Salvers with Paridaeza Motifs." In *Persian Art*, 14:3029–41.

Riviére, Jean-Loup. "La carte, le corps, la mémoire." In *CFT*, 83–91.

Roberts, Sean. "Poet and 'World Painter': Francesco Berlinghieri's *Geographia* (1482)." *IM* 62, no. 2 (2010): 145–60.

Robinson, B. W. *Fifteenth-Century Persian Painting: Problems and Issues*. New York: New York University Press, 1991.

———. "The Turkman School to 1503." In *ABC*, 214–47.

Robinson, B. W., E. J. Grube, G. M. Meredith-Owens, and R. Skelton, eds. *The Keir Collection: Islamic Painting and the Arts of the Book*. London: Faber and Faber, 1976.

Rogers, J. M. "Itineraries and Town Views in Ottoman Histories." In *HC 2.1*, 228–55.

———. "Two Masterpieces from 'Süleyman the Magnificent'—A Loan Exhibition from Turkey at the British Museum," *Orientations* 19, no. 8 (August 1988): 12–17.

Romm, James S. *The Edges of the Earth in Ancient Thought: Geography, Exploration, and Fiction*. Princeton, NJ: Princeton University Press, 1992.

Rosenthal, Franz. *Knowledge Triumphant: The Concept of Knowledge in Medieval Islam*. Leiden: E. J. Brill, 1970.

Roshdi, Rashed, ed. *Encyclopedia of the History of Arabic Science*. 3 vols. London: Routledge, 1996.

Roxburgh, David J. *The Persian Album, 1400–1600: From Dispersal to Collection*. New Haven, CT: Yale University Press, 2005.

———. "Pilgrimage City." In *The City in the Islamic World*, 2 vols., edited by Salma K. Jayyusi, 2:753–74. Leiden: Brill, 2008.

Rushdie, Salman. *Haroun and the Sea of Stories*. London: Granta Books, 1990.

Safier, Neil. *Measuring the New World: Enlightenment Science and South America*. Chicago: University of Chicago Press, 2008.

Saliba, George. *Islamic Science and the Making of the European Renaissance*. Cambridge, MA: MIT Press, 2007.

Salinari, Marina Emiliani. "An Atlas of the 15th Century Preserved in the Library of the Former Serail in Constantinople." *IM* 8 (1951): 101–2.

Salway, Benet. "Putting the World in Order: Mapping in Roman Texts." In *AP*, 193–234.
Sarna, N. M. *Understanding Genesis*. New York: Schoken Books, 1986.
Savage-Smith, Emilie. "The *Book of Curiosities*: An Eleventh-Century Egyptian View of the Land of the Infidels." In *GE*, 291–310.
———. "Celestial Mapping." In *HC 2.1*, 12–70.
———. *Islamicate Celestial Globes*. Washington, DC: Smithsonian Institution, 1984.
Saxl, Fritz. "The Zodiac of Quṣayr ʿAmra." Translated by Ruth Wind and Arthur Beer. *Early Muslim Architecture*, 2 vols., edited by K. A. C. Creswell, 1:289–303. Oxford: Clarendon Press, 1932.
Scafi, Alessandro. "Defining Mappaemundi." In *HWM*, 345–54. London: The British Library, 2006.
———. *Mapping Paradise: A History of Heaven on Earth*. Chicago: University of Chicago Press, 2006.
———. *Maps of Paradise*. London: The British Library, 2013.
Schama, Simon. "The Domestication of Majesty: Royal Family Portraiture." *Journal of Interdisciplinary History* 17, no. 1 (Summer 1986): 155–83.
———. *The Embarrassment of Riches: An Interpretation of Dutch Culture in the Golden Age*. New York: Alfred A. Knopf, 1987.
———. *Landscape and Memory*. New York: Vintage Books, 1995.
Schapiro, Meyer. *Late Antique, Early Christian and Medieval Art, Selected Papers*. New York: George Braziller, 1979.
Schick, Irvin Cemil. *Uğur Derman: 65 Yaş Armağanı*. Istanbul: Sabanci Üniversitesi, 2000.
Schmidt, K. "Göbekli Tepe, Southeastern Turkey: A Preliminary Report on the 1995–1999 Excavations." *Palèorient* 26, no. 1 (2000): 45–54.
———. "Göbekli Tepe—the Stone Age Sanctuaries: New Results of Ongoing Excavations with a Special Focus on Sculptures and High Reliefs." *Documenta Praehistorica* 37 (2010), 239–56.
Schmitz, Barbara, ed. *Islamic and Indian Manuscripts and Paintings in the Pierpont Morgan Library*. New York: The Pierpont Morgan Library, 1998.
Schröder, Stefan. "Kartographische Entwürfe iberischer Provenienz. Zu Raum- und Ordnungsvorstellungen auf der Iberischen Halbinsel in Karten des 9. Bis 12. Jahrhunderts." In *Von Mozarabern zu Mozarabismen*, edited by Matthias Maser, Klaus Herbers, Michele C. Ferrari, and Harmut Bobzin, 257–77. Münster: Aschendorf Verlag GmbH & Co. KG, 2014.
Schwartzberg, Joseph E. "Cosmographical Mapping." In *HC 2.1*, 332–87.
———. "Introduction to South Asian Cartography." In *HC 2.1*, 295–331.
Scott, James. "Perceptions of the World in the *Book of Jubilees*." In *GE*, 182–96.
Seed, Patricia. *Oxford Map Companion: One Hundred Sources in World History*. New York: Oxford University Press, 2014.
Segarizzi, Arn. "Jacopo Languschi, Rimatore Veneziano del Secolo XV." *Atti Della I. R. Accademia degli Agiati in Rovereto* 10–11 (1904): 179–85.
Sezgin, Fuat. *The Contribution of the Arabic-Islamic Geographers to the Formation of the World Map*. Frankfurt: Institut für Geschichte der Arabisch-Islamischen Wissenschaften an der Johann Wolfgang Goethe-Universität, 1987.
———. *Islamic Cartography*. 318 vols. Frankfurt: Institut für Geschichte der Arabisch-

Islamischen Wissenschaften an der Johann Wolfgang Goethe-Universität Frankfurt am Main, 2000–present.
———. *Islam Uygarliginda Astronomi Cografya ve Denizcili* [Astronomy, Geography, and Navigation in Islamic Civilization]. Istanbul: Boyut Yayıncılık, 2009.
———. *Mathematical Geography and Cartography in Islam and Their Continuation in the Occident*. 3 vols. Translated by Guy Moore and Geoff Sammon. Frankfurt: Institut für Geschichte der Arabisch-Islamischen Wissenschaften an der Johan Wolfgang Goethe-Universität, 2000–2007. Originally published as *Geschichte Des Arabischen Schrifttums: Mathematische Geographie und Kartographie im Islam und Ihr Fortleben im Abendland. Historische Darstellung* (Frankfurt: Institut für Geschichte der Arabisch-Islamischen Wissenschaften an der Johann Wolfgang Goethe-Universität Frankfurt am Main, 2000).
Shaban, M. A. *Islamic History: A New Interpretation*. 2 vols. Cambridge: Cambridge University Press, 1976.
Silverstein, Adam J. "Arabo-Islamic Geography." In *The Oxford Companion to World Exploration*, edited by David Buisseret, 59–62. Oxford: Oxford University Press, 2007.
———. "The Medieval Islamic Worldview: Arabic Geography in Its Historical Context." In *GE*, 273–90.
———. *Postal Systems in the Pre-modern Islamic World*. Cambridge: Cambridge University Press, 2007.
Simms, E. "The Turks and Illustrated Historical Texts." *ICTA* 5 (1978): 747–72.
Simpson, Mariana Shreve. *Arab and Persian Painting in the Fogg Art Museum*. Cambridge: Fogg Art Museum, 1980.
Simpson, William Kelly, ed. *The Literature of Ancient Egypt: An Anthology of Stories, Instructions, and Poetry*. New Haven, CT: Yale University Press, 1972.
Smith, Jonathan Z. *Map Is Not Territory: Studies in the History of Religions*. Leiden: E. J Brill, 1978. Reprint, Chicago: University of Chicago Press, 1993.
Smith, Richard J. *Mapping China and Managing the World: Culture, Cartography and Cosmology in Late Imperial Times*. Oxon, UK: Routledge, 2013.
Soja, Edward W. *Postmodern Geographies: The Reassertion of Space in Critical Theory*. London: Verso Press, 1989.
———. *Thirdspace: Journeys to Los Angeles and Other Real-and-Imagined Places*. Oxford: Basil Blackwell, 1996.
Soucek, Priscilla P. "The Arts of Calligraphy." In *ABC*, 7–34.
Soucek, Svat. *Piri Reis and Turkish Mapmaking after Columbus*. London: The Nour Foundation, 1996.
Sourdel, Dominique. *Medieval Islam*. Translated by J. Montgomery Watt. London: Routledge and Kegan Paul, 1979. Originally published as *L'islam medieval* (Paris: Presses Universitaires de France, 1979).
Spaulding, Jay. "Precolonial Islam in the Eastern Sudan." In *The History of Islam in Africa*, edited by Nehemia Levitzion and Randall L. Pouwels, 117–18. Athens: Ohio University Press, 2000.
"Special Issue on Islamic Heritage" In *Arts & the Islamic World* 4, no. 3 (Spring–Summer 1987): 50–59.
Spence, Jonathan D. *The Memory Palace of Matteo Ricci*. New York: Penguin Books, 1985.

Sprenger, Aloys. *Die Post- und Reiserouten des Orients*. Leipzig: F. A. Brockhaus, 1864.
Stchoukine, Ivan. "Un Manuscrit Illustré de la Bibliotheque de Mohammad II Fatih." *Arts Asiatiques* 19 (1969): 3–9.
———. "Miniatures Turques du temps de Mohammad II." *Arts Asiatiques* 15 (1967): 47–49.
———. *La peinture Turque d'après les manuscrits Illustrés*. 2 vols. Paris: Librairie orientaliste Paul Geuthner, 1966–1971.
Taeschner, Franz. "Die Geographische Literatur der Osmanen." *Zeitschrift der Deutschen Morgenländischen Gesellschaft* 77 (1923): 31–80.
Talbert, Richard. "Greek and Roman Mapping: Twenty-First Century Perspectives." In *CAMA*, 9–27.
———. "P.Artemid.: The Map." In *Images and Texts on the "Artemidorus Papyrus": Working Papers on P.Artemid. (St. John's College Oxford, 2008)*, edited by Kai Brodersen and Jaś Elsner, 57–64. Stuttgart: Franz Steiner Verlag, 2009.
———. *Rome's World: The Peutinger Map Reconsidered*. Cambridge: Cambridge University Press, 2010.
———. "Urbs Roma *to* Orbis Romanus: Roman Mapping on a Grand Scale." In *AP*, 163–91.
Taussig, Michael. *Mimesis and Alterity: A Particular History of the Senses*. New York: Routledge, 1993.
Thurman, Robert A. F. *Inside Tibetan Buddhism: Rituals and Symbols Revealed*. Edited by Barbara Roether. San Francisco, CA: Collins, 1995.
Tibbetts, Gerald R. *Arabic Texts Containing Material on South-East Asia*. Leiden: E. J. Brill, 1979.
———. *Arab Navigation in the Indian Ocean before the Coming of the Portuguese*. London: Royal Asiatic Society of Great Britain and Ireland, 1971. Reprint, 1981.
———. "Arab Navigation in the Red Sea." *Geographical Journal* 127, no. 3 (1961): 322–34.
———. "The Balkhī School of Geographers." *HC 2.1*, 108–36.
———. "The Beginnings of a Cartographic Tradition." *HC 2.1*, 90–107.
———. "Later Cartographic Developments." *HC 2.1*, 137–55.
Titley, Nora M. *Miniatures from Persian Manuscripts: A Catalogue and Subject Index of Paintings from Persia, India, and Turkey in the British Museum*. London: British Library, 1977.
———. *Miniatures from Turkish Manuscripts: A Catalogue and Subject Index of Paintings in the British Library and British Museum*. London: British Library, 1981.
———. *Persian Miniature Painting an its Influence on the Painting of Turkey and India*. London: British Library, 1983.
Togan, Zeki Velidi, ed. *Bīrūnī's Picture of the World*. Delhi: Archaeological Survey of India, 1941.
———. *On the Miniatures in Istanbul Libraries*. Publications of the Faculty of Letters of the University of Istanbul. Istanbul: Baha Matbaasi, 1963.
Tolmacheva, Marina. "Geography." In *Medieval Islamic Civilization: An Encyclopedia*, edited by Josef W. Meri, 1:284–88. New York: Routledge, 2006.
———. "The Medieval Arabic Geographers and the Beginnings of Modern Orientalism." *IJMES* 27, no. 2 (1995): 143–56.
———. "On the Arab System of Nautical Orientation." *Arabica* 27, no. 2 (1980): 180–92.
Tooley, R. V. *Maps and Map-Makers*. 6th ed. London: B. T. Batsford, 1978.
Touati, Houari. *Islam and Travel in the Middle Ages*. Translated by Lydia G. Cochrane. Chicago:

University of Chicago Press, 2010. Originally published as *Islam et voyage au Moyen Âge: Histoire et anthropologie d'une praqtique lettrée* (Paris: Éditions du Seuil, 2000).
———. Review of *An Eleventh-Century Egyptian Guide to the Universe: The Book of Curiosities*, edited and translated by Yossef Rapoport and Emilie Savage-Smith. *Studia Islamica* 109, no. 2 (2014): 344–47.
Toussaint, Auguste. *History of the Indian Ocean*. Translated by June Guicharnaud. London: Routledge and Kegan Paul, 1966. Originally published as *Histoire de l'océan Indien* (Paris: Universitaires de France, 1961).
Trimble, Jennifer. "Process and Transformation on the Severan Marble Plan of Rome." In *CAMA*, 67–97.
Trimingham, J. Spencer. *A History of Islam in West Africa*. Oxford: Oxford University Press, 1970.
Tuan, Yi-Fu. *Space and Place: The Perspective of Experience*. Minneapolis: University of Minnesota Press, 1977.
———. *Topophilia: A Study of Environmental Perception, Attitudes, and Values*. Englewood Cliffs, NJ: Prentice-Hall, 1974.
Türkay, Cevdet. *Istanbul Kütübhanelerinde Osmanli'lar Devrine Aid Türkçe—Arabca—Farsça Yazma Ve Basma Coğrafya Eserleri Bibliyoğrafyas*. Istanbul: Maarif Basimevi, 1958.
Turnbull, David. *Maps Are Territories: Science Is an Atlas*. Chicago: University of Chicago Press, 1993.
Unger, Eckhard. "From the Cosmos Picture to the World Map." *IM* 2 (1937): 1–7.
Unno, Kazutaka. "Cartography in Japan." In *HC* 2.2, 346–477.
Ünver, A. Süheyl. *Istanbul Risaleleri*. 6 vols. Istanbul: Büyüksehir Belediyesi Kültür Isleri Daire Baskanlıgı Yayınları, 1995.
Vanhove, Martine. "The Beja Language Today in Sudan: The State of the Art in Linguistics." In *Proceedings of the 7th International Sudan Studies Conference*, 1–15. University of Bergren, Norway, April 6–8, 2006. http://halshs.archives-ouvertes.fr/docs/00/06/52/11/PDF/Beja_State_of_the_Art_Bergen.pdf.
Vasaly, Ann. *Representations: Images of the World in Ciceronian Oratory*. Berkeley: University of California Press, 1993.
Verdet, Jean-Pierre. "Hiérarchies." In *CFT*, 76–82.
———. "La place de la terre." In *CFT*, 70–75.
Volk, Tyler. *Metapatterns across Space, Time, and Mind*. New York: Columbia University Press, 1995.
Von den Brincken, Anna-Dorothee. "Jerusalem on Medieval Mappaemundi." In *HWM*, 355–79.
von Folsach, Kjeld. *Art from the World of Islam in the David Collection*. Copenhagen: Davis Samling, 2001.
von Hees, Syrinx. "The Astonishing: A Critique and Re-reading of 'Ağā'ib Literature." *Middle Eastern Literatures* 8, no. 2 (July 2005): 101–20.
Vryonis, S. *The Decline of Medieval Hellenism in Asia Minor*. Berkley: University of California Press, 1971.
Wahl, François. "Le desir d'espace." In *CFT*, 41–46.
Ward, Rachel. *Islamic Metalwork*. New York: Thames and Hudson, 1993.
Warnock, Mary. *Imagination and Time*. Cambridge: Blackwell, 1994.

Warren, W. F. *The Earliest Cosmologies*. New York: Eaton & Mains, 1909.
Weitzmann, Kurt. *The Fresco Cycle of S. Maria di Castelsprio*. Princeton, NJ: Princeton University Press, 1951.
———. *Late Antique and Early Christian Book Illustration*. New York: 1977.
Welch, Stuart Cary. *Wonders of the Age: Masterpieces of Early Safavid Painting, 1501–1576*. Cambridge, MA: Fogg Art Museum, Harvard University, 1979.
Wendell, C. "Baghdad: Imago Mundi and Other Foundation Lore." *IJMES* 2 (1971): 99–128.
Wensinck, A. J. "The Ocean in the Literature of the Western Semites." *Afdeeling Letterkunde* 19, no. 2 (1918): 1–67.
Westrem, Scott. *The Hereford Map*. Turnhout, Belgium: Brepols, 2001.
Weyl, Hermann. *Symmetry*. Princeton, NJ: Princeton University Press, 1952. Reprint, 1980.
White, Hayden. *The Content of the Form: Narrative Discourse and Historical Representation*. Baltimore: Johns Hopkins University Press, 1987. Reprint, 1989.
———. *Figural Realism: Studies in the Mimesis Effect*. Baltimore: Johns Hopkins University Press, 1999.
Whitfield, Peter. *The Image of the World*. London: British Library, 1994.
Whittaker, C. R. "Mental Maps: Seeing Like a Roman." In *Thinking like a Lawyer: Essays on Legal History and General History for John Crook on His Eightieth Birthday*, edited by P. McKechnie, 81–112. Leiden: Brill, 2002.
———. *Rome and Its Frontiers: The Dynamics of Empire*. Oxon, UK: Routledge, 2004.
Whittington, Karl. *Body-Worlds: Opicinus de Canistris and the Medieval Cartographic Imagination*. Toronto: Pontifical Institute of Medieval Studies, 2014.
———. "The Psalter Map: A Case Study in Forming a Cartographic Canon for Art History." *Kunstlicht* 34, no. 4 (2013): 19–25.
Wilford, Francis. "An Essay on the Sacred Isles in the West, with Other Essays Connected with That Work." *Asiatick Researches* 8 (1808): 245–376.
Willcock, Malcolm M. *A Companion to the "Iliad."* Chicago: University of Chicago Press, 1976.
Williams, John. *The Illustrated Beatus*. 5 vols. London: Harvey Miller, 1994–2003.
———. "Isidore, Orosius, and the Beatus Map." *IM* 49 (1997), 7–32.
Winterson, Jeanette. *Sexing the Cherry*. New York: Vintage International, 1991.
Wintle, Michael. *The Image of Europe: Visualizing Europe in Cartography and Iconography throughout the Ages*. Cambridge: Cambridge University Press, 2009.
Witkam, Jan Just. "Arabic Manuscripts in Distress: The Frankfurt facsimile series." *Manuscripts of the Middle East* 4 (1989): 175–80.
Wittek, Paul. "De la défaite d'Ankara à la prise de Constantinople." *Revue des etudes islamiques* 12 (1938), 1–34.
———. *The Rise of the Ottoman Empire*. London: Royal Asiatic Society, 1938.
Woldering, Irmgard. *The Art of Egypt: The Time of the Pharoahs*. New York: Greystone Press, 1963.
Wood, Denis. *The Power of Maps*. New York: Guilford Press, 1992. Rev. ed., *Rethinking the Power of Maps*. New York: Guilford Press, 2010.
Woods, John. E. *The Aqquyunlu: Clan, Confederation, Empire; a Study in 15th/9th Century Turko-Iranian Politics*. Minneapolis: Bibliotheca Islamica, 1976.
Woodward, David, ed. *Art and Cartography*. Chicago: University of Chicago Press, 1987.

———. *Five Centuries of Map Printing*. Chicago: University of Chicago Press for the Newberry Library, 1975.
———. "Medieval *Mappaemundi*." In *HC 1*, 286–370.
———. "Reality, Symbolism, Time, and Space in Medieval World Maps." *Annals of the Association of American Geographers* 75 (1985): 510–21.
Wortzel, Adrianne. "Sayonara Diorama: Acting Out the World as a Stage in Medieval Cartography and Cyberspace." In *HWM*, 415–21.
Wright, John Kirkland. *The Geographical Lore at the Time of the Crusades: A Study in the History of Medieval Science and Tradition in Western Europe*. New York: Dover, 1965.
Yee, Cordell D. K. "Taking the World's Measure: Chinese Maps between Observation and Text." In *HC* 2.2, 96–127.
Young, John. "Beja: Local Conflict, Marginalization, and the Threat to Regional Security." *Researching Local Conflicts and Regional Security* 2 (2007): 1–19.
Zadeh, Travis. *Mapping Frontiers across Medieval Islam: Geography, Translation, and the ʿAbbasid Empire*. New York: I. B. Tauris, 2011.
Zagórski, Bogusław R. "Sea Names of the Arab World as a System." *Onomastica* 57 (2013): 205–28.
Zakariya, Mohamed. "The Hilye of the Prophet Muhammad." *Seasons*, Autumn–Winter 2003–2004, 13–22.
Zandvliet, Kees. "Les livres et les murs." In *CFT*, 436–41.
Zimmer, Heinrich. *Myths and Symbols in Indian Art and Civilization*. Princeton, NJ: Princeton University Press, 1972. Reprint, 1974.

Index

Page numbers in italics refer to figures.